GEOMATRIX CONSULTANTS, INC.
LIBRARY

Ground Improvement

As needs increase for redeveloping land in urban areas, major developments have resulted in the field of ground improvement, a process that is continuing and expanding. In particular, the specialist grouting technique of soilfracturing has been used widely for the control of ground movements associated with tunnelling, whilst soil mixing technology has found an increasing number of applications in Europe and North America.

The second edition of this well established book continues to provide an international overview of the major techniques in use. Comprehensively updated in line with recent developments, each chapter is written by an acknowledged expert in the field.

Ground Improvement is written for geotechnical and civil engineers, and for contractors working in grouting, ground improvement, piling and environmental engineering. Advanced students will find the book a valuable source of reference.

Mike Moseley was the staff director of Business Development for Keller Group plc until 2000. He has worked extensively over the past 30 years within the field of ground improvement, both at a national and international level.

Klaus Kirsch is chairman of the supervisory board of Keller Grundbau in Germany and also of the Keller Group Technology Committee, and is also a consultant to the board of Keller Group plc. He has been involved with some fundamental breakthroughs in soil engineering, including the first vibro stone project in the USA, and has authored many papers.

Ground Improvement

Second Edition

Edited by

M.P. Moseley and K. Kirsch

LONDON AND NEW YORK

First published 1993
by Taylor & Francis
Reprinted 1994

Published in the USA and Canada by
CRC Press, Inc.

Second edition published 2004
by Taylor & Francis
2 Park Square, Milton Park, Abingdon, Oxon, OX14 4RN

Simultaneously published in the USA and Canada
by Taylor & Francis
270 Madison Avenue, New York, NY 10016

Transferred to Digital Printing 2006

Taylor & Francis is an imprint of the Taylor & Francis Group

Typeset in Sabon by
Integra Software Services Pvt. Ltd, Pondicherry, India

British Library Cataloguing in Publication Data
A catalogue record for this book is available
from the British Library

Library of Congress Cataloging in Publication Data
A catalog record for this book has been requested

ISBN 0–415–27455–9

Printed and bound by CPI Antony Rowe, Eastbourne

Contents

Contributors

Sam Bandimere, 7898 West 119th Place, Broomfield, CO, 80020, USA.

Bengt B. Broms, M.S., Ph.D., Professor, Strindbergsgatan 36, S-115 31 Stockholm, Sweden.

Bob D. Essler, M.A., M.Sc., DIC, CEng., MICE, MASCE, 7 Riverside Walk, Airton, Skipton, North Yorkshire, BD23 4AF, UK.

Eduard Falk, Dipl.-Ing. Dr.techn., Keller Grundbau Ges.m.b.H., A-1151 Wien, Mariahilfer straße 129, Austria.

Sven Hansbo, Professor Emeritus, Lyckovagen 2, 182 74 Stocksund, Sweden.

Robert M. Rubright, B.S.C.E., Hayward Baker Inc., 1130 Annapolis Road, Odenton, MD, 21113, USA.

Barry Slocombe, B.Sc., M.Sc., GMICE., MASCE., Keller Ground Engineering, Oxford Road, Ryton-on-Dunsmore, Coventry, UK, CV8 3EG.

Wolfgang Sondermann, Dr.-Ing., Keller Holding GmbH, D-63006 Offenbach, Kaiserleistraße 44, Germany.

Gert Stadler, O. Univ.-Prof. Dipl.-Ing. Dr.mont., Graz University of Technology, A-8010 Graz, Lessingstraße 25/II, Austria.

Michał Topolnicki, Prof. Dr.hab. inż., Gdańsk University of Technology, Pl-80-952 Gdańsk, ul. Narutowicza 11/12 and Keller Polska SP. zoo, Pl-05-850 Ozarow Mazowiecki, ul. Poznańska 172, Poland.

Jimmy Wehr, Dr.-Ing. M.Sc., Keller Grundbau GmbH, D-63006 Offenbach, Kaiserleistraße 44, Germany.

Mitsuhiro Shibazaki, B.Sc., Construction Engineering, Chemical Grouting Company, Anzen Bldg. 1-6-4, Motoakasaka, Minato-Ku, Tokyo 107, Japan.

Introduction

K. Kirsch and M.P. Moseley

This second edition of *Ground Improvement* will provide the reader with a sound basis for the study and understanding of the most important ground improvement techniques. Considerable developments have occurred since publication of the first edition in 1993, not only in technical matters but also in plant and equipment, and rate of production. The last decade has seen an increasing demand for *in-situ* deep soil mixing work in Europe and North America, which is reflected in broader coverage in this edition.

All ground improvement techniques seek to improve those soil characteristics that match the desired results of a project, such as an increase in density and shear strength to aid problems of stability, the reduction of soil compressibility, influencing permeability to reduce and control ground water flow or to increase the rate of consolidation, or to improve soil homogeneity. These considerations are addressed by well-known international experts. Their contributions provide an overview of the development of each specific technology as well as details of plant and equipment required for their execution. Theoretical considerations on design aspects together with fields of application and limitations of the methods are also given. Case histories from around the world with aspects of testing, monitoring and process controls complete the description of the various ground improvement methods.

In ground improvement, distinction is made between methods of compaction or densification and methods of soil reinforcement through the introduction of additional material into the ground. This distinction offers the opportunity to divide the topics into several groups, which are covered by the chapters of this book, each of which can be read and studied separately. The extensive references given with each chapter enable further reading on each technology.

Chapter 1 of the book deals with methods of static compaction by preloading with and without consolidation aid. An updated overview of consolidation theory together with experiences gained from recent field data provide valuable information and practical guidance. The most common methods are described and illustrated by examples.

Chapter 2 describes vibratory compaction by depth vibrators, probably the most used dynamic method of soil improvement worldwide. It also covers vibro stone column methods which achieve their improvement effects from a combination of soil displacement and reinforcement through the introduction of granular or cementicious material into the soil.

Dynamic compaction uses the energy from the impact of heavy drop weights on the ground to achieve ground compaction. A comprehensive description of the technology and current plant and equipment is given in Chapter 3.

The important ground improvement method of permeation grouting is presented in Chapter 4, with an emphasis on recent advances in cement grouting. This emphasis gives tribute to the fact that the injection of chemicals into the ground as presented in the first edition is increasingly prohibited for environmental reasons and indeed has largely been replaced by jet grouting as described in Chapter 5.

Compaction grouting as covered in Chapter 6 is a soil improvement method that utilises both the compaction effect of grouting pressure and the reinforcing effect of the low strength grout material that is introduced into the ground during the process. This technology has its origins in the United States and is now being applied in Europe.

Soil fracturing or compensation grouting is described in Chapter 7 and uses similar soil improvement features to compaction grouting for different engineering purposes. This technique was initially used to control building settlement and to lift structures and has been used recently to control ground and structure movement during tunnelling operations, a factor of particular importance in urban renewal schemes.

Important non-displacing reinforcement methods are the lime/cement stabilisation method of Chapter 8 and the *in-situ* deep soil mixing method described in Chapter 9. Both methods create stiffer columns in soft ground consisting of a mixture of the natural soil with stabilising additives which are introduced as dry or wet components into and mechanically mixed with the soil. The lime/cement stabilisation methods originated in Sweden and deep soil mixing techniques in Japan, and both technologies are finding increasing application in Europe and North America, a factor that is reflected in the greater attention and detail given to these methods in this edition.

Ground improvement techniques continue to make considerable progress, both quantitatively and qualitatively, as a result of not only technology developments but also of an increasing awareness of the environmental and economic advantages of modern ground improvement methods. The selection of the correct ground improvement technique at an early stage in design can have an important effect on foundation choice and can often lead to more economical solutions when compared to traditional approaches. The expansion in the use of ground improvement techniques is further assisted by the increasing need to develop marginal land.

This development is reflected in the increasing degree of standardisation of the various methods in codes and similar technical recommendations that cover geotechnical design.

We are grateful and pleased to be able to present contributions from a group of authors who are recognised experts in their fields of ground improvement and who provide an international viewpoint in this important field of ground engineering.

Chapter 1

Band drains

S. Hansbo

1.1 Introduction

1.1.1 Background to the use of band drains

According to the fundamental concepts of soil mechanics the placement of an external load on a low-permeable soil layer will induce excess pore water pressure, causing a consolidation process in which pore water is squeezed out of the soil, accompanied by a gradual increase in effective stress and a corresponding decrease in excess pore water pressure. The consolidation process will continue until the excess pore water pressure has dissipated, a process whose duration depends on the consolidation characteristics of the soil and the drainage paths (the longer the drainage paths, the longer the consolidation process). The idea behind the installation of vertical band drains is to reduce the length of the drainage paths and thereby reduce the time of consolidation.

The use of vertical sand drains was first proposed in 1925, and patented in 1926, by Daniel D. Moran. He also suggested the first practical application of sand drains as a means of stabilizing of mud soil beneath a roadway approach to the San Francisco Oakland Bay Bridge (Johnson, 1970). This led to some successful laboratory and field experiments followed by the installation of the first drain system in 1934. Porter (1936) described these trials and contributed to the further use and development of the system.

The sand drains originally installed had generally a relatively large diameter, 0.4–0.6 m. Later on small-diameter sand drains came into use, for example 'sand wicks', 0.05 m in diameter, and 'fabridrains' – also called 'sand pack drains' – 0.12 m in diameter. The sand in these drains is packed into a synthetic fibre net-type tube, which prevents the drains from necking. Sand drains with a diameter of 0.18 m were utilized in the oldest and best-documented test field existing, the one situated at Skå Edeby, Sweden, established in 1957 (Hansbo, 1960). This test field is still under continuous observation.

Another type of circular-cylindrical drains, the so-called 'soil drain' was developed by, among others, Technique Louis Ménard. This consists of an open pre-fabricated tubular plastic core provided with perforations to admit inflow of pore water.

A range of techniques has been utilized for installation of sand drains. These include so-called non-displacement methods, such as shell and auger drilling, powered auger drilling, water-jetting, flight augering and wash-boring and displacement methods, typically by the use of a driven closed-end mandrel.

The first type of band drains, also named wick drains, introduced on the market was invented in Sweden by Walter Kjellman and his co-workers at the Swedish Geotechnical Institute. These drains, named cardboard wicks (Kjellman, 1947), were made of two cardboard sheets glued together with an external cross-section of 100 mm by 3 mm and included ten longitudinal internal channels, 3 mm in width and 1 mm in thickness (Figure 1.1). A special machine with a capacity for installing the cardboard wicks to a depth of 14 m was also developed (Figure 1.1). The efficiency of cardboard wicks was first investigated in a full-scale test at Lilla Mellösa, 20 km north of Stockholm, constructed between 1945 and 1947 as preliminary measure of a planned new Stockholm International Airport at the site in question. The test field comprised two test areas, 30 m by 30 m, one with cardboard wicks to a depth of 5 m, the other without drains, with an initial surface load of 2.5 m of gravel (40 kN/m^2) on 14 m deep, very soft, organic high-plasticity clay. The results of this investigation, which proved the expected functioning of the cardboard wick, were reported by Chang (1981).

The cardboard wick has served as a prototype for the various band drains existing on the market. It is an interesting fact that cardboard wicks were first installed for stabilization purposes only 5 years later than sand drains. Band drains and sand drains are thus of very nearly the same age. After the method of using cardboard wicks for the purpose of speeding up consolidation was introduced in Japan, a considerable number of installations were made. An installation to a depth of 17 m in Canada in the early seventies showed both economic and technical superiority of the cardboard wick in comparison with conventional sand drains (Flodin, 1973).

The market of band drains did not really open until the introduction of Geodrain, developed at the Swedish Geotechnical Institute in cooperation with Perstorp AB and Örebro Pappersbruk AB (Boman, 1973). It consists of a core of plastic material surrounded by a filter sleeve with an external cross-section of 95 mm by 4 mm, i.e. nearly the same dimensions as the cardboard wick. Both sides of the core are provided with 27 longitudinal grooves whose widths and depths vary with different makes. The filter sleeve was originally made of a special make of paper but was later changed into synthetic material.

Figure 1.1 The cardboard wick and the installation equipment (maximum depth of installation 14 m).

1.1.2 *Proprietary drains*

After some successful applications of Geodrains, a great number of band drains, having more or less similar characteristics but different drainage efficiencies, have been developed, some of which are shown in Figure 1.2.

Some of the different proprietary drains are summarized in Table 1.1. Although most band drains have a central core enclosed in a filter sleeve, drain types without filter sleeve also exist on the market. These generally consist of porous material, which allows water into the drains. A somewhat different type of band drain, the fibre drain, was developed in Singapore (Lee *et al.*, 1995). It consists of one layer of thin, closely knit jute burlap laid inside another layer of thick but coarsely knit burlap. Four coir strands, 3–6 mm in diameter, pass longitudinally through the inner core formed by the two layers of burlap.

It is interesting to note that Barron (1947), with reference to a contribution by Kjellman (1947), expresses his opinion that 'should wick material and installation machines become available in the United States, sand wells may be outmoded'. Nowadays very efficient installation machines have come into use and a large number of various band drains exists on the market. Barron's prophecy has certainly become true.

1.2 Construction and plant

1.2.1 *Construction*

A new Eurocode on vertical drainage will be published in the near future under the auspices of EFFC (European Federation of Foundation Contractors). This code includes detailed requirements and recommendations regarding installation and use of vertical drains.

Sites in need of vertical drainage often have a low bearing capacity. Consequently, there may be a need to reinforce the ground surface, for example by means of geotextile, in order to enable the installation machinery to enter the site without risk of soil failure. Usually, the drainage blanket, which is placed in advance on the ground surface before drain installation and serves as a working platform, is thick enough – at least 0.3–0.5 m – to safeguard the installation work.

The material in the drainage blanket must have excellent drainage properties. Otherwise it will be necessary to connect the vertical drains with a horizontal drainage system in order to avoid a build-up of backpressure in the drains that would delay the consolidation process (Figure 1.3).

The drains are usually installed in equilateral triangular or square pattern. Irrespective of which drain pattern is chosen, the number of drains required to achieve a certain rate of consolidation will be more or less the same. However, the equilateral triangular pattern is most optimal and preferable.

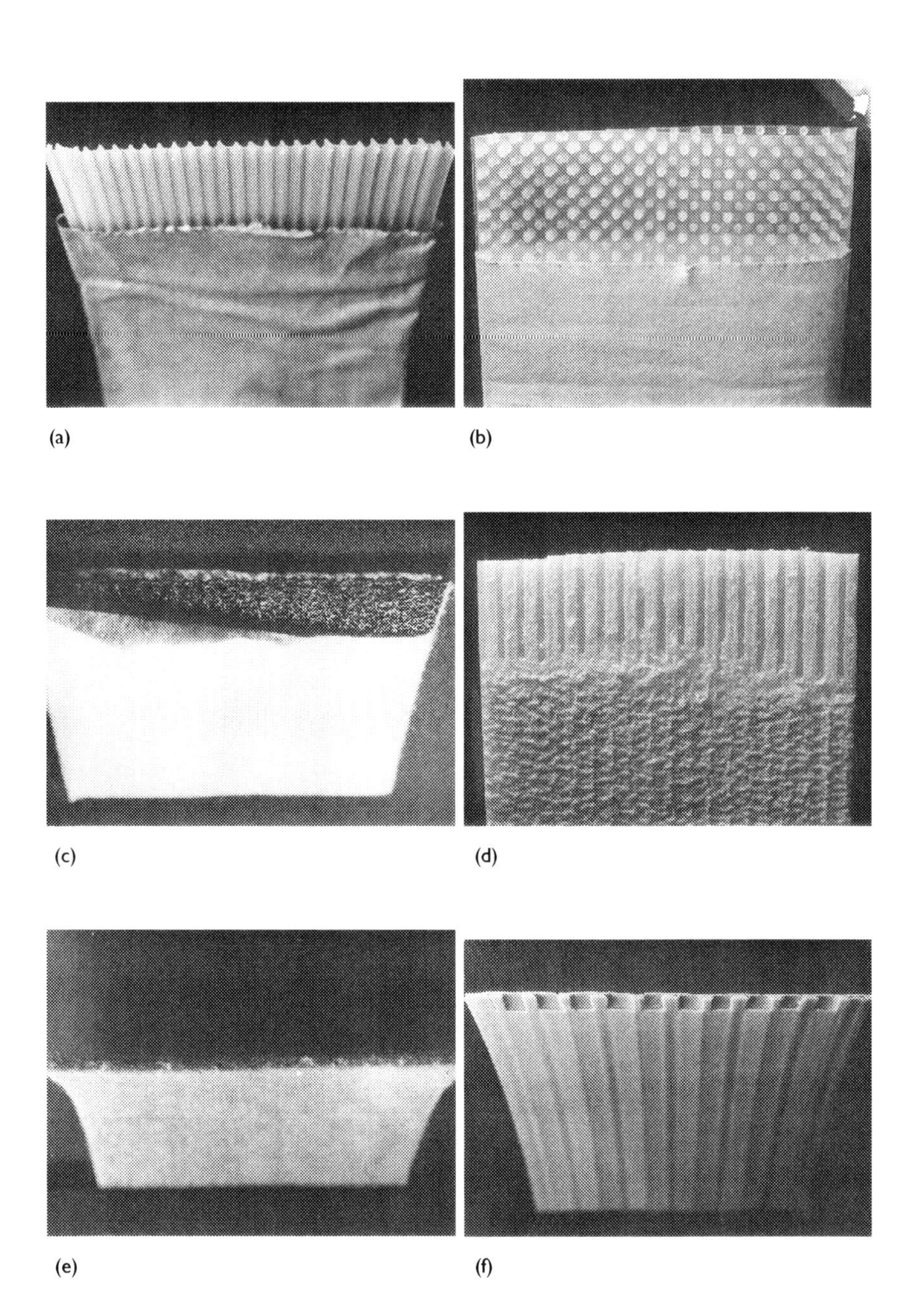

Figure 1.2 Examples of different makes of band drains: (a), (b), (c), (d) drains with central core surrounded by filter sleeve; (e), (f) drains without filter sleeve.

Table 1.1 Proprietors of various makes of band drains

Drain designation	*Proprietor*
Cardboard wick	No longer manufactured
Alidrain	Burcan Industries, Canada
Castleboard	Kinjo, Japan
Colbond	Arcadis, Holland
Desol	Funyick Trading Cö., Hong Kong
Flodrain	Nylex Shah Alam, Malaysia
Geodrain	Marubeni Construction Sales Inc., Japan
Mebradrain	Cofra, The Netherlands
OV-drain	Japan Vilene Company
PVC	Ohbayashi-Gumi, Japan
Tafnel	Toyokiso, Japan

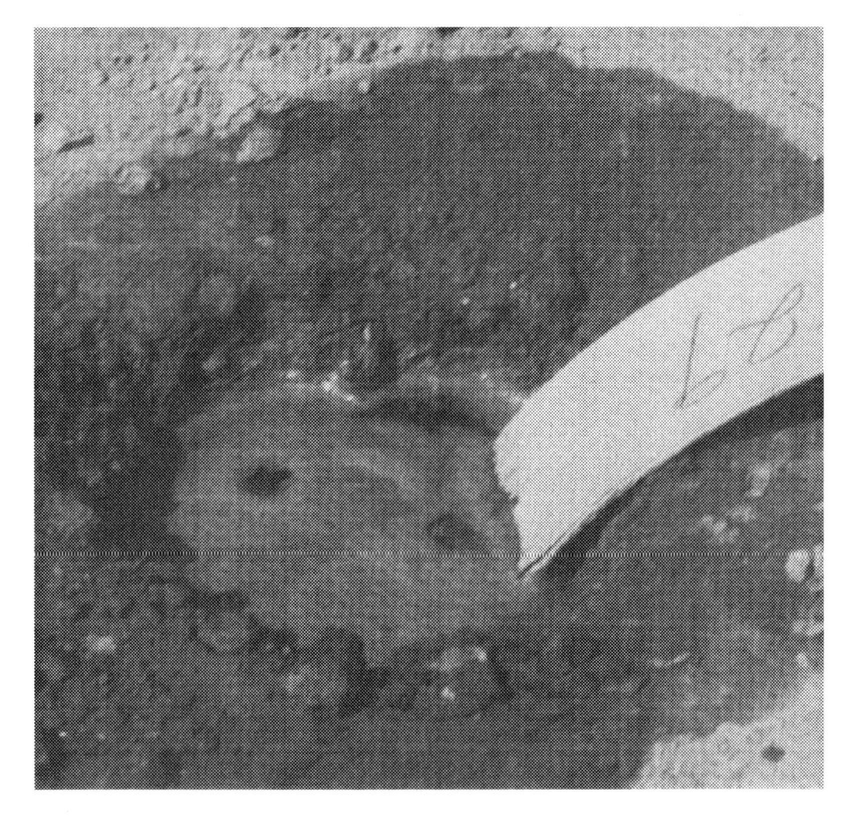

Figure 1.3 This is a case where a build-up of excess pore water backpressure in the drains can be expected because of insufficient permeability of the drainage blanket.

After completion of the drain installation, the drained area is loaded with a fill embankment in accordance with the loading schedule. In places where the subsoil has low undrained shear strength, stability along the edge of the fill embankment may be at stake. Stabilizing loading berms may then be required to avoid failure or excessive lateral movement in the soil. When considering the need of loading berms it must be taken into account that the drain installation process itself may entail an overall decrease of the undrained shear strength of the subsoil.

The cost of loading berms whenever needed may add considerably to the total cost of a vertical drainage project. Loading berms are no longer required by using the so-called vacuum method, which is an alternative to pre-loading by the use of a fill embankment. According to the vacuum method an airtight seal of the drainage blanket has first to be created. Then an underpressure is created in the drainage blanket, and consequently also in the drains, by means of a vacuum pump. Normally, 80 per cent of full vacuum can be achieved. The main difficulty with the vacuum method is to obtain and maintain a safe airtight seal.

The vacuum method is of special interest where soft soil under deep water has to be consolidated. Full vacuum would then have the same effect as a surcharge represented by atmospheric pressure plus the water pressure at the sea bed. For example, at a water depth of 10 m, 80 per cent vacuum would represent a surcharge of 180 kN/m^2 (about 16 m of fill placed on the sea bed).

1.2.2 Plant

Contractors working in the field of vertical drainage have generally developed their own type of equipment for drain installation. A common feature is that the drains are installed inside a steel mandrel, which protects the drain from being damaged. Two main principles of installation can be distinguished, the so-called static and dynamic installation methods. In the first case, the mandrel with the drain inside is pushed into the soil by static pressure, while in the second case, it is driven into the soil by means of a gravity hammer or a vibratory driver. Modern rigs for static and dynamic installations are shown in Figures 1.4 and 1.5, respectively. Floating rigs exist by which the drains can be installed from open water (Figure 1.6).

The method of installation – static or dynamic – does not seem to affect the efficiency of the drainage system. Dynamic methods should, however, be avoided wherever disturbance effects, usually evidenced by excess pore water pressure being built up during installation, may affect stability.

Before the drains are inserted into the soil they must be provided with an anchor, which keeps the drains in position when the mandrel is withdrawn (Figure 1.7). The anchor also prevents soil from intruding into the mandrel during installation (which may lock the drain to the mandrel by friction). Different contractors use different types of anchors.

After the mandrel is withdrawn, the drain should be cut in such a way that a good connection with the drainage blanket is ascertained (Figure 1.8). Rigs exist that are provided with a cutting device by which, for example in a drain installation from open water, the drains can be cut just above the drainage blanket placed on the bottom. In deep water this technique can save a considerable length of drains.

In order to reduce the disturbance effects caused by drain installation, the mandrels utilized are normally quite slender. Therefore, when they are inserted into the soil they may deviate considerably from the vertical, particularly in the case of deep installations. Since it is essential for a well-functioning drain system that the prescribed drain spacing is maintained throughout the drained soil layer, such a deviation from the vertical can have a negative and unpredictable effect on the consolidation process at great depth. Therefore, the mandrel should either be equipped with an inclinometer that gives information about the horizontal position of the drain at various depths, or the mandrel should be stiff enough to ensure verticality of the drains.

1.3 Analytical approach

1.3.1 Assumptions based on Darcian flow

Regarding the historical development of vertical drain analysis, special interest must be devoted to the contributions by Barron, which form a starting point in the understanding of the result to be expected by vertical drain installations. During the winter of 1941–1942, the Providence

(a)

Figure 1.4 Rigs for static installation of band drains: (a) Porto Tolle, Italy (installation to a depth of 30 m); (b) Arlanda International Airport, Stockholm.

(b)

Figure 1.4 (continued)

District incorporated drain wells in plans for reconstruction of a portion of Riverfront Dike, Hartford, Connecticut. This entailed the necessity of having a more exact analysis of the influence of vertical drains on the consolidation process. The analysis first published by Barron (1944) was based on existing solutions for one-dimensional vertical consolidation (Terzaghi, 1925) and radial flow of heat. Barron's analysis was based on the following assumptions (Figure 1.9):

- Darcy's flow law is valid.
- The soil is water saturated and homogeneous.
- Displacements due to consolidation take place in the vertical direction only.
- Excess pore water pressure at the drain well surface is zero.
- The cylindrical boundary of the soil mass is impervious, i.e. $\partial u/\partial \rho = 0$ at $\rho = R$.
- Excess pore water pressure at the upper boundary of the soil mass ($z = 0$) is zero.
- No vertical flow at the central cross-section of the soil mass, i.e. $\partial u/\partial z = 0$ at $z = l$.

(a)

(b)

Figure 1.5 Rigs for dynamic installation of band drains: (a) Changi Airport, Singapore (installation to a depth of 43 m); (b) installation in the Netherlands.

Figure 1.6 Floating rig for installation of four band drains in one operation from open water, Japan.

The differential equation governing the consolidation process is then given by the expression:

$$\frac{k_h}{\gamma_w}\left(\frac{1}{\rho}\frac{\partial u}{\partial \rho}+\frac{\partial^2 u}{\partial \rho^2}\right)+\frac{k_v}{\gamma_w}\frac{\partial^2 u}{\partial z^2}=\frac{a_v}{1+e}\frac{\partial u}{\partial t} \qquad (1.1)$$

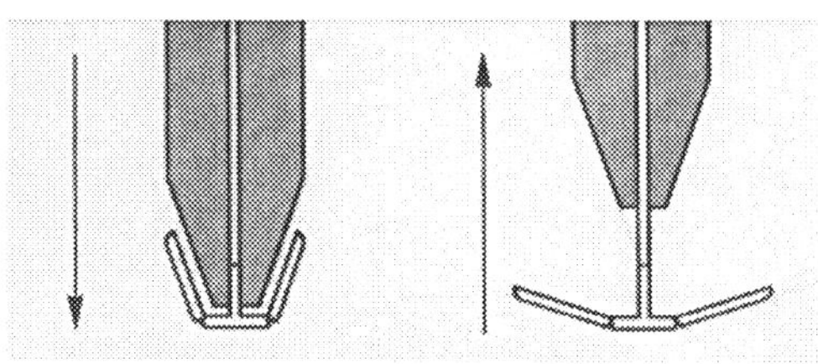

Figure 1.7 Example of drain anchor. The anchor, which is fixed to the drain tip, prevents soil from intruding into the mandrel during drain installation and keeps the drain in place when the mandrel is withdrawn.

where k_h and k_v = the permeability in the horizontal and the vertical directions, respectively, γ_w = the unit weight of water ($= g\rho_w$, where g is the acceleration of gravity and ρ_w is the density of water), ρ and z = the cylindrical coordinates in the radial and the vertical directions, respectively, u = excess pore water pressure, $a_v = -\Delta e/\Delta\sigma'$, the coefficient of theoretical compressibility, e = void ratio, $a_v/(1+e) = m_v = 1/M$ = coefficient of volume compressibility (M = oedometer modulus) and t = consolidation time.

Barron proposes that the total degree of consolidation, including the effect of combined radial and vertical outflow of water, be solved according to Carillo (1942) by the expression:

$$u_{\rho z} = \frac{u_\rho u_z}{u_0} \tag{1.2}$$

where $u_{\rho z}$ = remaining total excess pore water pressure after time t, u_ρ = remaining excess pore water pressure after time t due to radial drainage, u_z = remaining excess pore water pressure after time t due to vertical drainage and u_0 = excess pore water pressure at time $t = 0$.

Expressed in degree of consolidation, $U = 1 - u/u_0$, this yields:

$$U_{\rho z} = U_\rho + U_z - U_\rho U_z \tag{1.3}$$

where U_ρ = degree of consolidation due to radial outflow of pore water to the drains and U_z = degree of consolidation due to vertical outflow of pore water outside the drains.

As an alternative to $U = 1 - u/u_0$ we can also use the definition $\overline{U} = s/s_p$, where $\overline{U}$ = average consolidation, s = settlement at time t and s_p = total primary settlement. Then the settlement s_{hd} at time t, achieved by the effect of radial drainage only, can be written as

$$s_{hd} = \frac{s - s_{vd}}{1 - \dfrac{s_{vd}}{s_\rho}} \tag{1.4}$$

where s_{vd} = settlement caused by one-dimensional vertical consolidation.

Figure 1.8 Cutting of the band drains is made just above the drainage blanket and afterwards the drain is provided with an anchor.

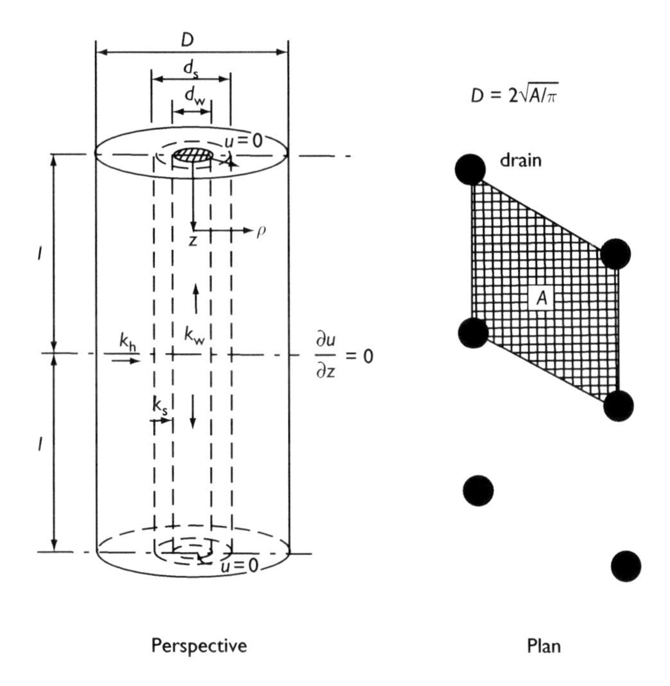

Figure 1.9 Terms used in the analysis of vertical drains: D = diameter of soil cylinder dewatered by a drain, d_s = diameter of the zone of smear, d_w = drain diameter, z = depth coordinate, l = length of drain when closed at bottom ($2l$ = length of drain when open at bottom), q_w = specific discharge capacity of the drain (vertical hydraulic gradient inside the drain $i = 1$).

Barron assumed two different cases to take place: the case of *free strains* and the case of *equal strains*.

In the *free strain hypothesis* Barron pre-supposes that no arching takes place and that shear strains caused by differential settlement do not redistribute the load-induced stresses within the soil at any time during consolidation.

In his original free strain analysis, Barron (1944) assumed that the installation of the drains did not affect the properties of the soil and that the permeability of the drain well was high enough for well resistance to be neglected. He later on included disturbance effects due to installation, a *zone of smear* (Barron, 1947) with reduced permeability k_s. In the *equal strain hypothesis* Barron (1947) presumes arching to redistribute the load so that the vertical strains at a certain depth z become equal irrespective of the radial distance ρ, and, consequently, no differential settlement will take place. This may seem a rather serious condition but is supported by field observations in areas provided with vertical drains.

The average degree of consolidation obtained for ideal drains according to Barron's free strain analysis is very nearly equal to that obtained according to Barron's equal strain analysis, equation (1.5), as shown in Figure 1.10. Therefore, the equal strain hypothesis has become the basis for routine design of vertical drain systems.

In the equal strain hypothesis Barron also includes the effect of *well resistance* on the consolidation process. Thus, in reality the drains may have a limited capacity of transporting the pore water entering into the drains during the consolidation process. Assuming complete drainage at $z = 0$ and $z = 2l$ the average degree of consolidation obtained at depth z by radial (horizontal) drainage $\overline{U}_{hz} = 1 - \overline{u}_{hz}/u_0$ is given by the correlation

$$\overline{U}_{hz} = 1 - \exp\left\{-\frac{8c_h t}{\nu D^2}\left[\frac{\exp[\beta(z - 2l)] + \exp(-\beta z)}{1 + \exp(-2\beta l)}\right]\right\} \tag{1.5}$$

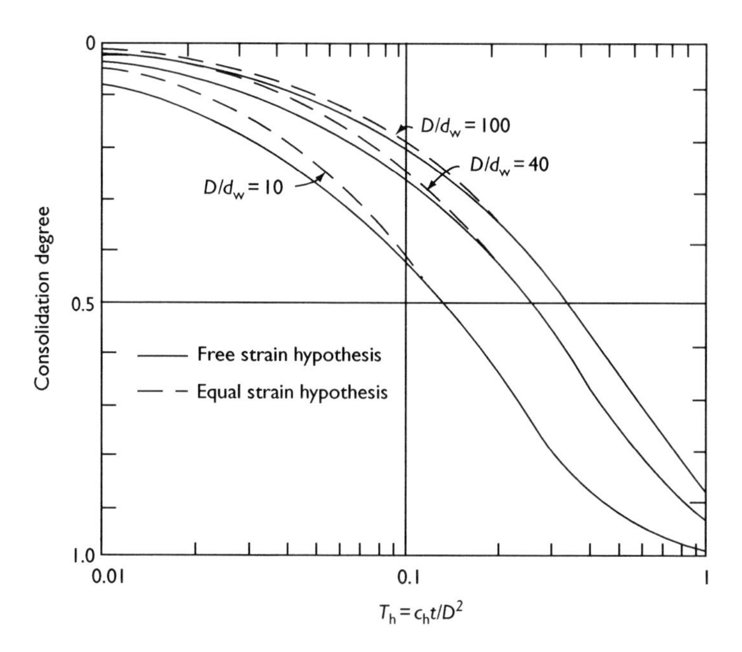

Figure 1.10 Comparison of average consolidation rates by radial drainage only for various values of D/d_w under conditions of equal vertical strains at any given time and no arching of overburden (free strain hypothesis). Ideal drain wells. No effect of smear. After Barron (1947).

where

$$\nu = \frac{D^2}{D^2 - d^2} \ln\left(\frac{D}{d_s}\right) - \frac{3}{4} + \frac{d_s^2}{4D^2} + \frac{k_h}{k_s}\left(\frac{D^2 - d_s^2}{D^2}\right) \ln\left(\frac{d_v}{d_w}\right),$$

$k_h, k_s, c_h \ldots$ as above,

$$\beta = \sqrt{\frac{8k_h\left(1 - \frac{d_s^2}{D^2}\right)}{d_w^2 k_w \nu}} = \sqrt{\frac{2\pi k_h\left(1 - \frac{d_s^2}{D^2}\right)}{\nu q_w}},$$

$$q_w = k_w \pi \frac{d_w^2}{4} = \text{specific discharge capacity of the drain.}$$

Another approach to the equal strain hypothesis in the simple case of no peripheral smear or well resistance, very similar to Barron's approach, was presented already in 1937 (Kjellman, 1947). Kjellman's approach was extended by the author (Hansbo, 1979, 1981) to include the effect of smear and well resistance. In this case, the average degree of consolidation is given by the relation:

$$\overline{U}_{hz} = 1 - \exp\left(-\frac{8c_h t}{\mu D^2}\right) \tag{1.6}$$

where

$$\mu = \frac{D^2}{D^2 - d_w^2}\left[\ln\left(\frac{D}{d_s}\right) + \frac{k_h}{k_s} \ln\left(\frac{d_s}{d_w}\right) - \frac{3}{4}\right] + \frac{d_s^2}{D^2 - d_w^2}\left(1 - \frac{d_s^2}{4D^2}\right)$$
$$- \frac{k_h(d_s^2 - d_w^2)}{k_s(D^2 - d_w^2)}\left(1 - \frac{d_s^2 + d_w^2}{4D^2}\right) + \frac{k_h \pi z(2l - z)\left[1 - \left(\frac{d_w}{D}\right)^2\right]}{q_w}$$

Omitting terms that for band drains are normally of minor significance we find:

$$\mu = \ln\left(\frac{D}{d_s}\right) + \frac{k_h}{k_s} \ln\left(\frac{d_s}{d_w}\right) - \frac{3}{4} + \frac{k_h \pi z(2l - z)}{q_w}$$

The average degree of consolidation $\overline{U}_{h,av}$ of the whole layer is obtained by exchanging the value of μ for

$$\mu_{av} = \ln\left(\frac{D}{d_s}\right) + \frac{k_h}{k_s}\ln\left(\frac{d_s}{d_w}\right) - \frac{3}{4} + \frac{2k_h\pi l^2}{3q_w}$$

The contribution to the consolidation process by pore water escape in the vertical direction can be solved according to Terzaghi (1925). Accordingly, the average degree of vertical consolidation $\overline{U}_{v,av}$, for values of $\overline{U}_{v,av} \leq 50$ per cent, is obtained by the relation

$$\overline{U}_{v,av} = 2\sqrt{\frac{c_v t}{\pi h^2}} \tag{1.7}$$

where $2h$ = the thickness of the soil layer, drained at top and bottom.

The average degree of consolidation achieved by one-dimensional vertical pore water flow (undrained condition) is generally below 50 per cent. Therefore, the total average degree of consolidation for fully penetrating drains, taking into account both undrained (c_v value assumed to be constant) and drained conditions, can be expressed by the relation

$$\overline{U}_{av} = 1 - \left(1 - \frac{2}{l}\sqrt{\frac{c_v t}{\pi}}\right)\exp\left(-\frac{8c_h t}{\mu_{av} D^2}\right) \tag{1.8}$$

where $2l$ = thickness of the clay layer when drained at top and bottom (Figure 1.9), corresponding to the length of the drains.

The consolidation equations presented above are based on the assumption that the value of c_h does not change during the consolidation process. An attempt to take into account a successive decrease of c_h with time of consolidation, expressed through a decrease in the coefficient of permeability, was made by Shiffman (1958). Shiffman's concept is based on a linear correlation between permeability coefficient and excess pore water pressure. For constant load, varying permeability (oedometer modulus assumed to remain constant during the consolidation process) and radial drainage only, the solution obtained for ideal drain wells (no well resistance, no smear) becomes:

$$\overline{U}_h = 1 - \frac{k_f}{k_f + k_0\left[\exp\left(\frac{8T_f}{v}\right) - 1\right]} \tag{1.9}$$

where k_f = final value of k_h, k_0 = initial value of k_h, $T_f = c_f t/D^2$, $c_f = k_f M/\gamma_w$ and $\upsilon = \left(D^2/(D^2 - d_w^2)\right) \ln(D/d_w) - 3/4 + d_w^2/4D^2$. The parameter υ is equal to the parameter ν in equation (1.5) when $d_s = d_w$.

The effect of well resistance was also taken into account by Yoshikuni and Nakanado (1974). Their solution, which includes both vertical and horizontal pore water flows (upper and lower boundary surfaces assumed to be drained) but does not include the effect of smear, ends up in a rather complex expression. Yoshikuni (1979) presented a very comprehensive analytical study of the effect on the consolidation process of various boundary and loading conditions and drain properties.

With the advances of the finite element and the finite difference methods the consolidation process achieved for any type of loading and drainage condition can be solved theoretically on the basis of given consolidation and drain parameters (e.g. Onoue, 1988; Zeng and Xie, 1989; Lo, 1991). Among these can be mentioned the finite element programme ILLICON developed at University of Illinois at Urbana-Champaign, USA, which is based on the following basic correlations (Lo, 1991):

$$\left(\frac{\partial e}{\partial \sigma'_v}\right)_t \frac{\partial \sigma'_v}{\partial t} + \left(\frac{\partial e}{\partial t}\right)_{\sigma'_v} = \frac{(1+e_0)^2}{\gamma_w(1+e)}\left[\frac{\partial k_v}{\partial z}\frac{\partial u}{\partial z} + k_v\left(\frac{\partial^2 u}{\partial z^2} - \frac{1}{1+e}\frac{\partial u}{\partial z}\frac{\partial e}{\partial z}\right)\right]$$

$$+ \frac{1+e}{\gamma_w}\left[\frac{\partial k_h}{\partial \rho}\frac{\partial u}{\partial \rho} + k_h\left(\frac{1}{\rho}\frac{\partial u}{\partial \rho} + \frac{\partial^2 u}{\partial \rho^2}\right)\right] \qquad (1.10)$$

The excess pore water pressure is given by the relation:

$$\frac{\partial \sigma_v}{\partial t} - \frac{\partial u}{\partial t} = \frac{\dfrac{\partial e}{\partial t} - \left(\dfrac{\partial e}{\partial t}\right)_{\sigma'_v}}{\left(\dfrac{\partial e}{\partial \sigma'_v}\right)_t} \qquad (1.11)$$

where $(\partial e/\partial t)_{\sigma'_v}$ = change in void ratio with time at a given effective stress and $(\partial e/\partial \sigma'_v)_t$ = change in void ratio owing to change in effective stress.

The results obtained by different design methods for drains with well resistance are compared in Figure 1.11.

The left-hand term in equation (1.10) is equivalent to the constitutive relation presented by Taylor and Merchant (1940).

$$\frac{de}{dt} = \left(\frac{\partial e}{\partial \sigma'_v}\right)_t \frac{\partial \sigma'_v}{\partial t} + \left(\frac{\partial e}{\partial t}\right)_{\sigma'_v}$$

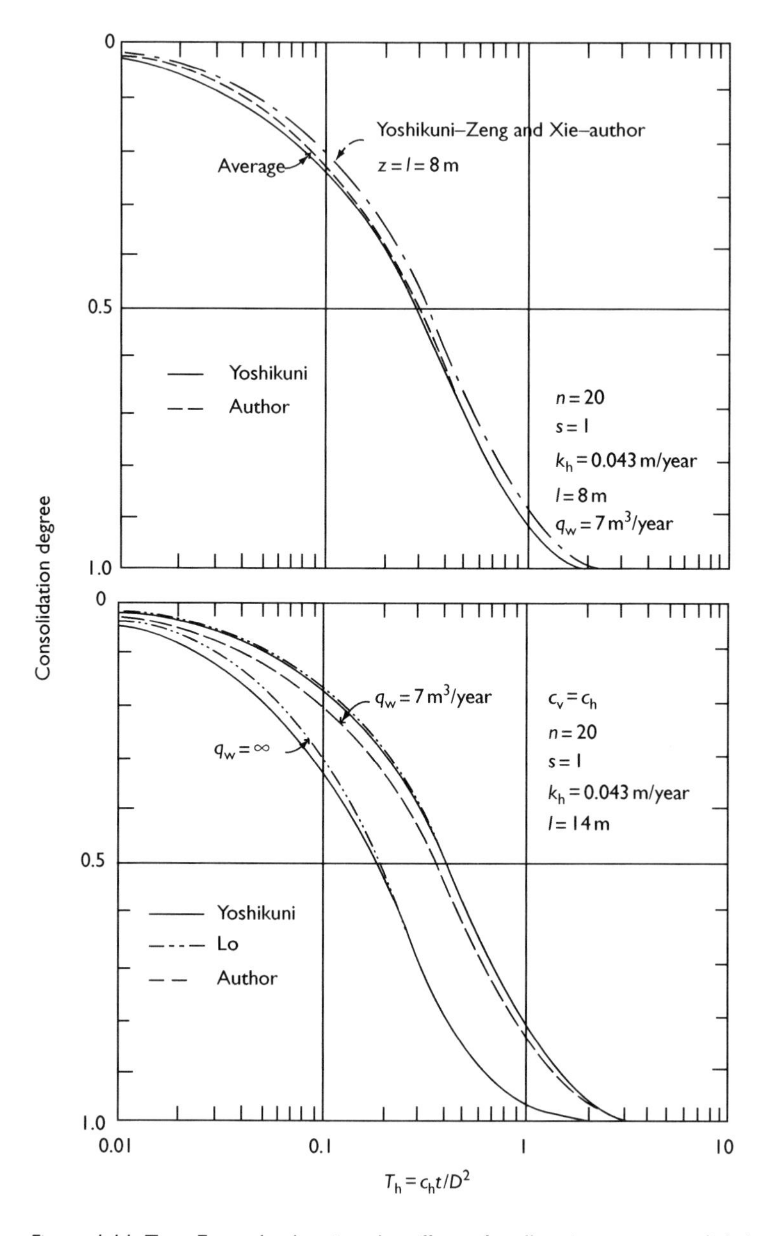

Figure 1.11 Top: Example showing the effect of well resistance on radial drainage (average and for $z = l$) according to Barron, equation (1.5), Yoshikuni, Zeng and Xie, and the author, equation (1.6). Bottom: Example showing the result of combined vertical and radial drainage without and with well resistance according to Yoshikuni, Lo (ILLICON), equation (1.10), and the author, equation (1.8).

In the case shown in Figure 1.11 the agreement between all the exemplified theories is very good. However, as was shown by Lo (1991), Barron's equal strain solution underestimates the average degree of consolidation when the discharge capacity of the drains is considerably smaller than in Figure 1.11. Choosing, for example, $q_w = 0.7\,m^3/year$ instead of $q_w = 7\,m^3/year$ (all other parameters unchanged), the degrees of consolidation obtained at depth $z = 8$ m by radial drainage only for the time factors $T_h = 0.1$, 1.0 and 5.0 become $U_h = 0.03$, 0.23 and 0.73, respectively, according to Barron's solution (equation (1.5)) and $U_h = 0.05$, 0.42 and 0.94, respectively, according to the author's solution (equation (1.6)). The solution given by equation (1.6) is in good agreement with, for example, the solution given by Lo, equations (1.10 and 1.11).

1.3.2 Assumptions based on non-validity of Darcy's Law

In the course of consolidation, the permeability in particular will be subjected to gradual reduction. However, case studies and experimental evidence have also shown that the coefficient of consolidation increases with increasing magnitude of the load that produces consolidation. For the determination of the coefficient of consolidation Terzaghi and Peck (1948) therefore recommended that the load increment 'applied to the sample after a pressure equal to the overburden pressure has been reached should be of the same order of magnitude as the load per unit area of the base of the structure'. A possible explanation to this phenomenon can be an exponential correlation between pore water flow and hydraulic gradient.

Results of permeability tests on clay samples presented by different researchers (e.g. Silfverberg, 1947; Hansbo, 1960; Miller and Low, 1963; Dubin and Moulin, 1986; Zou, 1996) have indicated that the pore water flow v caused by a hydraulic gradient i may deviate from Darcy's law $v = ki$ where k is the coefficient of permeability. Silfverberg and Miller, and Low drew the conclusion that there is a threshold gradient i_0 below which no flow will take place, yielding $v = k(i - i_0)$, while the author (Hansbo, 1960) proposed the following relations (Figure 1.12):

$$v = \kappa i^n \quad \text{when } i \le i_l \tag{1.12}$$

$$v = \kappa n i_l^{n-1}(i - i_0) \quad \text{when } i \ge i_l \tag{1.13}$$

In the author's opinion, $i_l = i_0 n/(n-1)$ represents the gradient required to overcome the maximum binding energy of mobile pore water (the physical background to non-linear conductivity behaviour is discussed in detail by Hansbo (1960)).

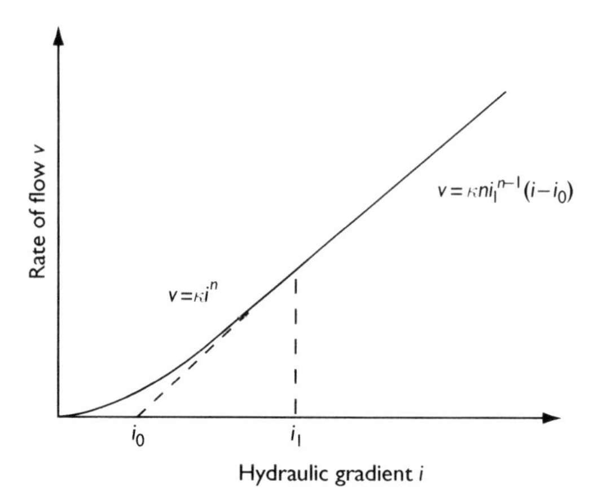

Figure 1.12 Hypothetical deviation from Darcy's law based on experimental evidence from results of permeability tests (Hansbo, 1960).

The relations (1.12–1.13) proposed by the author were also chosen by Dubin and Moulin (1986) in the analysis of Terzaghi's one-dimensional consolidation theory denoting $\kappa = ak$ where $a = i_l^{1-n}/n$ and $k = v/(i - i_0)$. The values of i_l have been found to vary from 4 to 10 (Hansbo, 1960) and 8 to 35 (Dubin and Moulin, 1986). Using the non-linear flow law given by equation (1.12) the consolidation equation, taking both smear and well resistance into account, can be written as:

$$\overline{U}_{hz} = 1 - \left[1 + \frac{\lambda t}{\alpha D^2}\left(\frac{\Delta h_0}{D}\right)^{n-1}\right]^{-\frac{1}{(n-1)}} \tag{1.14}$$

where $\Delta h_0 = \overline{u}_0/\gamma_w$ = the average increase in piezometric head caused by the placement of the load, $\lambda = \kappa_h M/\gamma_w$ = the coefficient of consolidation and $\alpha = n^{2n}\beta^n/4(n-1)^{n+1}$, with

$$\beta = \frac{1}{3n-1} - \frac{n-1}{n(3n-1)(5n-1)} - \frac{(n-1)^2}{2n^2(5n-1)(7n-1)}$$
$$+ \frac{1}{2n}\left[\left(\frac{\kappa_h}{\kappa_s} - 1\right)\left(\frac{D}{d_s}\right)^{\frac{1}{(n-1)}} - \frac{\kappa_h}{\kappa_s}\left(\frac{D}{d_w}\right)^{\frac{1}{(n-1)}}\right]$$
$$+ \frac{\left(1 - \frac{1}{n}\right)\left(\frac{d_w}{D}\right)^{1-\frac{1}{n}}\left(1 - \frac{d_w^2}{D^2}\right)^{\frac{1}{n}}\kappa_h \pi z(2l - z)}{2q_w}$$

The average degree of consolidation $\overline{U}_{h,av}$ for the whole layer is obtained by exchanging the last term in the β expression for

$$\frac{\left(1-\frac{1}{n}\right)\left(\frac{d_w}{D}\right)^{1-\frac{1}{n}}\left(1-\frac{d_w^2}{D^2}\right)^{\frac{1}{n}}k_h\pi l^2}{3q_w}$$

When the exponent $n \to 1$ (e.g. $n = 1.0001$), equation (1.14) yields the same result as equation (1.6) assuming $\lambda = c_h$ and $\kappa_h = k_h$ and $\kappa_s = k_s$. Thus equation (1.14) is generally applicable to consolidation effects produced by band drains and can, therefore, replace equation (1.6).

The hydraulic gradient i outside the zone of smear $(D/2 \geq \rho \geq d_s/2)$ becomes

$$i = \frac{\Delta h_0}{D}(1-\overline{U}_{hz})\left[\frac{1}{4\alpha(n-1)}\left(\frac{D}{2\rho}-\frac{2\rho}{D}\right)\right]^{\frac{1}{n}} \tag{1.15}$$

The variation of hydraulic head outside the zone of smear during the consolidation process becomes

$$\Delta h = \frac{\Delta h_0}{2^n\beta n^2}(1-\overline{U}_h)\left\{\begin{aligned}&F\left(\frac{2\rho}{D}\right)-F\left(\frac{d_s}{D}\right)+\frac{\kappa_h}{\kappa_s}\left[F\left(\frac{d_s}{D}\right)-F\left(\frac{d_w}{D}\right)\right]\\&+\frac{\kappa_h}{q_w}\pi z(2l-z)\left(1-\frac{1}{n}\right)\left(1-\frac{d_w^2}{D^2}\right)^{\frac{1}{n}}\left(\frac{d_w}{D}\right)^{1-\frac{1}{n}}\end{aligned}\right\} \tag{1.16}$$

where $F(x) = x^{(1-1/n)}[1-((1-1/n)/(3n-1))x^2-((1-1/n)^2/2(5n-1))x^4-((1-1/n)^2(2n-1)/6n(7n-1))x^6-\cdots]$ with the variable x representing $2\rho/D$, d_s/D and d_w/D.

The best agreement between theory and observations has been obtained for $n = 1.5$ (Hansbo, 1960, 1997a,b) which yields:

$$\overline{U}_{hz} = 1-\left(1+\frac{\lambda t}{\alpha D^2}\sqrt{\frac{\Delta h_0}{D}}\right)^{-2} \tag{1.17}$$

where $\alpha = 4.77\beta\sqrt{\beta}$ and, omitting terms of minor significance,

$$\beta = 0.270 + \frac{1}{3}\left[\left(\frac{\kappa_h}{\kappa_s} - 1\right)\left(\frac{d_w}{d_s}\right)^{\frac{1}{3}} - \frac{\kappa_h\left(\frac{d_w}{D}\right)^{\frac{1}{3}}}{\kappa_s} + \frac{\left(\frac{d_w}{D}\right)^{\frac{1}{3}}\kappa_h \pi z(2l - z)}{2q_w}\right]$$

The average degree of consolidation $\overline{U}_{h,av}$ for the whole layer is obtained by exchanging β for

$$\beta_{av} = 0.270 + \frac{1}{3}\left[\left(\frac{\kappa_h}{\kappa_s} - 1\right)\left(\frac{d_w}{ds}\right)^{\frac{1}{3}} - \frac{\kappa_h\left(\frac{d_w}{D}\right)^{\frac{1}{3}}}{\kappa_s} + \frac{\left(\frac{d_w}{D}\right)^{\frac{1}{3}}\kappa_h \pi l^2}{3q_w}\right]$$

Inserting β_{av}, the total average degree of consolidation, taking into account both undrained and drained conditions, can be expressed by the relation

$$\overline{U}_{av} = 1 - \left(1 - \frac{2}{l}\sqrt{\frac{c_v t}{\pi}}\right)\left(1 + \frac{\lambda t}{\alpha_{av} D^2}\sqrt{\frac{\Delta h_0}{D}}\right)^{-2} \tag{1.18}$$

where $2l$ = thickness of the clay layer when drained at top and bottom (Figure 1.9), corresponding to the length of the drains.

One may question the possibility of combining the consolidation theory based on Darcian flow for one-dimensional consolidation with non-Darcian flow for the effect of vertical drains. However, the effect of non-Darcian flow on one-dimensional consolidation is negligible in the beginning of the consolidation process, cf. Figure 1.13.

The influence of various magnitudes of well resistance on the results obtained according to equation (1.17) is exemplified in Figure 1.14.

In field conditions the hydraulic gradient i in most cases is very small in comparison with laboratory conditions. Choosing as an example a case with $n = 1.5$, $\Delta h_0 = 5$ m, $D = 1.05$ m, $d_w = 0.066$ m and $d_s = 0.15$ m (no well resistance), equation (1.15) in the initial state ($\overline{U}_{hz} = 0$) yields $i < 10$ in 85 per cent of the total drained cylinder. At an average degree of consolidation of 50 per cent, $i < 10$ in 98 per cent of the total drained cylinder. If $\Delta h = 8$ m the corresponding figures become 65 and 91 per cent, respectively. Thus, if we have to deal with non-Darcian flow with values of i_l about 10, equation (1.18) can replace equation (1.8) in most cases, except when the quotient $\Delta h/D$ is very large. In the latter case, one had better apply equation (1.8).

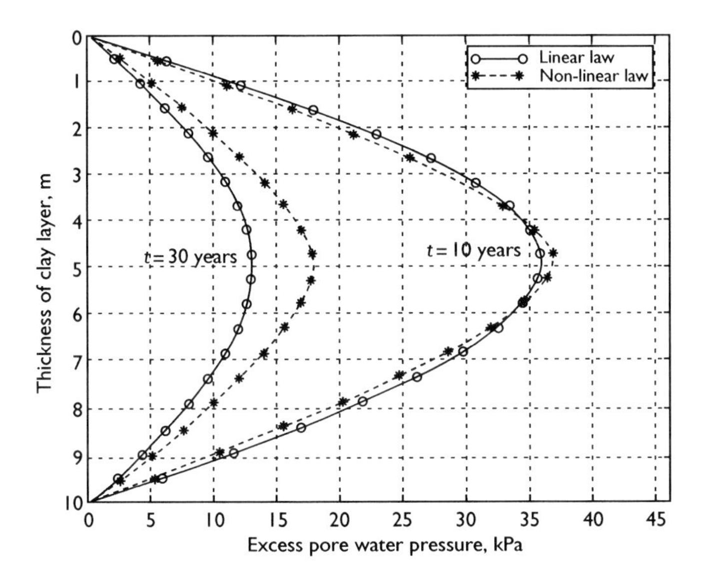

Figure 1.13 Comparison of the effect of Darcian and non-Darcian flow on one-dimensional consolidation of a 10 m thick clay layer, drained at top and bottom. Remaining excess pore water pressure in a clay layer drained at top and bottom after 10 and 30 years of loading. Initial excess pore water pressure 55 kPa at the top of the clay layer and 40 kPa at the bottom. Input parameters: $M = 200$ kPa, $\gamma_w = 10$ kN/m^3, while in Darcian flow $k = 0.025$ m/year and in non-Darcian flow $\kappa = 0.015$ m/year, $n = 1.5$ and $i_l = 4$.

It is interesting to note that Shiffman's approach with a gradual decrease of the coefficient of permeability in the course of consolidation has an effect on the consolidation rate similar to that based on non-Darcian flow. However, Shiffman's concept is difficult to apply in practical design and can also be questioned from a physical point of view. The assumption of a linear correlation between permeability coefficient and excess pore water pressure is not verified. Moreover, the coefficient of consolidation determined by oedometer tests tends to increase with increasing effective pressure, which contradicts the concept put forward by Shiffman.

1.4 Choice of parameters

1.4.1 Equivalent diameter of band drains

The first type of band drains, the so-called cardboard wick, which was invented and introduced on the market by the Swedish Geotechnical Institute, was assumed by Kjellman (1947) to have an equivalent diameter of

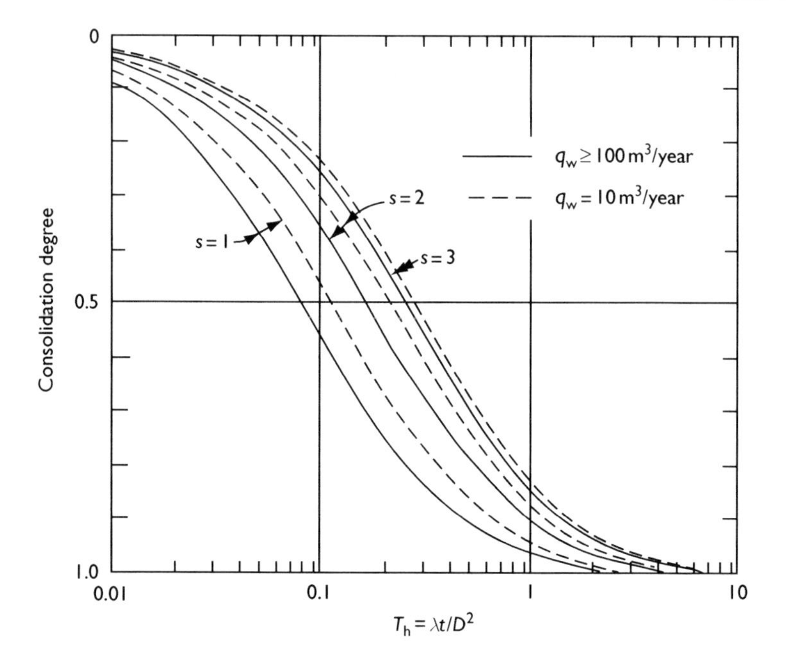

Figure 1.14 Example showing the effect of smear and well resistance according to equation (1.14) for $D/d_w = 20$, $z = l = 10$ m, $\kappa_h = 0.03$ m/year, $\kappa_h/\kappa_s = 4$ and $\Delta h/D = 2$.

50 mm. The author (Hansbo, 1979) showed that the process of consolidation for a circular drain and a band drain is very nearly the same if the circular drain is assumed to have a circumference equal to that of the band drain, (i.e. Figure 1.15).

$$d_w = \frac{2(b+t)}{\pi} \tag{1.19}$$

where b = the width of the drain and t = the thickness of the drain.

According to Atkinson and Eldred (1981), the diameter given by equation (1.19) should be reduced for the effect of convergence of flow lines towards the corners of the wick drain; they proposed:

$$d_w = \frac{(b+t)}{2} \tag{1.20}$$

The magnitude D/d_w in the latter case will increase by about 27 per cent. Which of the relation (1.19) or (1.20) is to be used can be estimated from the comparison of the degree of consolidation obtained in the two cases,

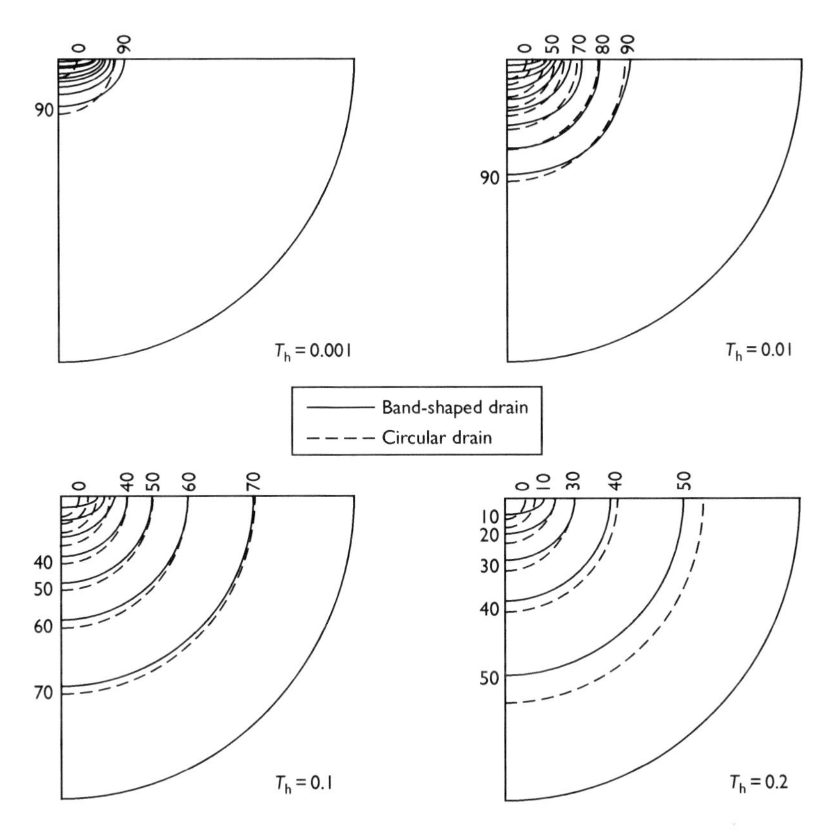

Figure 1.15 Comparison of consolidation effects (remaining excess pore water pressure Δu in percentage of initial excess pore water pressure Δu_0) caused by a band drain (100 mm in width and 4 mm in thickness) and a circular drain with the same circumference ($d = 66.2$ mm).

band drains and circular drains, shown in Figure 1.15. This comparison indicates that the equivalent diameter ought to be put somewhat larger than according to equation (1.19). However, the difference in result between the two assumptions is insignificant in comparison with the influence on the result exerted by the choice of other consolidation parameters to be applied in the design.

The equivalent diameter of various makes of band drains according to equation (1.19) is shown in Table 1.2.

1.4.2 The zone of smear

The effect on the consolidation parameters of disturbance caused by the installation of drains, expressed in terms of zone of smear, depends very

Table 1.2 Characteristics of various band drains

Drain-make	*Core width* (mm)	*Core thickness* (mm)	*Filter sleeve*	$d_{w,eq.}$ (mm)
Alidrain	100	6	Yes	67
Amerdrain	92	10	Yes	65
Bando chemical	96	2.9	Yes	63
Cardboard wick	100	3	No	65
Castleboard	94 ± 2	2.6 ± 0.5	Yes	62
Colbond CX 1000	100	5	Yes	67
Desol	95	2	No	62
Fibre drain	80–100	8–10	Yes	63
Flodrain	95	4	Yes	63
Geodrain, L-type	95.8 ± 2.0	3.4 ± 0.5	Yes	63
Geodrain, M-type	95.8 ± 2.0	4.2 ± 0.5	Yes	64
Mebradrain	100	3–4	Yes	66
OV-drain	103	2.5	No	67
PVC	100	1.6 ± 0.2	No	65
Tafnel	102	6.9	No	69

much on the method of drain installation, the size and shape of the mandrel, and the soil structure (Sing and Hattab, 1979; Bergado *et al.*, 1993). Two problems exist: to find the correct diameter value d_s of the zone of smear and to evaluate the effect of smear on the permeability.

The extent of the zone of smear and the disturbance effects depend on the type of soil and the geometrical dimensions of the installation mandrel. Remoulding will take place inside a volume equal to the volume displaced by the mandrel. Outside the remoulded zone, disturbance of the subsoil will also occur owing to distortion. The extent of the zone of distortion is a function of the stiffness of the soil. The stiffer the soil, the larger the zone of influence, and vice versa. Investigations of the extent of the zone of smear in the case of displacement-type circular drains indicate that the diameter d_s of the zone of smear can be assumed to be equal to 2 times the diameter of the drain (Holtz and Holm, 1973; Akagi, 1976; Bergado *et al.*, 1992). Recent investigations on a laboratory scale (Indraratna and Redana, 1998) indicate that the extent of the smear zone can be put equal to 3–4 times the cross-sectional area of the band drain. If the installation mandrel is non-circular, which is the normal case, the diameter d_s according to this result would yield an area corresponding to about 4 times the cross-sectional area of the mandrel. In many cases the cross-sectional area of the installation mandrel is typically about $7000\,mm^2$ in size, which yields $d_s \approx 0.19\,m$, i.e. about 3 times the equivalent diameter of the band drain.

Several authors have treated the other problem, the choice of permeability in the zone of smear. Of course, the permeability in the zone of smear will vary from a minimum nearest to the drain to a maximum at the outer border of the zone. The most conservative solution to the problem is to assume that horizontal layers in the undisturbed soil are turned vertical in the zone of smear, resulting in the quotient k_h/k_s being equal to the quotient $k_h/k_v = c_h/c_v$ (see also Bergado *et al.*, 1992).

Onoue *et al.* (1991) and Madhav *et al.* (1993) divide the zone of smear into two subzones: an inner, highly disturbed zone and an outer transition zone in which the disturbance decreases with increasing distance from the drain. Madhav *et al.* conclude that the author's solution based on axi-symmetric smear conditions and the assumption of only one smear zone is 'reasonably accurate for all practical purposes'. Chai *et al.* (1997) use a linear variation of the permeability in the zone of smear on the one hand and a bilinear variation on the other and conclude that the assumption of one single average value of permeability will underevaluate the effect of smear. Hird and Moseley (2000) conclude, on the basis of laboratory tests, that the ratio k_h/k_s for layered soil, assuming $d_s = 2d_w$, can be much larger than that mentioned in the literature but that the assumption of $k_s = k_v$ in these cases is much too severe.

1.4.3 Effect of well resistance

Because drains nowadays are frequently installed to great depths, well resistance has become a matter of increasing interest. This is understandable since well resistance in such cases can cause a serious delay in the consolidation process.

Before a certain drain-make is accepted for a job, evidence ought to be presented that the drain fulfils the requirements on discharge capacity assumed in the design. Generally, appropriate laboratory testing can give sufficient evidence of the discharge capacity to be expected under field conditions (Hansbo, 1993), but in case of a drain-make never before used in practice, full-scale field tests are recommended. Full-scale field tests also serve the purpose of showing that the drains are strong enough to resist the strains subjected to them during installation. Moreover, compression of the soil in the course of consolidation settlement entails folding of the drain, which may bring about clogging of the channel system. The latter effect is difficult to discern in a laboratory test (cf. Lawrence and Koerner, 1988).

As a general rule investigations on a laboratory scale of the discharge capacity of a drain ought to be carried out in a way that simulates field conditions as closely as possible. The drain should be placed in the soil and be subjected to a consolidation pressure similar to that expected in practice. Unfortunately it is very difficult to reproduce on a laboratory scale the deformation conditions prevailing in reality in the field (this is more or less the same as in the oedometer test). One possible method is to construct an oedometer for samples of large diameter and height in which the band drain can be installed centrally. The discharge capacity of the drain can then be investigated for different consolidation pressures by measuring the water flow through the drain at a hydraulic gradient $i = 1$. Such tests have been carried out in Italy and Japan, in the latter for investigating the effect of folding on the discharge capacity of Colbond and Geodrain. Otherwise, laboratory methods of testing different band drains have been based on small-scale tests with the drain placed centrally in a cylindrical soil sample. In these tests the influence on the discharge capacity of the lateral pressure against the drains has been investigated. The results obtained reveal a great influence on the discharge capacity of the lateral consolidation pressure (Figure 1.16).

There are several reasons why the discharge capacity of a drain may become low: siltation of the channels in the core of band drains, unsatisfactory drain-makes with too low a discharge capacity, necking of drains, etc. Back-calculated values of discharge capacity of drains under field conditions have been reported to be quite low for certain makes of band drains without filter (Hansbo, 1986; Chai *et al.*, 1996). However, most of the band drains marketed today have a high enough discharge capacity ($q_w > 150\,m^3$/year) as to become negligible in the design (cf. Hansbo, 1986, 1994). Moreover, the influence of well resistance decreases with increasing time of consolidation.

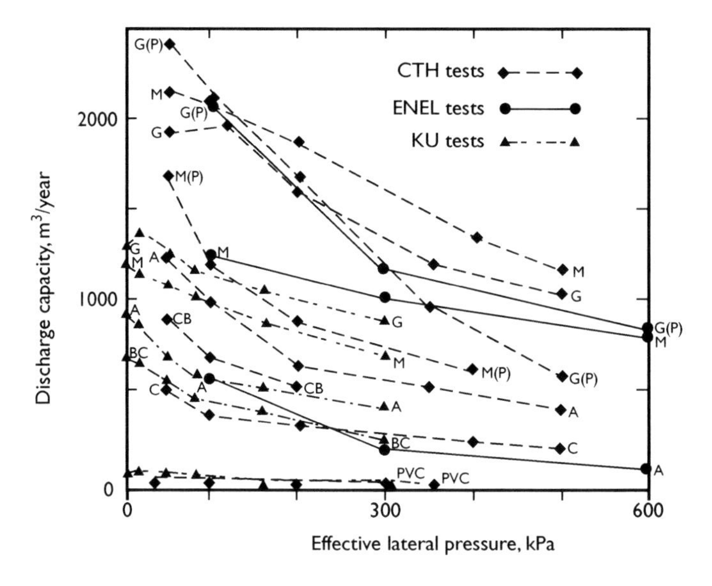

Figure 1.16 Results of discharge capacity tests for different band drains carried out on a laboratory scale. Drains enclosed in soil. In the ENEL tests (Jamiolkowski *et al.*, 1983) the drains were tested in full scale, in the CTH tests (Hansbo, 1983) and in the KU tests (Kamon, 1984) with reduced width (40 and 30 mm, respectively). Legend: A = Alidrain, BC = Bando chemical, CB = Castle board, C = Colbond, G = Geodrain, M = Mebradrain and (P) indicates filter sleeve of paper.

The new Eurocode on vertical drainage, previously mentioned, includes detailed description of how to evaluate the discharge capacity of pre-fabricated band drains under various soil conditions and soil temperature.

1.4.4 Multi-layer system

The effect of well resistance according to equations (1.6) and (1.14) is calculated on the assumption that we have to deal with homogeneous soil conditions (k_h and κ_h assumed constant).

If the soil consists of layers with different characteristics, this can be taken into account in a simple way as suggested by Onoue (1988). The consolidation process is calculated on the assumption that the whole soil profile is homogeneous and has the consolidation properties of each of the respective layers. The distribution of excess pore pressure in the respective layer is extracted and plotted as shown in Figure 1.17.

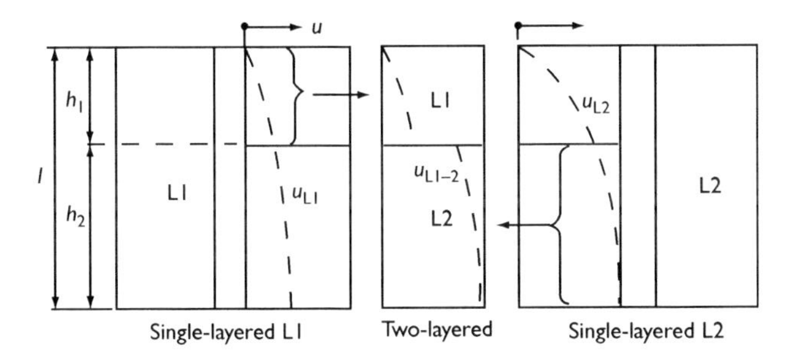

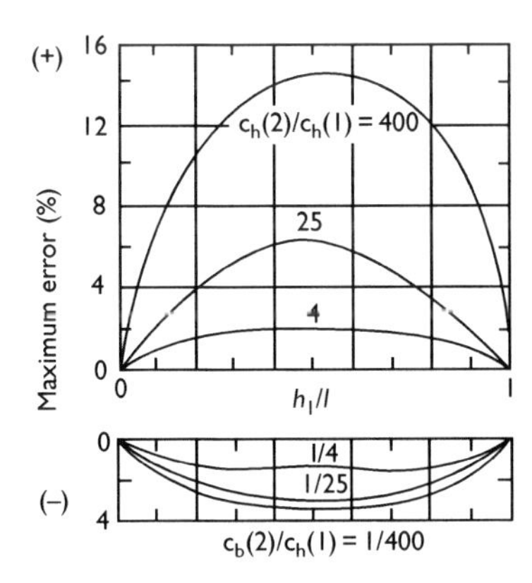

Figure 1.17 Consolidation of anisotropic soil by vertical drains with well resistance exemplified for a two-layered soil deposit with different characteristics. Simplified solution (top) and consequential error according to Onoue (1988).

1.4.5 Filter requirements

Most of the band drains are made up of a central core with longitudinal channels surrounded by a filter of synthetic material. The first requirement on these materials is that they must be strong enough to resist the tension and the wear and tear which takes place during drain installation.

Much attention has been devoted to filter criteria. Among the problems mentioned the risk of siltation and the strength of the filters have to be taken into account. The risk of blinding owing to too low a permeability of the filter has almost negligible effect on the consolidation behaviour. The filter and the low-permeable cake of soil particles, which may be formed outside the filter and cause so-called blinding, will have a fairly small thickness. The consequence of this blinding is easily recognized if the filter is considered as a zone of smear. Assuming, for example, that the filter/filter cake has a thickness of as much as 2 mm (corresponding to $d_s = d_w + 0.004$ m) and that its permeability becomes only 20 per cent of the permeability of the surrounding soil ($k_s/k_h = 0.2$), the average degree of consolidation, using band drains with an equivalent diameter $d_w = 0.066$ m, will differ from the ideal case by a maximum of only 2–3 per cent, a negligible difference in result.

The filter material has also been considered important. When the first modern pre-fabricated band drain, the Geodrain, was introduced on the market, the filter was made of specially prepared paper material. Although the effectiveness of these drains was demonstrated by the results of a large number of drain installations, the use of paper as filter material was severely questioned. The main reason for questioning the use of paper was the risk of filter deterioration caused by fungi or bacteria. This risk has been proven by full-scale experiments to be overstated (Figure 1.18). Moreover, there are cases where clogging of the drains would be desirable after full consolidation under the design load has been attained, which generally requires a pre-consolidation time of about one year.

1.4.6 Correlation between λ and c_h

The ratio of λ to c_h will depend on the hydraulic gradient prevailing in the horizontal direction during the consolidation process. This value can be estimated on the basis of the expression for i, given by equation (1.15). Since the parameters M and γ_w are independent of the flow conditions, we have $c_h/\lambda = k_h/\kappa_h$. Equalizing the areas created below the flow and gradient curves in the two cases, non-Darcian and Darcian flow, we find the correlation:

$$\frac{\lambda}{c_h} = \frac{n+1}{2i^{n-1}} \quad \text{when } i \leq i_l \tag{1.21}$$

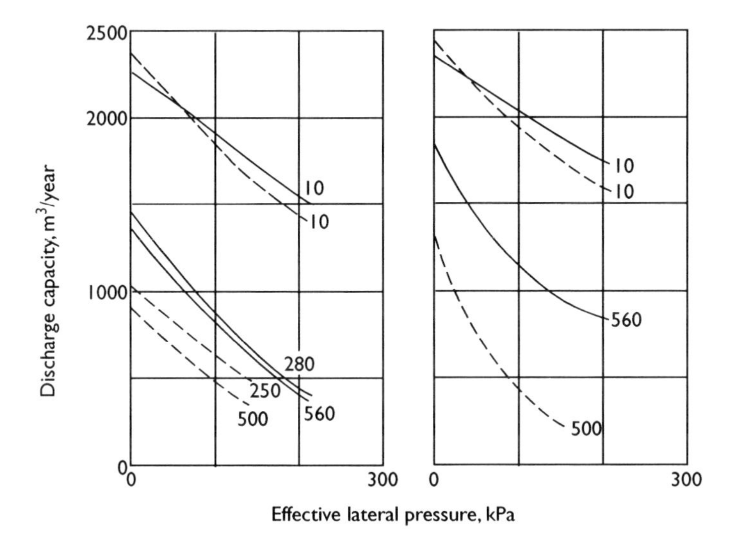

Figure 1.18 Influence on discharge capacity of filter deterioration (Koda *et al.*, 1986). Tests on band drains (type Geodrain) with filter sleeves of paper (broken lines) and synthetic material (full lines) which were pulled out of peat (to the left) and gyttja (to the right) after different lengths of time after installation (number of days given in figure).

and

$$\frac{\lambda}{c_\mathrm{h}} \approx \frac{i^2}{2}\left[\frac{i_\mathrm{l}^{n+1}}{n+1} + n i_\mathrm{l}^{n-1}(i - i_\mathrm{l})\left(\frac{i - i_\mathrm{l}}{2} + \frac{i_\mathrm{l}}{n}\right)\right]^{-1} \quad \text{when } i \geq i_\mathrm{l} \qquad (1.22)$$

Assuming, for example, that the maximum gradients reached during the consolidation process are respectively 2, 5, 15, 25 and 75 and that the exponent $n = 1.5$ and the limiting gradient $i_\mathrm{l} = 8$, we find in due order $\lambda/c_\mathrm{h} = \kappa_\mathrm{h}/k_\mathrm{h} \approx 0.88$, 0.56, 0.34, 0.29 and 0.25. Thus, the higher the value of Δh and the smaller the drain spacing, the lower the ratio of λ to c_h.

1.5 Settlement analysis

Among the main problems to resolve remain the pre-determination of the primary consolidation settlement and the influence of secondary consolidation. This pre-determination is generally based on the results of oedometer tests, which may give quite misleading information about the deformation properties of the soil owing to sample disturbance. Empirical correlations are also utilized as a means of establishing the deformation characteristics.

A misinterpretation of the pre-consolidation pressure or of the consolidation parameters (compression modulus, coefficient of consolidation or permeability) may give a completely wrong picture of the final settlement and the consolidation rate to be expected.

1.5.1 Determination of the pre-consolidation pressure

The pre-consolidation pressure σ'_c is usually determined according to the well-known Casagrande procedure from the shape of the oedometer curve presented in the $\log \sigma'/\varepsilon$ (or $\log \sigma'/e$) diagram. This method, however, may give quite a false picture of the pre-consolidation pressure in the case of disturbed soil samples. Therefore, it is always necessary to check the shape of the oedometer curve in linear scales σ'/ε (or σ'/e). As demonstrated in Figure 1.19, the use of log σ'/ε diagrams may give the impression of the existence of a pre-consolidation pressure where it cannot be noticed because of sample disturbance.

The pre-consolidation pressure will be affected by the disturbance caused by drain installations. Lightly overconsolidated clay may turn into normally consolidated clay after the drains have been installed. This has to be taken

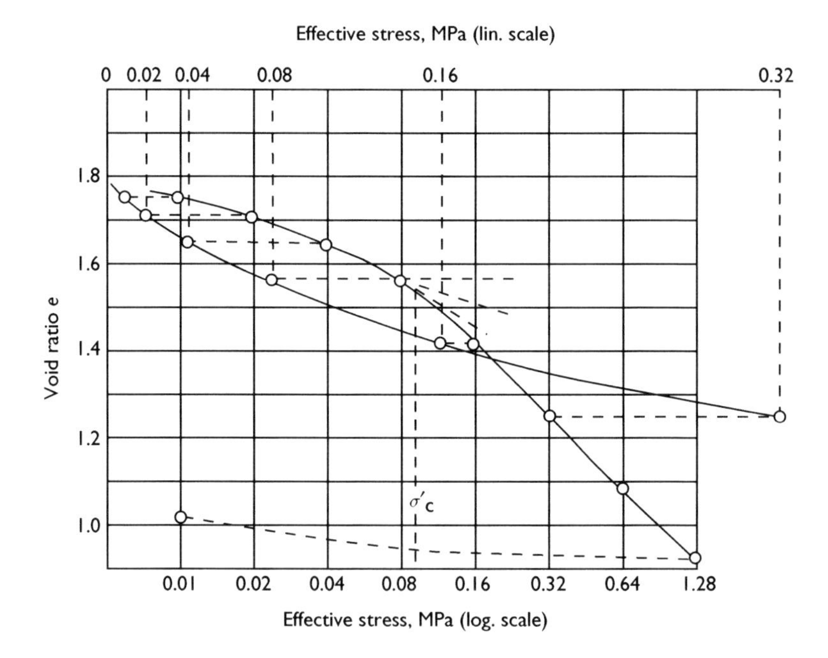

Figure 1.19 Result of oedometer test presented by a consultant in a semi-logarithmic diagram. The pre-consolidation pressure was determined according to Casagrande's well-known method. The linear presentation shows no sign of a pre-consolidation pressure.

into account when calculating the settlement to be expected under an embankment or an area provided with vertical drains. Thus, a certain underestimation of the pre-consolidation pressure in this case may be justified. On the other hand, one has to take care not to mislead the building proprietor in installing vertical drains where drains are not required.

1.5.2 Primary settlement

In order to cope with the influence on the course of settlement of a time-consuming stepwise placement of the load, each break in the rate of loading can be analysed either on the basis of a direct use of the oedometer curves or by the use of oedometer parameters. In the analysis the traditional approach is to use the virgin compression ratio $CR = C_c/(1+e_0)$, which represents the relative compression $\varepsilon = \Delta h/h_0$ achieved along the virgin oedometer curve ($C_c = \Delta e/\log[(\sigma'_0 + \Delta\sigma')/\sigma'_0]$, e_0 = initial void ratio and h_0 = the initial height of the oedometer sample). The primary consolidation settlement s is then obtained by the relation,

$$s = \sum_{j=1}^{m} \Delta h_j \left[(RR)_j \log\left(\frac{\sigma'_{cj}}{\sigma'_{0j}}\right) + (CR)_j \log\left(\frac{\sigma'_j - \sigma'_{cj}}{\sigma'_{cj}}\right)\right] \tag{1.23}$$

where Δh_j is the thickness of layer j, $(RR)_j$ is the recompression ratio of layer j (overconsolidated state), σ'_{0j} is the effective overburden pressure of layer j, $(CR)_j$ is the virgin compression ratio of layer j, σ'_{cj} is the pre-consolidation pressure of layer j, σ'_j is the effective vertical stress of layer j at the end of the primary consolidation period and m is the number of layers.

A more modern approach to the settlement analysis based on oedometer tests is based on the variation of the oedometer modulus shown in Figure 1.20.

In this case the settlement s caused by loading is obtained by the relation,

$$s = \sum_{j=1}^{m} \Delta h_j \left[\frac{\sigma'_{cj} - \sigma'_{0j}}{M_{0j}} + \frac{\sigma'_{Lj} - \sigma'_{cj}}{M_{Lj}} + \frac{1}{M'_j} \ln\left(1 + \frac{M'_j[\sigma'_j - (\sigma'_{Lj} - \sigma'_{cj}) - s(\gamma - \gamma')_j]}{M_{Lj}}\right)\right] \tag{1.24}$$

where σ'_{0j}, σ'_{cj}, σ'_j and m are as above, $M_j = \Delta\sigma'_j/\Delta\varepsilon_j$ is the compression modulus of layer j determined by oedometer test (see Figure 1.20), $M'_j = \Delta M_j/\Delta\sigma'_j$, $\sigma'_j \geq \sigma'_{Lj} - \sigma'_{0j}$, $\sigma'_{Lj} - \sigma'_{cj}$ is the stress interval above the pre-consolidation pressure with constant modulus M_{Lj}, and $s(\gamma - \gamma')_j$

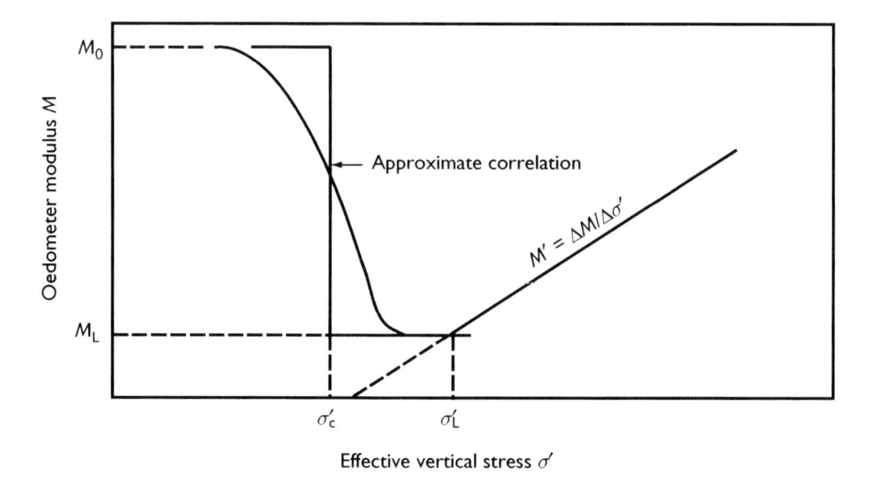

Figure 1.20 Oedometer parameters used in settlement analysis.

represents the reduction in total stress increase of layer j caused by the fact that part of the load becomes submerged in the course of settlement.

In the analysis one has to take into account the influence on the compression characteristics of the disturbance taking place during drain installations. These disturbance effects are not compensated for by the disturbance due to sampling. Thus, sample disturbance results in a lower value of the compression index, whereas disturbance of the soil in nature results in a higher value of the compression index.

Owing to the uncertainties involved in the calculation of the total primary consolidation settlement, test areas are recommended wherever possible. In this case the primary consolidation settlement can be determined according to Asaoka (1978). Asaoka's method is based on the following procedure. The settlement observed at different equal time intervals Δt is plotted in a diagram with s_{i-1} as ordinate and s_i as abscissa where indices $i-1$ and i refer to times $t-\Delta t$ and t. The primary consolidation settlement is obtained when $s_{i-1} \rightarrow s_i$.

1.5.3 Secondary settlement

Secondary consolidation refers to the change in void ratio (relative compression) taking place with time at a given effective vertical stress (cf. equation (1.10)). Thus, secondary consolidation has to be taken into account during the whole of the consolidation process. In the case of vertical drainage, however, the primary consolidation period is generally short enough for the influence of secondary consolidation to be ignored. Secondary consolidation settlement can therefore be analysed in the traditional way, i.e. starting at the end of the primary consolidation period.

The secondary compression during time Δt can be obtained by the relation:

$$\varepsilon = \alpha_s \log\left[\frac{(t_p + \Delta t)}{t_p}\right] \quad (1.25)$$

where t_p is the time of primary consolidation.

The secondary compression ratio $\alpha_s = d\varepsilon/d(\log t)$ represents the inclination of the rectilinear tail of the oedometer curve in the $\log(t/\varepsilon)$ diagram (Buisman, 1936). This requires that the oedometer test be carried out by means of a stepwise load increase. Nowadays, however, oedometer tests are often performed as CRS (constant rate of strain) tests which makes impossible a direct judgement of the secondary compression indexes. Therefore, one often has to rely on half-empirical correlations. According to Mesri and Godlewski (1977) the most typical values of C_α/C_c, where $C_\alpha = \alpha_s(1 + e_0)$ represents the secondary compression index and C_c the primary compression index, are 0.04 ± 0.01 for inorganic soft clays and 0.05 ± 0.1 for organic soft clays. The α_s value increases with increasing water content.

1.6 Monitoring of vertical drain projects

Monitoring of vertical drain projects is more or less a must since the consolidation characteristics determined by oedometer tests may be misleading. An early follow-up of the results obtained will form the basis for a correct estimate of the result to be expected, so-called active design.

The monitoring systems utilized for the control of vertical drain projects usually consist of vertical settlement meters of various types and piezometers placed at different depths in the soil. In the case of pilot tests the size of the test area is often limited in relation to the thickness of the soil layer subjected to consolidation. Therefore, in such cases the influence on the vertical settlement of horizontal displacements has to be taken into account. This purpose is usually achieved by the installation of inclinometers along the border of the test area.

Considering the derivation of the consolidation theory, the follow-up of the consolidation process nearest at hand is to check the course of excess pore pressure dissipation. However, experience shows that the interpretation of the consolidation process on the basis of pore pressure measurements may be quite intricate. The main problem in the case of vertically drained areas consists in uncertainty about the exact position of the filter tip of the piezometer in relation to the surrounding drains. Therefore, the observations can give a misleading conception of the average excess pore water pressure dissipation. To find the average degree of consolidation on the basis of pore pressure observations, the piezometer tip should be placed about halfway between the outer border of the drained cylinder and the drain (cf. Figure 1.9). However, in practice the aim is generally to have it placed at the outer border of the drained cylinder (halfway between the drains). Other difficulties arise from the fact that the piezometer tip, owing to frictional forces against the piezometer tube by settlement of overlying soil layers, may be penetrating the underlying soil, thereby creating additional excess pore pressure. Phenomena such as influence of pore gas, erroneous pore pressure readings, collapse of soil structure, structural viscosity, secondary consolidation, and the fact that the ground water level may not revert to its original position, may also contribute to discrepancies observed between the consolidation degree based on settlement and that based on pore pressure observations. One must also bear in mind that the drain installation in itself causes excess pore water pressure, which may extend even far outside the drained area (cf. Hansbo, 1960).

The aim of pre-loading in combination with vertical drainage is usually to eliminate unacceptable settlement under future loading conditions. The preconsolidation pressure in the soil has to be increased up to, or preferably above, the effective stress level induced by the future load. Settlement observations of the soil surface may be strongly influenced by vertical consolidation U_v and thus lead to the impression that the acceleration of the consolidation process caused by the drains is faster than in reality. In active design, this can be checked theoretically by inserting the values of c_v and c_h, respectively λ, found by trial and error to yield acceptable agreement with the course of surface settlement. Then the degree of consolidation obtained by the aid of the drains can be checked by inserting $c_v = 0$ and the value of c_h or λ found. However, if there are layers with more unfavourable consolidation characteristics than on the whole, these layers will be decisive.

1.7 Case records

It is interesting to check by case records whether the theory based on non-Darcian flow gives results in better agreement with real behaviour than the theory based on the validity of Darcy's law. As already mentioned, the author in the study of the full-scale tests at Skå-Edeby (Hansbo, 1960) found that the best agreement between theory and practice was obtained by assuming non-Darcian flow with the exponent $n = 1.5$; i.e. according to the flow law $v = \kappa i\sqrt{i}$. A better agreement has been found in several cases between case records and theoretical analysis, based on non-Darcian flow with the exponent $n = 1.5$, than theoretical analysis based on Darcian flow (Robertson *et al.*, 1988; Hansbo, 1994, 1997a,b). In reality

$v = 1.5\kappa\sqrt{i}(i - i_l/3)$ when $i \geq i_l$ but in vertical drain projects the gradient i is largely inferior to the value of i_l during the whole consolidation process (Hansbo, 2001).

In a case where the monitoring system is based on settlement observations of the soil surface (which is most common), the settlement observed vs total settlement refers to the average degree of consolidation. The settlement due to one-dimensional vertical consolidation can be obtained utilizing the diagrams of average one-dimensional consolidation vs time factor T_v found in most textbooks on soil mechanics. If, on the other hand, the stress increase varies non-linearly with depth below the ground surface, or if the consolidation characteristics vary with depth, then the effect of one-dimensional vertical consolidation can be calculated by finite difference methods as suggested by Helenelund (see Hansbo, 1994) or by means of finite element methods. As shown in Figure 1.13 the difference in result between Darcian and non-Darcian flow will be insignificant in the beginning of the consolidation process and, therefore, Terzaghi's solution can be applied when judging the influence of one-dimensional vertical outflow of water. This concept forms the basis of equation (1.18). Alternatively, the contribution to the settlement obtained by one-dimensional consolidation can be studied in a dummy area without drains and with similar loading conditions. However, because of disturbance effects caused by drain installation, the consolidation characteristics of the soil in an area with drains will be different from those in an area without drains.

1.7.1 The Skå-Edeby test field

The test field arranged at Skå-Edeby, situated some 25 km west of Stockholm, is one of the oldest and most well-documented test fields in the world. It was established by the Swedish Government in 1957 for the purpose of examining the effectiveness of vertical sand drains in a then planned soil improvement project for a new International Airport (for full details about the test field, see Hansbo, 1960).

The soil conditions in the test field can be summarized as follows. Below a 1.5 m thick dry crust, the soil consists of normally consolidated, high-plasticity clay to a depth of 9–15 m (average about 12 m). From the results of oedometer tests the following consolidation characteristics were found: coefficients of consolidation $c_v = 0.17\ m^2/year$ (standard deviation $0.03\ m^2/year$) and $c_h = 0.7\ m^2/year$ (only one test).

Comparisons between theoretical and measured consolidation rates earlier presented by the author (Hansbo, 1997b), for test areas I (sand drains; 0.9, 1.5 and 2.2 m spacing), II and III (sand drains; 1.5 m spacing; different loading conditions) at Skå-Edeby, have shown that the consolidation theory based on non-Darcian flow with the exponent n equal to 1.5 gives better agreement with observations than the consolidation theory based on Darcian flow.

The results obtained in the *undrained* Test Area IV are of interest in the study of the combined effect of one-dimensional consolidation and the effect of vertical drainage. A follow-up of the course of settlement in Test Area IV according to Asaoka (1978) yields a total primary settlement of 1.12 m while the estimated primary settlement was 1.3 m (Hansbo, 1960).

The excess pore water pressure distribution in the undrained Test Area IV is still under continuous observation (the latest observation was made in 1982, i.e. 25 years after loading). The value of c_v determined by oedometer tests is considerably lower than the value found by trial and error. Thus, a good correlation between theoretical and observed course of settlement is obtained by assuming an average value of $c_v = 0.62\,m^2$/year, i.e. more than 3 times higher than the value determined by oedometer tests. In Figure 1.21 the excess pore water pressure distribution after 14 and 25 years of loading, obtained by assuming $c_v = 0.55\,m^2$/year (Darcian flow) and $\lambda_v = 0.35\,m^2$/year (non-Darcian flow), is shown.

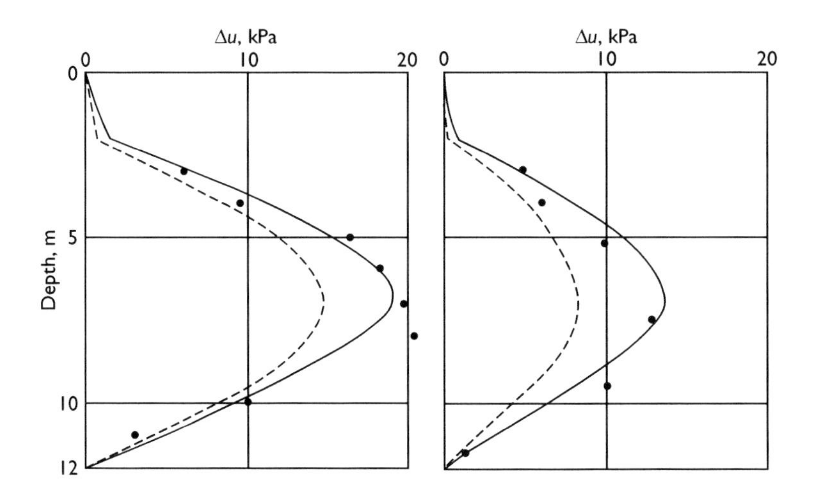

Figure 1.21 Comparison between calculated and observed pore pressure distribution in Test Area IV (undrained) at Skå-Edeby, Sweden, 14 and 25 years after loading with 1.5 m of gravel (load $= 27\,kN/m^2$). Broken line designates Darcian flow, unbroken line non-Darcian flow. Assumed average consolidation coefficient below the dry crust (from 2 to 12 m depth): in Darcian flow $c_v = 0.54\,m^2$/year (based on settlement observations), in non-Darcian flow $\lambda_v = 0.36\,m^2$/year with $n = 1.5$ and $i_l = 5$. In Test Area IV the initial excess pore water pressure was observed equal to 24 kPa (corresponding to Skempton's pore pressure coefficients $B = 1$ and $A = 0.77$).

These values of the coefficient of consolidation yield theoretically an average excess pore water pressure in the 5 m thick layer from 2.5 to 7.5 m depth of about 20, 19, 17 and 16 kPa (equal results in both Darcian and non-Darcian flow) after respectively 2, 3½, 5 and 6 years of consolidation, corresponding to respectively $\overline{U}_v \approx 18$ per cent, 22 per cent, 30 per cent and 34 per cent. The difference in result after 14 and 25 years of loading between Darcian and non-Darcian flow supports the assumptions of non-Darcian flow.

In practice, the influence of one-dimensional vertical consolidation exerted on the consolidation process at normal drain spacing and thickness of the drained layer is unimportant for the evaluation of the drainage project. It may have an important influence, however, if the drain spacing relatively speaking is large as compared to the thickness of the drained layer.

In Test Area V (30 m diameter), band drains, type Geodrain with paper filter, were installed with 0.9 m spacing. The initial load, 27 kN/m^2, was doubled after 3½ years of consolidation. In a case like this, the following procedure of settlement analysis is followed. The degree of consolidation $\overline{U}_1$ and the settlement s_1 achieved under load q_1 at time t_1 when the additional load q_2 is being placed is calculated first. The part of the load q_1 that is still producing primary consolidation settlement, i.e. $\Delta q = q_1(1 - \overline{U}_1)$, is added to q_2 and the primary settlement process to be added after time t_1 under load $\Delta q + q_2$ is added to the settlement s_1. As shown in Table 1.2, the equivalent diameter of Geodrain can be put equal to $d_w = 0.066$ m. Assuming a zone of smear $d_s = 0.19$ m and $k_h/k_s = \kappa_h/\kappa_s = 4$ the best fit between observations and theory is obtained by applying the values $c_h = 0.45$ m^2/year and $\lambda = 0.23$ m^2/year (Figure 1.22).

As can be seen, the values of the coefficients of consolidation in this case are smaller than the values arrived at in the undrained area, in spite of the fact that the coefficient of consolidation to be expected is normally higher in horizontal pore water flow than in vertical pore water flow due to the structural features of the soil (e.g. due to the existence of sand and silt layers).

This obviously depends on the fact that the drain installation causes disturbance effects leading to a reduction of the coefficient of consolidation. The disturbance is evidenced by the fact that the drain installation at Skå-Edeby in itself induced excess pore water pressure of 30–50 kPa.

1.7.2 The Bangkok test field

In connection with the planning of a new international airfield in Bangkok, Thailand, three test areas were arranged in order to form a basis for the design of soil improvement by pre-loading in combination with vertical drains. The results of the settlement observations in two of these test areas, TS 1 and TS 3 (Hansbo, 1997b), showed a better agreement with equation (1.18)

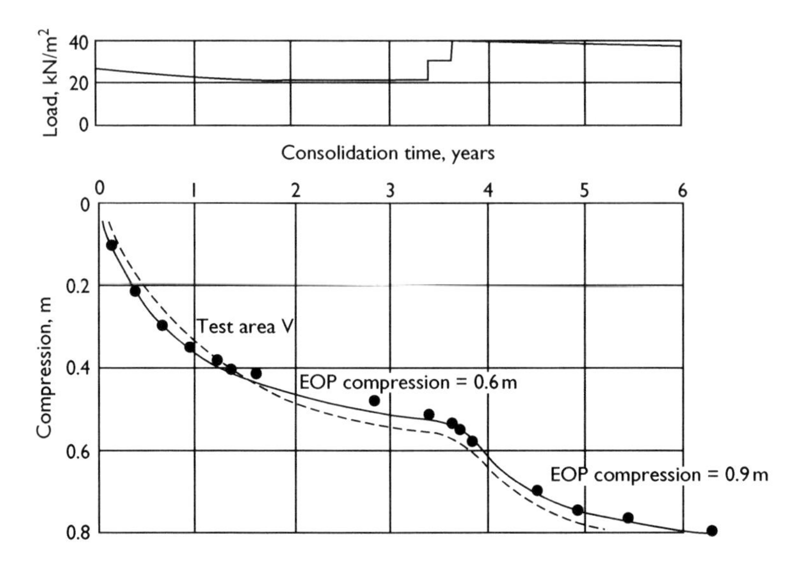

Figure 1.22 Settlement of ground surface obtained at Skå-Edeby in Test Area V: 0.9 m drain spacing ($D = 0.95$ m). Broken line: Darcian flow with $c_h = 0.45\ m^2$/year; unbroken line: non-Darcian flow with $\lambda = 0.23\ m^2$/year.

than with equation (1.8). In this paper, the results obtained in test area TS 3 will be examined in detail.

The crest width of TS 3 is 14.8 m (square) and the bottom width 40 m. It is provided with an approximately 10 m wide loading berm, 1.5 m thick. The fill placed on the area amounts to a maximum of about 4.2 m, corresponding to a load of about 80 kN/m^2. Owing to submergence of the fill during the course of settlement, the load, in kN/m^2, will be reduced successively by about 8s, where s = settlement, in metre, of the soil surface.

The drains, type Mebradrain, were installed to a depth of 12 m in a square pattern with a spacing of 1.0 m which yields $D = 1.13$ m. The equivalent drain diameter determined according to equation (1.19) becomes $d_w = 0.066$ m. The equivalent diameter of the mandrel becomes $d_m = 0.10$ m. The smear zone is estimated at $d_s = 0.20$ m. The permeability ratios k_h/k_s and κ_h/κ_s are assumed equal to the ratio c_h/c_v.

The consolidation characteristics of the clay deposit, determined by oedometer tests, can be summarized as follows (DMJM International, 1996; Kingdom of Thailand, Airports Authority of Thailand, 1996): average coefficient above the pre-consolidation pressure $c_v = 1.06\ m^2$/year (standard deviation = 0.061 m^2/year), $c_h = 1.37\ m^2$/year (standard deviation = 0.050 m^2/year). This yields $k_h/k_s = \kappa_h/\kappa_s = 1.3$ (in a paper previously published by the author (Hansbo, 1997b) this ratio was assumed equal to 2).

The virgin compression ratio CR varies from 0.3 to 0.55 (average 0.43; standard deviation 0.1) and the average recompression ratio $RR = 0.03$ (standard deviation = 0.007). The clay penetrated by the vertical drains is slightly overconsolidated with a pre-consolidation pressure about 15–50 kPa higher than the effective overburden pressure. The clay below the tip of the drains is heavily overconsolidated. In this case the influence on the consolidation process of one-dimensional vertical consolidation will be ignored owing to difficulties in assessing the drainage conditions.

The monitoring system consisted of vertical settlement meters placed on the soil surface at different depths and inclinometers to study the horizontal displacements. Unfortunately, the results of the settlement observations at various depths are contradictory and, therefore, only the surface settlement observations can be trusted. The contribution to the vertical settlement of horizontal deformations is analysed on the basis of the inclinometers placed at 7.8 m from the centre of the test area. Denoting the area created by horizontal deformation vs depth by A, the vertical settlement s owing to the horizontal deformations is calculated as the mean of the two values $4A/14.8$ and $\pi A/14.8$. The total settlement observed, including the settlement caused by horizontal deformations, and the thus corrected settlement curve, representing merely the effect of consolidation, are shown in Figure 1.23.

A follow-up of the course of consolidation settlement according to Asaoka's method, based on settlement observations at equal time intervals (cf. Section 1.5.2), results in the following relation $s_i = 0.2625 + 0.8195 s_{i-1}$ which yields the primary settlement $s_p = 1.45$ m.

The settlement analysis based on the λ method is carried out in four successive steps: load-step 1 with load $\Delta q_1 = 20\,\text{kN/m}^2$; load-step 2 with load $\Delta q_2 = 30\,\text{kN/m}^2$; load-step 3 with $\Delta q_3 = 10\,\text{kN/m}^2$ and load-step 4 with $\Delta q_4 = 20\,\text{kN/m}^2$. The primary consolidation settlement caused by a load intensity of $80\,\text{kN/m}^2$, determined on the basis of the compression characteristics, becomes equal to 1.4 m. By slightly modifying the compression characteristics to yield a final primary consolidation settlement of 1.45 m, we find $\Delta s_1 = 0.15$ m, $\Delta s_2 = 0.6$ m, $\Delta s_3 = 0.2$ m and $\Delta s_4 = 0.5$ m (in total 1.45 m).

The analysis of the consolidation process according to equation (1.18) has to be carried out in the following way. The degree of consolidation $\overline{U}_1$, inserting $\Delta h_1 = \Delta q_1/\gamma_w$, determines the course of settlement in the first load-step. When calculating the course of settlement in the second load-step we have to apply the value $\Delta h_2 = (1 - \overline{U}_1)\Delta q_1/\gamma_w + \Delta q_2/\gamma_w$ and the settlement at the end of the load-step is obtained from $\Delta s = \Delta s_1 \overline{U}_1 + [\Delta s_1(1 - \overline{U}_1) + \Delta s_2]\overline{U}_2$, and so on. Now, 50 days after the start of loading (consolidation time $t = 50 - 15 = 35$ days; $\Delta h = 2$ m; $\lambda = 0.37\,\text{m}^2/\text{year}$) we find $\overline{U} = 0.21$ which yields $s = 0.03$ m. In load-step 2 the load $\Delta q_2 = 30\,\text{kN/m}^2$ has to be increased by $0.79 \times 20 = 16\,\text{kN/m}^2$ corresponding to $\Delta h_{2,\text{corr}} = 4.6$ m and $\Delta s_{2,\text{corr}} = 0.6 + 0.12 = 0.72$ m. After 75 days when load-step 2 is

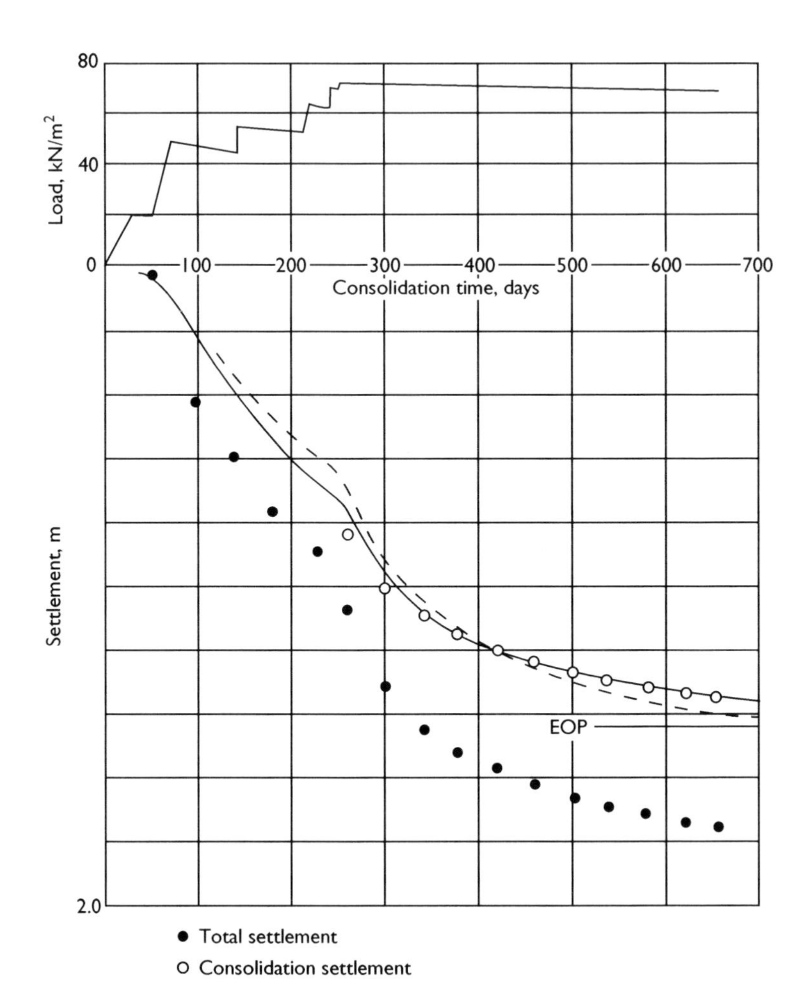

Figure 1.23 Settlement of ground surface in the Bangkok test field, Thailand. Test area TS 3: 1.0 m drain spacing ($D = 1.13$ m). Settlement corrected with regard to immediate and long-term horizontal displacements. EOP = end of primary consolidation settlement estimated according to Asaoka's method. Full lines: analytical results according to equation (3.18). Broken lines: analytical results according to equation (3.8).

completed we find $[t = (75 - 50)/2]\overline{U} = 0.12$ which yields $\Delta s = 0.09$ m and $s = 0.12$ m. After 140 days when load-step 3 is being applied we have ($t = 12.5 + 65 = 77.5$ days) $\overline{U} = 0.50$ from which $\Delta s = 0.36$ m and $s = 0.39$ m. This yields $\Delta h_{3,\mathrm{corr}} = 3.3$ m and $\Delta s_{3,\mathrm{corr}} = 0.2 + 0.36 = 0.56$ m. After 220 days when load-step 4 is being applied we find ($t = 80$ days) $\overline{U} = 0.46$ from which $\Delta s = 0.26$ m and $s = 0.26 + 0.39 = 0.65$ m. This yields

$\Delta h_{4,\mathrm{corr}} = 1.8 + 2.0 = 3.8$ m and $\Delta s_{4,\mathrm{corr}} = 0.5 + 0.3 = 0.8$ m. After 250 days when load-step 4 is completed we have ($t = 15$ days) $\overline{U} = 0.13$ which yields $\Delta s = 0.10$ m and $s = 0.75$ m. We have $\overline{U} = 0.59$, $\overline{U} = 0.74$ and $\overline{U} = 0.89$, 100, 200 and 400 days later from which $\Delta s = 0.47$ ($s = 0.47 + 0.65 = 1.12$ m), 0.67 ($s = 1.24$ m) and 0.71 m ($s = 1.36$ m), respectively.

The theoretical course of settlement determined in the conventional way is less complicated in that the total consolidation curve can be determined for each load-step separately and added to each other. Assuming $c_h = 0.93\,\mathrm{m^2/year}$ (in the paper previously mentioned (Hansbo, 1997b) the coefficient c_h was assumed to be equal to $1.2\,\mathrm{m^2/year}$ owing to the fact that the ratio k_h/k_s was put equal to 2 instead of 1.3 now applied) we find, to give an example, 400 days after the start of the loading process ($t_1 = 385$ days, $\overline{U} = 0.92$; $t_2 = 340$ days, $\overline{U} = 0.89$; $t_3 = 260$ days, $\overline{U} = 0.82$; $t_4 = 170$ days, $\overline{U} = 0.67$) the settlement $s = (0.92)(0.15) + (0.89)(0.6) + (0.82)(0.2) + (0.67)(0.5) = 1.17$ m.

The results obtained by the two methods of analysis are shown in Figure 1.23. Inserting the maximum value $\Delta h = 4.6$ m into equation (1.15), the values $c_h = 0.93\,\mathrm{m^2/year}$ and $\lambda = 0.37\,\mathrm{m^2/year}$, correspond to $i_l = 3.5$ and $i_{max} = 22.5$.

1.7.3 The Vagnhärad vacuum test

Torstensson (1984) reported an interesting full-scale test in which consolidation of the clay was achieved by the vacuum method. The subsoil at the test site consists of post-glacial clay to a depth of 3 m and below this of varved glacial clay to a depth of 9 m underlain by silt. The clay is slightly overconsolidated with a pre-consolidation pressure about 5–20 kPa higher than the effective overburden pressure. The coefficient of consolidation c_h was found equal to $0.95\,\mathrm{m^2/year}$ and the average virgin compression ratio CR equal to 0.7 (max. 1.0).

The vacuum area, 12 m square, was first covered by a sand/gravel layer 0.2 m in thickness, and then by a Baracuda membrane which was buried to 1.5 m depth along the border of the test area and sealed by means of a mixture of bentonite and silt. Mebradrains ($d_w = 0.066$ m) were installed in a square pattern with 1.0 m spacing to a depth of 10 m. The equivalent diameter of the mandrel $d_m = 0.096$ m. The average underpressure achieved by the vacuum pump was 85 kPa. After 67 days the vacuum process was stopped and then resumed after 6 months of rest. From the shape of the settlement curve (Figure 1.24) Asaoka's method yields the correlation $s_i = 0.0756 + 0.9075 s_{i-1}$ from which $s_p = 0.82$ m. This value is low with regard to the loading conditions and the compression characteristics. The main reason seems to be that the applied vacuum effect is not fully achieved in the drains. Thus, the primary settlement 0.82 m corresponds to a vacuum

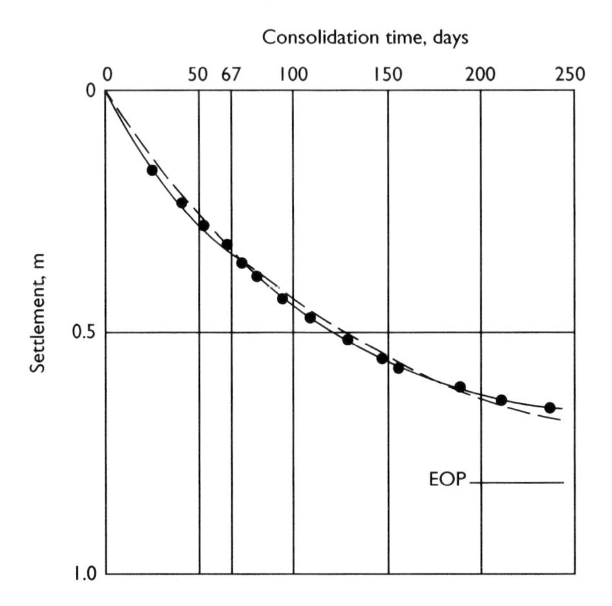

Figure 1.24 Results of settlement observations at Vagnhärad, Sweden. Consolidation by vacuum. 1.0 m drain spacing ($D = 1.13$ m). EOP = end of primary consolidation settlement estimated according to Asaoka's method. Full lines: analytical results according to non-Darcian flow, equation (1.18). Broken lines: analytical results according to Darcian flow, equation (1.8).

effect of about 35 kPa ($\Delta h = 3.5$ m). Another reason may be that the test area is too small as compared to the thickness of the clay layer.

The theoretical settlement curve in this case has to be determined in two steps, the first one up to a loading time of 67 days leading to a settlement $s_1 = \overline{U}_{h1} s_p$. In the next load step, starting again from the time of resumption of the application of vacuum, the remaining primary settlement is obtained from the relation $\Delta s = \overline{U}_h (s_p - s_1)$, i.e. the settlement $s_t = s_1 + \overline{U}_{ht}(s_p - s_1)$ where t starts from the time of resumption of the application of vacuum. In this case, where vacuum is applied to create underpressure in the drains, the effect of vertical one-dimensional consolidation is eliminated.

Inserting the values $D = 1.13$ m, $d_w = 0.066$ m, $d_s = 0.19$ m, $k_h/k_s = \kappa_h/\kappa_s = 4$ and $\Delta h = 3.5$ m into equations (1.8) and (1.18), the best agreement between theory and observations is found for $\lambda = 0.95$ m^2/year and $c_h = 2.4$ m^2/year (Figure 1.24). Even in this case the λ theory agrees better with observations than the classical theory.

Inserting the maximum value $\Delta h = 3.5$ m into equation (1.15) yields $i_{max} = 7.3$, and inserting the values $c_h = 2.4$ m^2/year and $i_{max} = 7.3$ into equation (1.21) yields $\lambda = 1.1$ m^2/year. $\lambda = 0.95$ m^2/year corresponds according to equation (1.21) to $i_{max} = 10$.

1.7.4 The Arlanda project

The extension of the international airfield at Arlanda, situated some 30 km north of Stockholm, entailed, among other things, the construction of a new runway at a site with very bad soil conditions. A detailed description of the Arlanda project and the soil conditions at the site is given by Eriksson *et al.* (2000). The soil at the site consists of up to 5 m of peat underlain by high-plasticity, very soft, normally consolidated clay with a maximum thickness of about 10 m. After excavation of the peat layer, pre-loading has been undertaken both in undrained condition and in combination with vertical drain installations, the latter wherever the thickness of the clay layer exceeds 5 m. The consolidation process is monitored by settlement and pore pressure observations. Mebradrains were installed in an equilateral triangular pattern with a drain spacing of 0.9 m. The core of Mebradrains is now equal to the core once used only in Geodrains (see Hansbo, 1986; Eriksson *et al.*, 2000). Two cases of observations will be presented: one (site K) where the overload consists of 19.5 m sand and gravel ($\Delta q = 390$ kN/m^2) and the other (site L) where the overload consists of 16.2 m sand and gravel ($\Delta q = 325$ kN/m^2).

The soil consists of clay with silt and sand seams, at site K to a depth of 9.7 m (with a sand layer from 1.6 to 1.8 m) and at site L to a depth of 7.8 m. The undrained shear strength of the clay is fairly constant, about 5–10 kPa, irrespective of depth. In order to cope with the influence on the course of settlement of a time-consuming stepwise placement of the load, each break in the rate of loading has been analysed separately on the basis of a direct use of the oedometer curves. The total settlement values calculated on the basis of the oedometer tests were checked by Asaoka's method. This resulted in the correlations $s_i = 0.5905 + 0.7755 s_{i-1}$ at site K, and $s_i = 0.2488 + 0.8487 s_{i-1}$ at site L. The primary settlements thus obtained become $s_p = 2.63$ m at site K and $s_p = 1.64$ m at site L. Based on the results of the oedometer tests, the following settlement values Δs were obtained at site K: load-step 0–80 kN/m^2, $\Delta s_1 = 1.63$ m; load-step 80–215 kN/m^2, $\Delta s_2 = 0.64$ m; load-step 215–390 kN/m^2, $\Delta s_3 = 0.36$ m. At site L the following settlement values Δs were obtained: load-step 0–80 kN/m^2, $\Delta s_1 = 1.02$ m; load step 80–325 kN/m^2, $\Delta s_2 = 0.62$ m.

The coefficient of consolidation c_v according to the oedometer tests varies from about 0.2–0.3 m^2/year just above the pre-consolidation pressure to about 0.5–1.0 m^2/year (maximum 2.5 m^2/year) at the end of primary consolidation under the applied overload. The coefficient of consolidation c_h was not determined.

The loading conditions and the settlement observations in the two cases are shown in Figure 1.25. Assuming $d_w = 0.066$ m and $d_s = 0.19$ m, $k_h/k_s = \kappa_h/\kappa_s = 3$, chosen because of the existence of silt and sand seams in the clay deposit, $l = 4.5$ m and $c_v = c_h/3$ the best fit between theory and observations is obtained for $\lambda = 0.7$ m^2/year and $c_h = 2.6$ m^2/year.

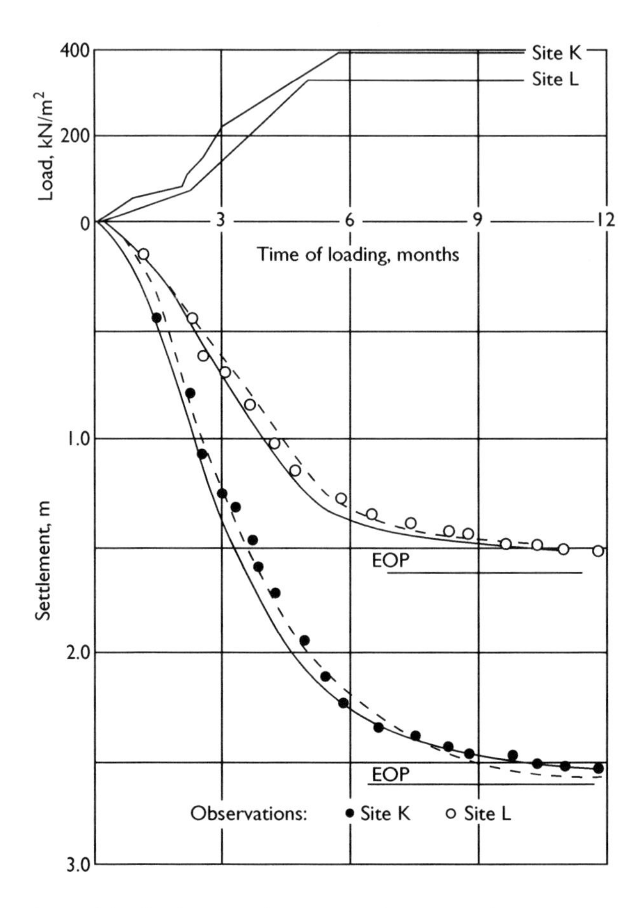

Figure 1.25 Settlement of ground surface under fill embankment for a new runway under construction at Arlanda Airport, Stockholm, sites K and L, drain spacing 0.9 m ($D = 0.95$ m). Unbroken line represents non-Darcian flow (exponent $n = 1.5$), broken line represents Darcian flow.

In this case, as was already demonstrated in Section 1.7.2, when a great deal of the consolidation process takes place during the loading period, the analysis of the consolidation process according to equation (1.18) has to be carried out in the following way. The degree of consolidation $\overline{U}_1$, inserting $\Delta h_1 = \Delta q_1/\gamma_w$, determines the course of settlement in the first load-step. Now, when calculating the course of settlement in the second load-step we have to apply the value $\Delta h_2 = (1 - \overline{U}_1)\Delta q_1/\gamma_w + \Delta q_2/\gamma_w$. At the end of the load-step we have $\Delta s = \Delta s_1 \overline{U}_1 + [\Delta s_1(1 - \overline{U}_1) + \Delta s_2]\overline{U}_2$, and so on. The settlement curves are then adjusted for the rate of loading according to the well-known graphical procedure suggested by Terzaghi.

Choosing site K as an example of the analysis that forms the basis of the settlement diagrams shown in Figure 1.25, we find, when load-step 1 is completed, inserting the consolidation time of 1 month (the length of time that corresponds to full loading condition), $\overline{U}_1 = 0.46$ according to equation (1.18). This yields $\Delta s = 0.46 \times 1.63 = 0.76$ m and a piezometric head $\Delta h_2 = 0.54 \times 8 + 135/10 = 17.8$ m. At the completion of load-step 2, inserting a time of consolidation of half a month (the time of full loading), we have $\overline{U}_2 = 0.38$ which yields $\Delta s = 0.38(2.27 - 0.76) + 0.76 = 1.33$ m. The piezometric head now becomes $\Delta h_3 = 0.62 \times 17.8 + 175/10 = 28.5$ m. Finally, at the completion of load-step 3, when the definite load has been reached, we have (time of consolidation under full load $1^1/_4$ month) $\overline{U}_3 = 0.70$ which yields $s = 0.70(2.63 - 1.33) + 1.33 = 2.24$ m. Three months later we find $\overline{U}_3 = 0.93$ from which $s = 2.45$ m.

Using instead equation (1.8) as a basis for the settlement analysis, we find at the completion of load-step 1, $\overline{U}_1 = 0.42$, and, consequently, $s = 0.42 \times 1.63 = 0.69$ m. At the completion of load-step 2 ($t_1 = 2$ months; $t_2 = 0.5$ month) we find $\overline{U}_1 = 0.65$ and $\overline{U}_2 = 0.25$ from which $s = 0.65 \times 1.63 + 0.25 \times 0.64 = 1.22$ m. Finally, at the completion of load-step 3 ($t_1 = 4.5$ months; $t_2 = 3$ months; $t_3 = 1.25$ months) we find $\overline{U}_1 = 0.90$, $\overline{U}_2 = 0.79$ and $\overline{U}_3 = 0.49$ from which $s = 0.90 \times 1.63 + 0.79 \times 0.64 + 0.49 \times 0.36 = 2.15$ m. Three months later we find $\overline{U}_1 = 0.98$, $\overline{U}_2 = 0.95$ and $\overline{U}_3 = 0.89$, i.e. $s = 2.52$ m.

The hydraulic head at the completion of load-step 2, $\Delta h_2 = 17.8$ m, and the degree of consolidation $\overline{U}_2 = 0.38$ result according to equation (1.15) in a maximum hydraulic gradient during the consolidation process of $i_{max} \approx 30$. This is considerably higher than the limiting hydraulic gradient according to equation (1.12), which for $\lambda = 0.7$ m^2/year and $c_h = 2.6$ m^2/year yields $i_l \approx 7$. In this case the agreement between theory and observations is equally good according to the classical theory based on Darcian flow, equation (1.8), and the theory based on non-Darcian flow, equation (1.18).

1.8 Summary

Results of permeability tests on clay indicating a deviation from Darcy's flow law are unequivocally confirmed by the results of full-scale investigations on consolidation rates obtained in vertical drain projects in different parts of the world. Thus, the consolidation theory developed on the assumption of an exponential correlation between flow velocity v and hydraulic gradient i, below a certain limiting value i_l, originally put forward by the author in 1960, undoubtedly agrees better with case records than the classical consolidation theory based on the validity of Darcy's flow law. The value of the exponent n in the exponential flow law $v = \kappa i^n$ can generally be put equal to 1.5 in accordance with the author's original proposal. Only when the maximum hydraulic gradient created by the overload is excessive in relation to the value of i_l, limiting the exponential correlation between flow and hydraulic gradient, may the classical consolidation theory give equally good (or possibly even better) agreement with observations.

It should be noticed that the equation based on an exponential correlation between flow rate and hydraulic gradient, governing the consolidation rate in a vertical drain project, is general and can be utilized also when the correlation is linear. The agreement with the classical solution becomes satisfactory if the exponent n is put equal to 1.0001.

In a vertical drain project, the effect on the consolidation process of one-dimensional vertical consolidation in undrained condition is relatively difficult to predict but can generally be neglected. Its contribution can be included by increasing the coefficient of consolidation to be applied in the vertical drain analysis. However, one has to consider that such a solution can be misleading with respect to the degree of consolidation obtained in the middle of the clay deposit where the effect of vertical one-dimensional consolidation is minimum. From a practical viewpoint, the design of a vertical drain system has to be based on the result achieved in the layer with the lowest coefficient of consolidation and at the depth where the influence of one-dimensional vertical consolidation is lowest.

Excess pore pressure observations may seem to be the most logical way of checking the degree of consolidation achieved in a vertical drain project. However, the pore pressures observed may be misleading for several reasons: the position of the piezometer in relation to the drains is uncertain; the pore water pressure may not revert to its original value; often obstructions around the piezometers entail a change in drain pattern (Eriksson *et al.*, 2000), and so forth.

Settlement observations are usually reliable but then the final primary consolidation settlement has to be known. In a test area this does not represent a serious problem since its value can be predicted by the aid of Asaoka's method. However, with regard to the influence of one-dimensional vertical consolidation on the rate of settlement, the designer of a vertical drain system should strive to find the effect on settlement caused by radial drainage only. This is important because of the reasons mentioned above. In practice, this can be done by the use of equations taking into account both radial and vertical pore water flow where the coefficients of consolidation c_h (λ) and c_v are determined by trial and error to give good agreement between theory and observations. Then the effect of radial drainage only is obtained by putting $c_v = 0$.

In cases where occasional overloading is utilized in order to avoid future settlement, the problem of when to remove the overload is of paramount interest. This problem is particularly important when the purpose is to avoid trouble from secondary settlements. Layers in the soil profile, not revealed in the soil investigation, may have more unfavourable consolidation characteristics than the soil profile as a whole, which can make it difficult on the basis of settlement observations to decide whether or not the set goal has been fulfilled.

Because of the complexity encountered in the evaluation of the consolidation properties of soil deposits, test areas ought to be arranged wherever possible. In cases where the widths of such test areas are small as compared to the depth of the consolidating layer, the vertical settlements will vary across the test areas and can also be strongly affected by horizontal, outward displacements taking place at various depths in the soil along the borderlines of the test areas. The latter phenomenon ought to be checked by installation of inclinometers at suitable points along the borderlines in order its effect be taken into account.

Bibliography

Akagi, T. (1976) Effect of displacement type sand drains on strength and compressibility of soft clays, PhD Thesis, University of Tokyo.

Asaoka, A. (1978) Observational procedure of settlement prediction, *Soils and Foundations*, Jap. Soc. Soil Mech. Found. Eng., Vol. 18, No. 4, pp. 87–101.

Atkinson, M.S. and Eldred, P.J.L. (1981) Consolidation of soil using vertical drains, *Géotechnique*, Vol. 31, No. 1, pp. 33–43.

Barron, R.A. (1944) *The influence of drain wells on the consolidation of fine-grained soils*, Diss., Providence, US Eng. Office.

Barron, R.A. (1947) Consolidation of fine-grained soils by drain wells, *Transactions ASCE*, Vol. 113, Paper No. 2346, pp. 718–742.

Bergado, D.T., Akasami, H., Alfaro, M.C. and Balasubramaniam, A.S. (1992) Smear effects of vertical drains on soft Bangkok clay, *Journal of Geotechnical Engineering*, Vol. 117, No. 10, pp. 1509–1530.

Bergado, D.T., Alfaro, M.C. and Balasubramaniam, A.S. (1993) Improvement of soft Bangkok clay using vertical drains, *Geotextiles and Geomembranes*, Vol. 12, No. 7, pp. 615–664.

Boman, P. (1973) Ny metod att dränera lera (A new method of draining clay), *Väg-och Vattenbyggaren*, No. 2, p. 192.

Buisman, A.S.K. (1936) Results of long duration settlement tests, *Proceedings of the 1st International Conference on Soil Mechanics and Foundation Engineering*, Cambridge, USA, Vol. 1, Paper No. F-7, pp. 103–106.

Carillo, N. (1942) Simple two and three dimensional cases in the theory of consolidation of soils, *Journal of Mathematics and Physics*, Vol. 21, No. 1, pp. 1–5.

Chai, J., Bergado, D.T., Miura, N. and Sakajo, S. (1996) Back calculated field effect of vertical drain, *2nd International Conference on Soft Soil Engineering*, Nanjing, pp. 270–275.

Chai, J.C., Miura, N. and Sakajo, S. (1997) A theoretical study on smear effect around vertical drain, *Proceedings of the 14th International Conference on Soil Mechanics and Foundation Engineering*, Hamburg, Vol. 3, pp. 1581–1584.

Chang, Y.C.E. (1981) *Long-term consolidation beneath the test fills at Väsby*. Swedish Geotechnical Institute, Report No. 13.
DMJM International – Scott Wilson Kirkpatrick, Norconsult International, SPAN, SEATEC (1996) Back-calculation of full-scale field tests (1993–1995). *Part of SBIA Preliminary Design Report*.
Dubin, B. and Moulin, G. (1986) Influence of critical gradient on the consolidation of clay. In: Yong/Townsend (eds), *Consolidation of Soils. Testing and Evaluation, ASTM* STP 892, pp. 354–377.
Eriksson, U., Hansbo, S. and Torstensson, B.A. (2000) Soil Improvement at Stockholm-Arlanda Airport, *Ground Improvement*, Vol. 4, pp. 73–80.
Flodin, N. (1973) Svenska pappdräner i Canada (Swedish cardboard drains in Canada), *Väg-och Vattenbyggaren*, No. 3, p. 120.
Hansbo, S. (1960) *Consolidation of clay, with special reference to the influence of vertical sand drains*, Diss. Chalmers Univ. of Technology. Swedish Geotechnical Institute, Proc. No. 18.
Hansbo, S. (1979) Consolidation of clay by band-shaped prefabricated drains, *Ground Engineering*, Vol. 12, No. 5, pp. 16–25.
Hansbo, S. (1981) Consolidation of fine-grained soils by prefabricated drains, *Proceedings of the 10th International Conference on Soil Mechanics and Foundation Engineering*, Stockholm, Vol. 3, Paper 12/22, pp. 677–682.
Hansbo, S. (1983) Discussion, *Proceedings of the 8th European Conference on Soil Mechanics and Foundation Engineering*, Helsinki, Vol. 3, Spec. Session 2, pp. 1148–1149.
Hansbo, S. (1986) Preconsolidation of soft compressible subsoil by the use of prefabricated vertical drains, *Tijdschrift der openbare werken van België, Annales des travaux publics de Belgique*, No. 6, pp. 553–562.
Hansbo, S. (1993) Band drains. In: Moseley, M.P. (ed.), *Ground Improvement*, Blackie Academic & Professional, CRC Press, Inc.
Hansbo, S. (1994) *Foundation Engineering*, Elsevier Science B.V., Developments in Geotechnical Engineering, 75.
Hansbo, S. (1997a) Practical aspects of vertical drain design, *Proceedings of the 14th International Conference on Soil Mechanics and Foundation Engineering*, Hamburg, Vol. 3, pp. 1749–1752.
Hansbo, S. (1997b) Aspects of vertical drain design – Darcian or non-Darcian flow, *Géotechnique*, Vol. 47, No. 5, pp. 983–992.
Hansbo, S. (2001) Consolidation equation valid for both Darcian and non-Darcian flow, *Géotechnique*, Vol. 51, No. 1, pp. 51–54.
Hird, C.C. and Moseley, V.J. (2000) Model study of seepage in smear zone around vertical drains in layered soil, *Géotechnique*, Vol. 50, pp. 89–97.
Holtz, R.D. and Holm, G. (1973) Excavation and sampling around some drains at Skå-Edeby, Sweden, *Proceedings of Nordic Geotechnical Meeting 1972*, Trondheim. Norwegian Geotechnical Institute.
Indraratna, B. and Redana, I.W. (1998) Laboratory determination of smear zone due to vertical drain installation, *Journal of Geotechnical Engineering, ASCE*, Vol. 123(5), pp. 447–448.
Jamiolkowski, M., Lancelotta, R. and Wolski, W. (1983) Summary of Discussion, *Proc. VIIIth ECSMFE*, Helsinki, Vol. 3, Spec. Session 6.

Johnson, S.J. (1970) Foundation precompression with vertical sand drains, *Journal of Soil Mechanics and Foundation Division, ASCE*, Vol. 96, SM 1, pp. 145–175.

Kamon, M. (1984) Function of band-shaped prefabricated plastic board drain, *Proceedings of the 19th Japanese National Conference on Soil Mechanics and Foundation Engineering*.

Kingdom of Thailand, Airports Authority of Thailand (1996) *The full-scale field test of prefabricated vertical drains for the second Bangkok International Airport. Final Report*. Submitted by Asian Institute of Technology in Association with Infinity Services Co. Ltd, Vol. III.

Kjellman, W. (1947) Consolidation of fine-grained soils by drain wells, *Trans. ASCE*, Vol. 113, pp. 748–751 (Contribution to the discussion on Paper 2346).

Koda, E., Szymanski, A. and Wolsky, W. (1986) Laboratory tests on Geodrains – Durability in organic soils. *Seminar on Laboratory Testing of Prefabricated Band-shaped Drains*, Milano, 22–23 April.

Lawrence, C.A. and Koerner, R.M. (1988) Flow behavior of kinked strip drains. In: R.D. Holtz (ed.), *Geosynthetics for Soil Improvement, Geotechnical Special Publications*, No. 18.

Lee, S.L., Karunaratne, G.P., Aziz, M.A. and Inoue, T. (1995) An environmentally friendly prefabricated vertical drain for soil improvement, *Proc. Bengt Broms Symp. on Geot. Eng.*, Singapore, pp. 243–261.

Lo, D.O.K. (1991) *Soil improvement by vertical drains*. Diss., Univ. of Illinois at Urbana-Champaign.

Madhav, M.R., Park, Y.-M. and Miura, N. (1993) Modelling and study of smear zones around band shaped drains, *Soils and Foundations*, Vol. 33, No. 4, pp. 133–147.

Mesri, G. and Godlewski, P.M. (1977) Time and stress-compressibility inter-relationship. *ASCE, Journal of Geotechnical Engineering Division*, GT 5, pp. 417–430.

Miller, R.J. and Low, P.F. (1963) Threshhold gradient for water flow in clay systems, *Proc. Soil Science Society of America*, November–December, pp. 605–609.

Onoue, A. (1988) Consolidation of multilayered anisotropic soils by vertical drains with well resistance, *Soils and Foundations*, Jap. Soc. Soil Mech. Found. Eng., Vol. 28, No. 3, pp. 75–90.

Onoue, A., Ting, N., Germaine, J.T. and Whitman, R.V. (1991) Permeability of disturbed soil around vertical drains. In: *ASCE Geot. Special Publ.* No. 27, pp. 879–890.

Porter, O.J. (1936) Studies of fill construction over mud flats including a description of experimental construction using vertical sand drains to hasten stabilization. *Proceedings of the 1st International Conference on Soil Mechanics and Foundation Engineering*, Vol. I, Cambridge, USA, Paper No. L-1, pp. 229–235.

Robertson, P.K., Campanella, R.G., Brown, P.T. and Robinson, K.E. (1988) Prediction of wick drain performance using piezocone data. *Canadian Geotech. J.*, 25, pp. 56–61.

Shiffman, R.L. (1958) Consolidation of soil under time-dependent loading and varying permeability, *Proceedings of the Highway Res. Board*, Vol. 37.

Silfverberg, L. (1947) In Statens Getekniska Institut, 1949. *Redogörelse för Statens Geotekniska instituts verksamhet under åren 1944–1948 (Report on the activities*

of the Swedish Geotechnical Institute during the years 1944–1948). Swedish Geotechnical Institute, Meddelande No. 2.
Sing and Hattab (1979) Sand drains, *Civil Engineering*, June, pp. 65–67.
Taylor, D.W. and Merchant, W. (1940) A theory of clay consolidation accounting for secondary compression, *Journal Mathematics and Physics*, No. 1.
Terzaghi, K. (1925) *Erdbaumechanik*, Leipzig u. Wien.
Terzaghi, K. and Peck, R. (1948) *Soil Mechanics in Engineering Practice*, John Wiley & Sons, Inc.
Torstensson, B.-A. (1984) Konsolidering av lera medelst vakuummetoden och/eller konstgjord grundvattensänkning i kombination med vertikaldränering. Resultat av fullskaleförsök i Vagnhärad. (Consolidation of clay by means of the vacuum method and/or artificial groundwater lowering in combination with vertical drainage. Results of full-scale tests at Vagnhärad). *Internal Report*.
Yoshikuni, H. (1979) Design and construction control of vertical drain methods, PhD Thesis Found. Engg. Series, Gihodo, Tokyo (in Japanese).
Yoshikuni, H. and Nakanado, H. (1974) Consolidation of soils by vertical drain wells with finite permeability, *Soils and Foundations*, Vol. 14, No. 2, pp. 35–45.
Zeng, G.X. and Xie, K.H. (1989) New development of the vertical drain theories. *Proceedings of the 12th International Conference on Soil Mechanics and Foundation Engineering*, Rio de Janeiro, Vol. 2, Paper 18/28, pp. 1435–1438.
Zou, Y. (1996) A non-linear permeability relation depending on the activation energy of pore liquid, *Géotechnique*, Vol. 46, No. 4, pp. 769–774.

Chapter 2

Deep vibro techniques

W. Sondermann and W. Wehr

2.1 Introduction and history

For over 60 years depth vibrators have been used to improve the bearing capacity and settlement characteristics of weak soils. Vibro compaction is probably the oldest dynamic deep compaction method in existence. It was introduced and developed to maturity by the Johann Keller Company in 1936, which enabled the compaction of non-cohesive soils to be performed with excellent results. A detailed description of the method from its beginnings up to the pre-war period is given by Schneider (1938) and by Greenwood (1976) and Kirsch (1993) for the period thereafter.

This original process, now referred to as vibro compaction, has since been applied successfully on numerous sites around the world. When carrying out compaction work using the vibro compaction method in water-saturated sands with high silt content, these sands, when lowering the depth vibrator and during subsequent compaction, are liquefied to such an extent that the compaction effect only occurs after a very long vibration period or it does not occur at all. In such soils, the vibro compaction method reaches its technical and economic limits (Kirsch and Sondermann, 2003).

To overcome the limitations of the vibro compaction method, a technique to insert the vibrator into the soil without the aid of simultaneously flushing in water was developed in 1956. After the vibrator is lifted, the temporarily stable cylindrical cavity is filled with coarse material, section by section. The coarse material is then compacted by repetitive use of the vibrator. This vibro replacement procedure came to be known as the conventional dry method. Such technical developments in dense stone column construction allowed for a greater range of treatable weak natural soils and man-made fills. Vibro replacement continues to be widely used in Europe to improve weak soil. It has a reputation for providing stable ground which allows for safe and economic construction of residential and light commercial and industrial structures.

The conventional dry method utilises the vibrator to displace the surrounding soil laterally, rather than for primary compaction of the original soil. The crushed stone is pressed laterally into the soil during both the cavity-filling stage and compaction stage. This produces stone columns that are tightly interlocked with the surrounding soil. Groups of columns created in this manner can be used to support large loads. The conventional dry method reliably produces stone columns to depths of 8 m in cohesive soils that have a shear strength of at least 20 kN/m^2.

Bottom feed vibrators, which introduce the stones through the vibrator tip during lift, are used to overcome the disadvantage of possible cavity collapse that can occur with the conventional dry method in cohesive soils with a high water content. During withdrawal of the vibrator, stone and compressed air are delivered through the vibrator tip, preventing cavity collapse. This method is known as dry vibro replacement. In 1972, a German patent on this method was applied for.

Reliable stone column production by vibro compaction in cohesive soils with a high water content is achievable with the aid of a heavy water jet. Water is jetted from the vibrator tip as the vibrator is lowered to the desired depth. Mud flushes loosened soil and rises to the surface, stabilising the cavity. This is known as the wet vibro replacement method.

After the bottom feed system had been developed, it was possible to install injected stone columns in 1976 by means of an injection of a cement–bentonite suspension near the bottom of the vibrator (Jebe and Bartels, 1983). The voids of the stone column skeleton are thereby filled with this suspension.

Finally vibro concrete columns were developed using a conventional concrete pump to deliver the concrete to the bottom of the vibrator via the tremie system.

In very soft nearly liquid soils vibro replacement is not applicable due to the lack of lateral support of the soil. A geotextile coating may be used around the column to ensure filter stability and to activate tensile forces to avoid lateral spreading of the column. This method was developed in 1992 and applied first in early 1993 for a dam project in Austria (Keller, 1993). A compilation of various projects with geotextile-coated columns may be found in Sidak and Strauch (2003).

Indeed, all the above techniques have been chosen for many major structures in the USA and Europe, endorsing their value in promoting safe and economic foundations to a wide range of buildings and soil conditions.

Probably the oldest recommendation on the use of vibro was issued by the German transport research society in 1979 (FGFS, 1979). Later the US department of transportation published the 'design and construction of stone columns' manual (USDT, 1983) followed by the British ICE 'specification for ground treatment' (ICE, 1987) and the BRE publication 'specifying vibro stone columns' (BRE, 2000). The latest effort has been made by the European community to standardise the execution of vibro works in 'ground treatment by deep vibration' (European Standard WG12, 2003).

2.2 Vibro processes

The operational sequence of the vibro compaction method is illustrated in Figure 2.1. During operation, the cylindrical, horizontally vibrating depth vibrator is usually suspended from a crane or like equipment. It weighs 15–40 kN, with a diameter of 30–50 cm and a length of 2–5 m. The vibrator reaches application depth by means of extension tubes.

The vibrator shell is constructed of steel pipe, forming a cylinder. Eccentric weight in the lower section is powered by a motor at the top end of a vertical shaft within the vibrator. Energy for the motor is supplied through the extension tubes. The rotational movement of the eccentric weights causes vibrations of the vibrator. The vibratory energy is transferred from the vibrator casing to the surrounding soil. This energy affects the surrounding soil without being dependent on the vibrator's depth of operation. A vibration damping device between the vibrator and extension tubes prevents the vibratory energy from being transmitted to the extension tubes. Supply pipes for water and air (optional) are also enclosed in the extension tubes. The pipes can deliver their payload through the vibrator tip as well as through special areas of the extension tubes to aid the ground penetration action of the vibrator.

During vibro compaction, the motor runs as the depth vibrator is inserted into the soil (Figure 2.1). The insertion is aided by water flushing. Field experience has shown that penetration is more effective when a larger volume of water is used, rather than a higher pressure. The water flow will

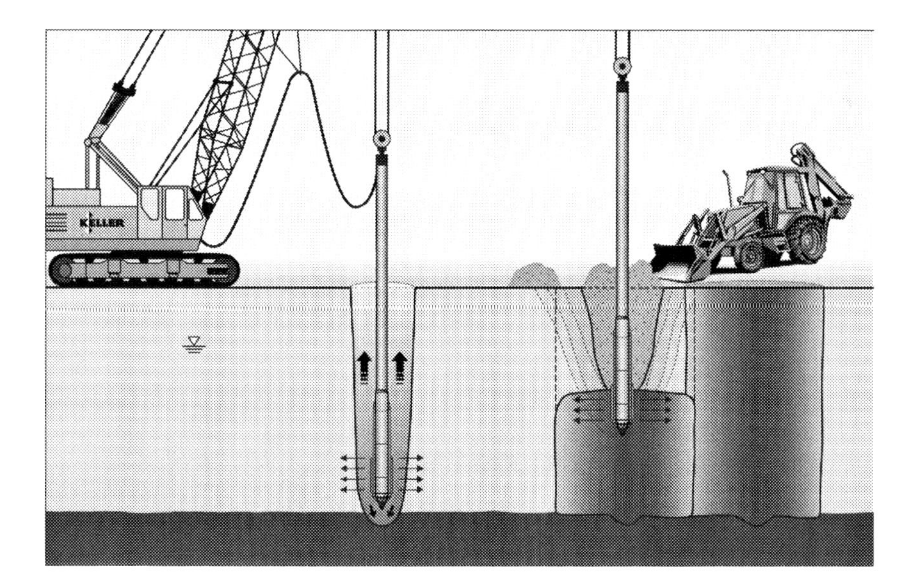

Figure 2.1 Vibro compaction method operating phases (by courtesy of Keller Group).

expel some loosened sand through the annulus around the vibrator. The granular soil targeted for compaction sees a fast reduction in temporary excess pore water pressure. At compaction depths greater than 25 m, additional flushing lines and compressed air may need to be utilised.

The water and air flows are normally stopped or reduced after the vibrator arrives at its specified depth and the compaction process stages have been initiated. Field experience has determined that lifting the vibrator in stages of 0.5 or 1.0 m, after 30–60 s of application tends to produce the best results. During the compaction process, granular material adjacent to the vibrator sees a reduction in pore volume, which is compensated for by introducing sand via the annulus. It is possible for settlement of the surface to range from 5 to 15 per cent of the compaction depth. This range depends on the density prior to compaction, as well as the targeted degree of compaction.

After the initial insertion and compaction processes have been completed at a particular location, the vibrator is moved to the next location and lowered to the depth specified for compaction.

Compacted soil elements with specified diameters can be created by performing the compaction procedures in grid patterns. Open-pit brown coal mining areas, such as those at Lusatia, have had vibro compaction performed at depths greater than 50 m (Degen, 1997b). Typically, the layout of compaction probe centres is based on an equilateral triangle. A distance between 2.5 and 4.5 m usually separates the centres. This distance is determined by grain crushability (shell content), required density, vibrator capacity and grain size distribution of the sand. The production stage of extensive projects can be greatly enhanced if a comprehensive soil study is done, with the added benefit of a test programme prior to going out to tender/bidding on the project. Guideline values for the strength properties of sand, which can aid the design of such projects, are displayed in Table 2.1. After completion

Table 2.1 Guideline values for the strength properties of sand (according to Kirsch, 1979)

Density	*Very loose*	*Loose*	*Medium dense*	*Dense*	*Very dense*
Relative density I_D [%]	<15	15–35	35–65	65–85	85–100
SPT [N/30 cm]	<4	4–10	10–30	30–50	>50
SCPT q_c [MN/m^2]	<5	5–10	10–15	15–20	>20
DCPT (light) [N/10 cm]	<10	10–20	20–30	30–40	>40
DCPT (heavy) [N/10 cm]	<5	5–10	10–15	15–20	>20
Dry density γ_d [kN/m^3]	<14	14–16	16–18	18–20	>20
Modulus of deformation [MN/m^2]	15–30	30–50	50–80	80–100	>100
Angle of internal friction [°]	<30	30–32.5	32.5–35	35–37.5	>37.5

of the vibro compaction work, it may be necessary to recompact the working surface down to a depth of about 0.5 m by using surface compactors.

Currently, depth vibrators are used to produce vibro stone columns in cohesive soils that exhibit low water content. For this production variant to be successful, the soil consistency must be able to hold the form of the entire cavity after the vibrator has been removed. This allows for the subsequent repeated delivery and compaction of stone column material to proceed, uninhibited by obstruction. With the dry or displacement method, the soil cavity is prevented from collapsing by the compressed air being released from the vibrator tip.

An alternative method to construct vibro stone columns in cohesive soils with high water content involves the use of a strong water jet that ejects water under high pressure from the vibrator tip. The cavity is stabilised by the mud that rises to the surface and flushes out loosened soil. The cavity is then filled in stages, through the annulus, with coarse fill, which surrounds the vibrator tip and is compacted into the stone column form as the vibrator is lifted. This is known as the wet/replacement method. A mud, or 'spoil', containing high quantities of soil particles is transported to specially designed settling tanks, or ponds, by way of trenches. This procedure is complicated, can be messy, but it is important to separate the water and mud from the operations area, where it is easily accessed when the time comes to discharge it (Kirsch and Chambosse, 1981).

Grain diameters of the stones and gravel which comprise the fill material for the wet method range from 30 to 80 mm. Stone column installation to depths as great as 26 m has been reported (Raju and Hoffman, 1996). The wet method guarantees stone column continuity for a wide range of soft soils.

Grain diameters of the stones or gravel which comprise the fill material when using a bottom feed vibrator typically range from 10 to 40 mm. The fill is delivered to the vibrator tip by means of a pipe. After the vibrator arrives at the specified depth, compressed air is used to help deliver the fill, as the vibrator is subsequently lifted in stages as it compacts the fill (Figure 2.2).

Carrier equipment typically consists of specially designed machines, known as vibrocats, which have vertical leaders. The vibrocats control the complex bottom feed vibrators, equipped with material lock and storage units, which deliver fill material to the vibrator by means of specialised mechanical or pneumatic feeding devices (Figures 2.3 and 2.4).

The installation of vibro mortar columns is similar to the dry vibro replacement method apart from the cement–bentonite suspension filling the voids of the stone skeleton inside the column. This results in a much stiffer column compared to a conventional vibro stone column.

For the installation of vibro concrete columns the tremmie system is connected to a mobile concrete pump. Before penetrating, the system is charged with concrete. The vibrator then penetrates the soil until the required depth has been achieved. The founding layer, if granular, is further compacted by

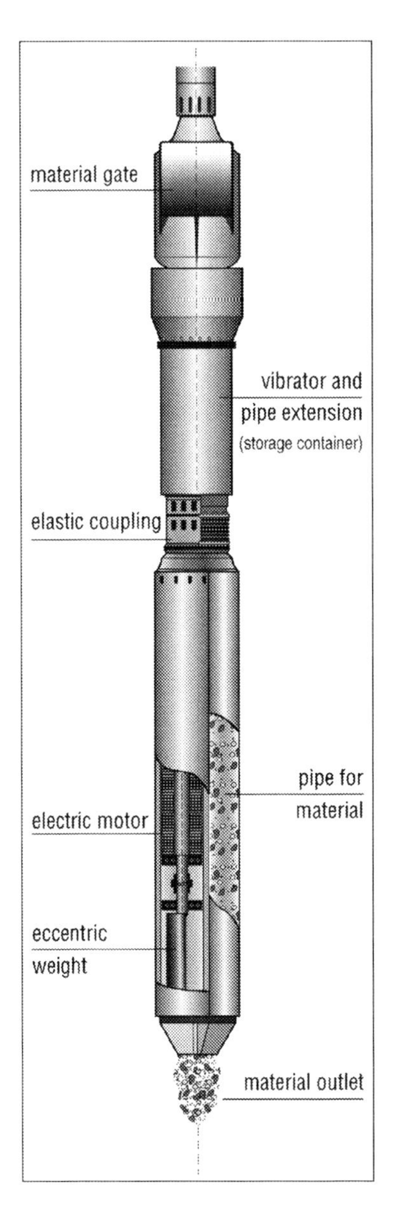

Figure 2.2 Details of a bottom feed vibrator (by courtesy of Keller Group).

the vibrator. Concrete is pumped out from the base of the tremmie at positive pressure. After raising the vibrator in steps, it re-enters the concrete shaft displacing it into a bulb until a set resistance has been achieved. Once the bulb end is formed, the vibrator is withdrawn at a controlled rate from the soil

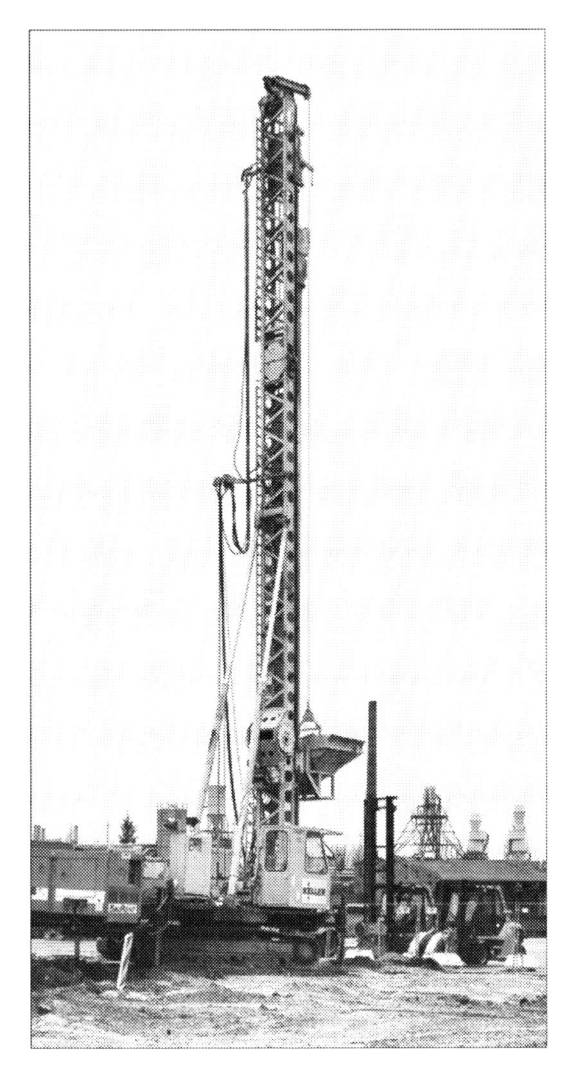

Figure 2.3 Vibrocat with bottom feed vibrator.

whilst concrete continues to be pumped out at positive pressure. Once completed, the column can be trimmed and reinforcement placed as required.

Vibro geotextile columns consist of a sand or stone core with a geotextile coating. The advantage of a vibro geotextile column to other geotextile columns (Schüßler, 2002) is the well-densified granular infill resulting only in small settlements of the soil-column system.

The installation is usually performed in several steps in order not to damage the geotextile. First a hole is created with the vibrator to the

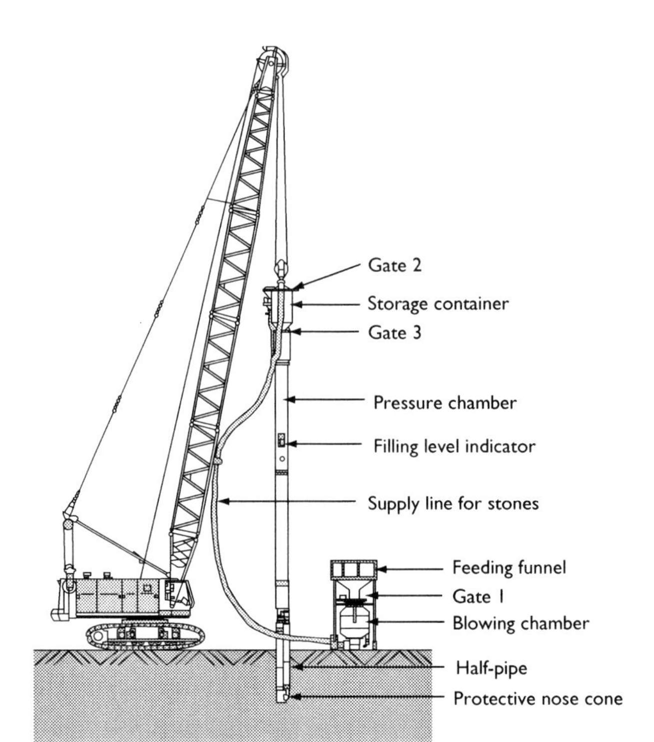

Figure 2.4 Bottom feed vibrator with pneumatic feeding device (Degen, 1997a).

required depth and the vibrator is extracted. In the next step the geotextile is mounted over the vibrator above the ground surface and subsequently the penetration is repeated with the geotextile to the same depth as before. On the way up, preferably stones are filled and densified inside the geotextile like in the usual dry bottom feed process.

If there is only one certain very soft layer it is possible to build a vibro stone column below this layer first, insert a vibro geotextile column or a vibro mortar column only in the very soft layer for economical reasons and finish the upper part of the column as an ordinary vibro stone column again.

2.3 Vibro plant and equipment

The equipment developed for the vibro compaction and vibro replacement processes comprises four basic elements:

1 the vibrator, which is elastically suspended from extension tubes with air or water jetting systems;

2 the crane or base machine, which supports the vibrator and extension tubes;
3 the stone delivery system used in vibro replacement;
4 the control and verification devices.

The principal piece of equipment used to achieve compaction is the vibrator (Figure 2.5).

The drive mechanism can be an electric motor or a hydraulic motor, with the associated generator or power pack usually positioned on the crawler rig in the form of a counter weight.

The typical power range in vibrators is 50–150 kW, and can go as high as 200 kW for the heaviest equipment. Rotational speeds of the eccentric weights in the cases of electric drives are determined by the frequency of the current and the polarity of the motor. For example, 3000 or 1500 rpm vibrating frequency are obtainable from a 50 Hz power source, and 3600 or 1800 rpm vibrating frequency from a 60 Hz power source, with a single or double-pole drive, respectively. A 5 per cent reduction in the frequency applied to the ground occurs, corresponding to the magnitude of the 'slip'

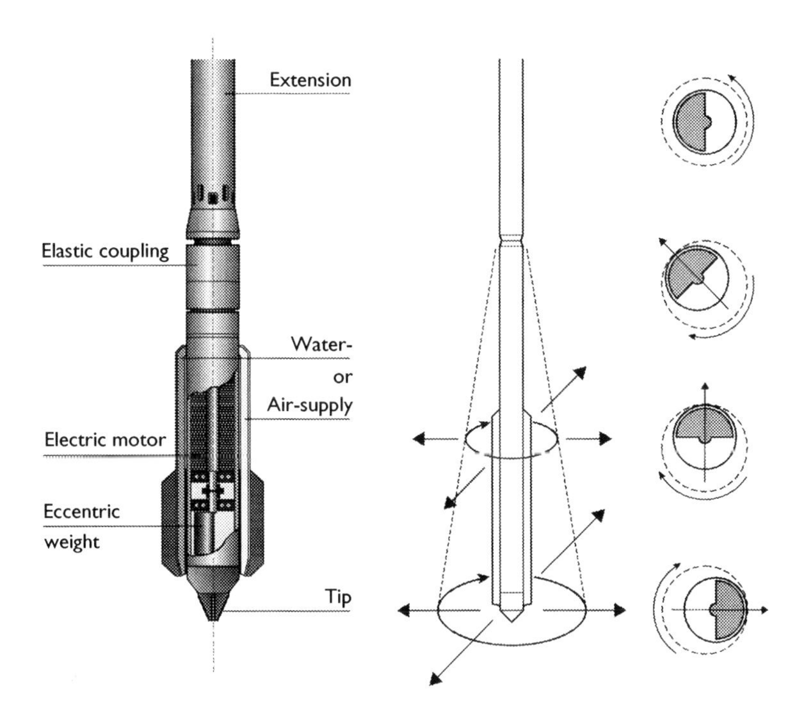

Figure 2.5 Depth vibrator and principle of vibro compaction (by courtesy of Keller Group).

experienced with asynchronous motors. The use of frequency converters has recently become economical as a result of modern control technology. The frequency converters enable limited variation of the operating frequency of the electric motors.

During rotation, the eccentric weight (Figure 2.5) generates horizontal force. This horizontal force is transmitted to the ground through the vibrator casing and (depending on the vibrator type) ranges from 150 to 700 kN. When the vibrator is freely suspended with a lack of lateral confinement, the vibration width (double amplitude) totals 10–50 mm. Acceleration values of up to 50 g are obtainable at the vibrator tip. It is practically complicated to measure crucial operational data during the compaction process. Therefore, any data given on vibrators apply to those which are freely suspended, lacking lateral confinement.

It is up to the designer to create a vibrator which is optimal for the specific application. One major challenge of design lies in keeping maintenance costs within standards that are economically tolerable. Based on field experience, the most effective compaction of sands and gravels is done by vibrating frequencies which approach the natural soil-vibrator system frequency, or 'resonance' for elastic systems, which ranges between 20 and 30 Hz.

Fellin (2000), who considered vibro compaction as a 'plasto-dynamic problem', has confirmed theoretically the knowledge gained from practical vibro operation conditions. Fellin's goal, by constant analysis of information obtained on the vibrator movement during compaction performance, was to create 'on-line compaction control'. His work's theoretical results confirm the observation that when using a constant impact force, the vibration's effect range increases as the vibrator frequency decreases, whereas compaction increases when the impact force increases.

The thickness of soil depths to be treated determines the overall length of vibrator, extension tubes and lifting equipment, which, in turn, determines the size of crane to be used.

Purpose-built tracked base machines (vibrocats) have been constructed to support vibrators: first, to ensure the columns are truly vertical and second to be able to apply the frequently required or desired vertical compressive force, which accelerates the introducing and compacting processes.

The construction of stone columns requires the importation and handling of substantial quantities of granular material. This stone is routinely handled with front end loaders, working from a stone pile and delivering stone to each compaction point.

To increase the performance of the vibro system multiple vibrators may be applied on one base machine. For example a barge with a 120–150 t crane was used for the Seabird project in India, with four vibrators (Keller, 2002). Alternatively a special frame was constructed on a barge suspending five vibrators (Keller, 1997).

2.4 Design and theoretical considerations

2.4.1 *Vibro compaction*

The purpose of vibro compaction is the densification of the existing soil. The feasibility of the technique depends mainly on the grain size distribution of the soil. The range of soil types treatable by vibro compaction and vibro replacement are given in Figure 2.6. The degree of improvement will depend on many more factors including soil conditions, type of equipment, procedures adopted and skills of the site staff. Such variables do not permit an optimum design to be established in advance but rather require the exercise of experience and judgement for their successful resolution.

For small projects, the design of vibro compaction work can be based on the experience of the contractor. For large projects it is preferable and advisable to conduct a trial in advance of contract works. A typical layout of vibro compaction probes for a trial is given in Figure 2.7. The trial allows for three sets of spacings between probes, together with pre- and post-compaction testing, often performed using cone penetration testing equipment. The degree of improvement achieved can be used to optimise the design, as shown in Figure 2.8.

The technical success of vibro compaction work is measured by the level of densification achieved against a specified target. The densification can be readily checked using standard penetration tests or cone penetration tests. Comparisons can be made between pre- and post-compaction testing, and care should be taken to ensure that the same techniques of testing are used

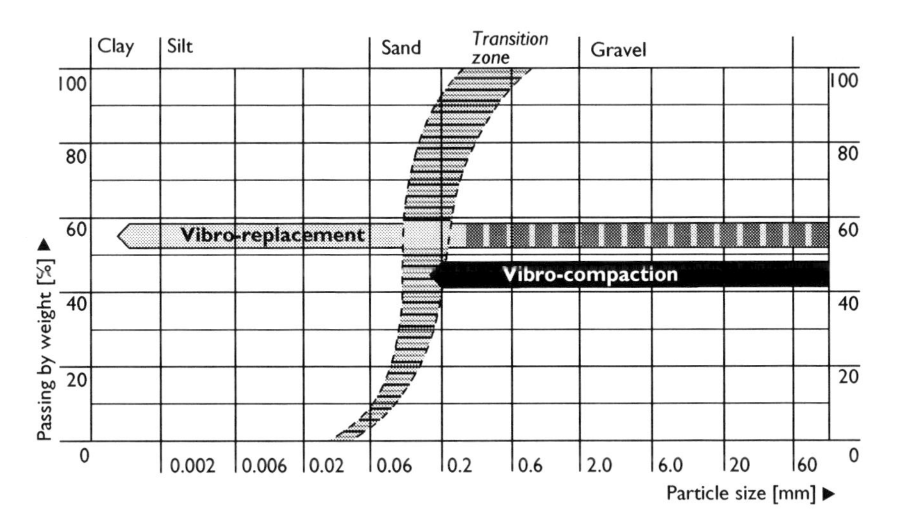

Figure 2.6 Range of soil types treatable by vibro compaction and vibro replacement (stone columns).

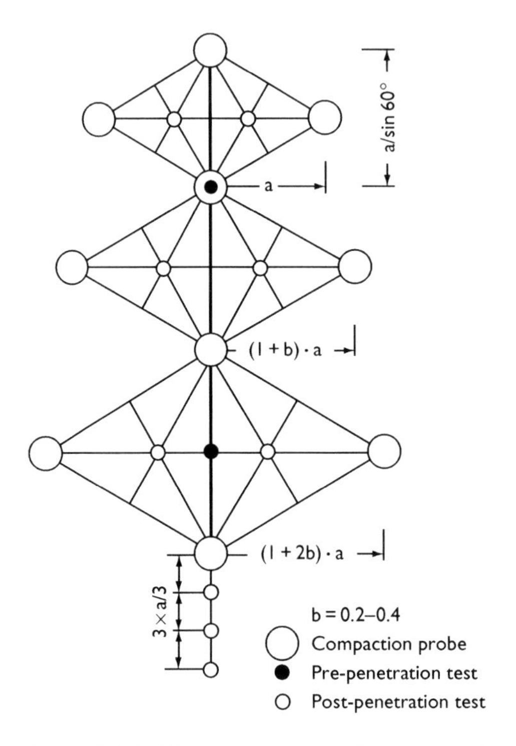

Figure 2.7 Trial arrangement for vibro compaction (Moseley and Priebe, 1993).

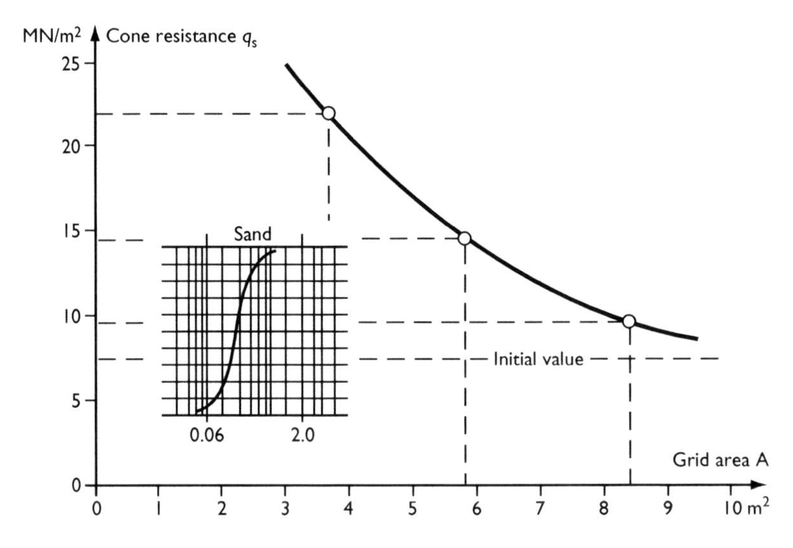

Figure 2.8 Results of vibro compaction trial (Moseley and Priebe, 1993).

in each situation. Control of performance is a further important element in carrying out vibro compaction work. This is best achieved by using a standardised procedure, established at the pre-contract trial, such as pre-determined lifts of the vibrator at pre-determined time intervals and/or pre-determined power consumptions. Only such a regular procedure can reveal whether variations in test results are due to the inherent inhomogeneity of the soils being treated or by insufficient compaction.

The soil being treated, the degree of densification required, the type of vibrator being used and production rates all have an influence on the spacing of vibro compaction probes. Areas treated per probe vary commonly between 6 and 20 m^2. Vibrator development over the past decade has allowed considerable increases in the area treated by each insertion of the vibrator. This development continues and will enable further expansion of the treatment envelope.

Sands and gravels bearing negligible cohesion are compatible with vibro compaction. The silt (grain size < 0.06 mm) percentage of such soils should be less than 10 per cent for ideal performance. Compaction is substantially hindered by clay particles (grain size < 0.002 mm) to the point that the procedure is unable to be performed without extra measures, including the introduction of coarse-grained fill. Reference to the grain size distribution diagram (Figure 2.6) usually determines application limits. However, application limits for material that is very coarse is typically determined empirically, taking into consideration the penetration effectiveness of the respective vibrator. Static cone penetration tests can also serve to estimate values of soil compactibility for compaction methods. Given that the local skin friction-to-point resistance (friction ratio) falls between 0 and 1, and the point resistance is a minimum of 3 MPa, the soil can be considered to be compactible (Massarsch, 1994).

The efficiency of compaction is also greatly influenced by the permeability of the soil. When permeability is too low ($<10^{-5}$ m/s), compaction effectiveness decreases as permeability decreases, whereas when permeability is too high ($>10^{-2}$ m/s), penetration of the soil by the vibrator becomes increasingly more difficult as the permeability increases (Greenwood and Kirsch, 1983).

The carbonate or shell content is important for the densification of highly compressible soils with low cone resistance and high friction ratio. Cemented soils are not considered here.

Correlations between the CPT cone resistance and the relative density are well established for silica sand. Unfortunately there are not many references concerning this correlation for calcareous sands and no systematic research has been undertaken. Vesic (1965) added 10 per cent of shells to quartz sand which resulted in a decrease of the CPT cone resistance by a factor of 2.3. Bellotti and Jamiolkowski (1991) compared CPT cone resistances $q_c(\text{silica})/q_c(\text{shells}) = 1 + 0.015(\text{Dr} - 20)$ yielding ratios between 1.3 and

2.2 increasing with relative density Dr. Almeida *et al.* (1992) compared normalised CPT cone resistances of calcareous Quiou sand and silica Ticino sand, which yielded ratios ranging from 1.8 to 2.2 proportionate to increasing relative density. Foray *et al.* (1999) compared pressuremeter limit pressure of silica sand and carbonate sands which resulted in ratios ranging from 2 to 3 proportionate to increasing initial vertical stress. Finally Cudmani (2001) looked at normalised cone resistances of seven sands yielding ratios between 1.4 and 3.5 depending on initial soil pressure and relative density.

2.4.2 *Vibro replacement stone columns*

The reduction of consolidation time and compressibility, and the increase of load-bearing capacity and shear strength, are what the effect of vibro replacement in soft fine-grained soils is determined by. The *in-situ* soil characteristics, the placement and geometry of the stone columns, and the soil-mechanical properties of the column composition, are what determine the scale of ground improvement achieved. Aside from the settlement rate increase (generated by the stone column's drainage effect), the reduction of overall settlement is the goal of the vibro column installation. Quite simply, stone columns are effective in reducing settlement since they are stiffer than the surrounding soil. Between stone columns and the ground, the effective stiffness ratio relies considerably on lateral support provided by the surrounding soil when the stone columns have loads put upon them. In order to mobilise the lateral support and generate the interaction between the soil and columns, a horizontal deformation is required. This deformation inevitably causes settlement at the ground surface. Bell (1915) relays the most simplistic relationship for calculating load-bearing behaviour. A maximum lateral support of $\sigma_h = \gamma z + 2c_u$ can be provided by the adjacent cohesive soil possessing a cohesion c_u at depth z. If it is assumed that the passive earth pressure coefficient $K_p = \tan^2(\pi/4 + \phi/2)$ is used, then the above supporting pressure allows a maximum vertical column stress of $\sigma_o = K_p(\gamma z + c_u)$, with ϕ being the angle of the internal friction of the column material (Figure 2.9). This equation, while underestimating the column's load-bearing capacity, still conveys the significance of column and ground interaction. The equation also reveals the differences in load-bearing behaviours of stone columns when compared to load-carrying elements of greater stiffness.

Minimum shear strength of ground proposed for improvement is frequently given in the form of a c_u value of 15 kN/m^2 (AUFS, 1979; Smoltczyk and Hilmer, 1994). It must be noted that no attention is given to the positive effects of the three-dimensional behaviour, the influences of adjacent columns, the dilatation of column material (Van Impe and Madhav, 1992) and most importantly, the rapid increase in the soil's shear strength owing to the stone column's drainage effect. As a consequence, the successful production

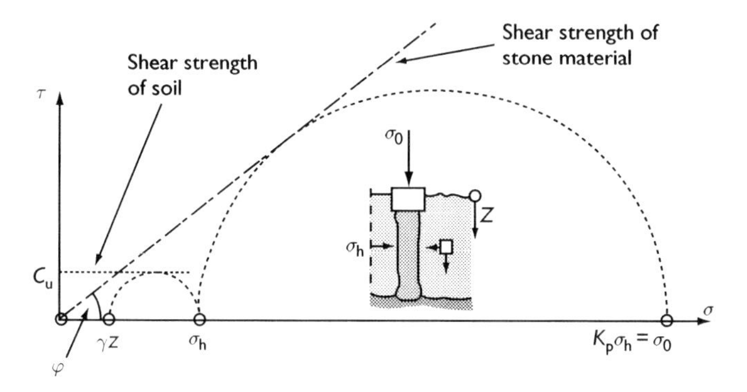

Figure 2.9 Influence of lateral support on column stress (Brauns, 1978).

of foundations in much softer soils via vibro replacement has been achievable (Raju and Hoffmann, 1996). Many model tests have been conducted in order to more clearly grasp the column/soil interactions and the influences of adjacent columns (Hu, 1995). In qualitative terms, these tests show the failure mechanism on one side and the group effect on the other (Figure 2.10).

With ultimate vertical load, the failure of stone columns is a result of relatively low lateral support in the upper third (bulging), or the column toe being punched into the underlying soil, such as with 'floating' foundations (Figure 2.11). However, such high rates of deformation precede the failure in every case that the column's serviceability is generally no longer provided. Therefore, we can conclude that the equations used to calculate the deformation, or 'serviceability state' of the discussed foundations are much more relevant than the outcome of limit load assessment of stone columns.

Soyez (1987) and Bergado *et al.* (1994) have conducted a thorough overview of the various design methods. The authors show the distinction between calculating single columns and calculating column grid patterns. In Europe, Priebe's (1995) design method for vibro replacement stone columns has gained acceptance as a valid method (Figure 2.12).

Thus, in Figure 2.12, the improvement factor, depending on angles of internal friction of the stone column, is related to the ratio of the stone column area and the area being treated by the column. The improvement factor indicates how many times the compression modulus increases for a grid of stone columns and, vice versa, to what extent the settlement of a raft foundation will be reduced.

The basic design curves assume the stone column material to be incompressible, and Figure 2.13 allows an adjustment to be made for this by plotting a fictitious area ratio, which has to be added to the actual area ratio, against the compression modulus ratio for soil and stone column material.

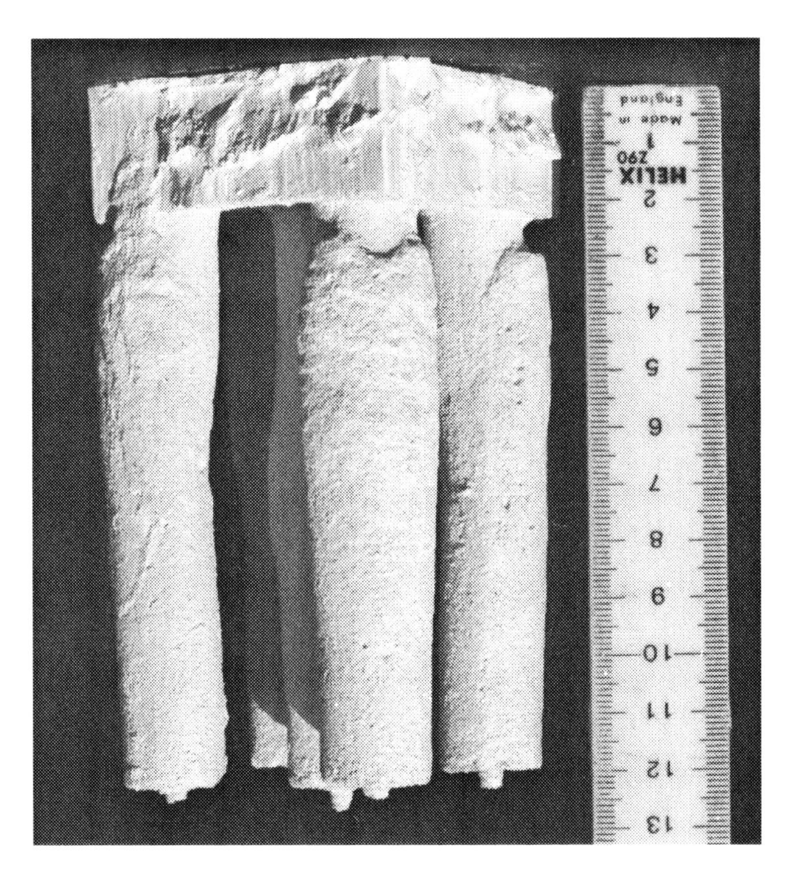

Figure 2.10 Failure mechanism of vibro replacement stone columns in the case of group effect (according to Hu, 1995).

With regard to settlement performance, theoretical approaches predominantly refer to an infinite grid of columns. Load tests executed in practice on footings resting on small numbers of columns do not fulfil the assumptions. Accordingly, evaluations of settlement performance of a footing on a limited number of stone columns are only approximations.

Practical design charts that consider load distribution as well as reduced lateral support on columns situated underneath footing edges have been presented by Priebe (1995). These charts allow the estimation of settlement of a rigid foundation on a limited number of stone columns as a function of the settlement of an infinite raft supported by an infinite grid of columns, outlined above.

The method pre-supposes that the footing area attributed to a stone column and the foundation pressure is identical. There exists an optimum

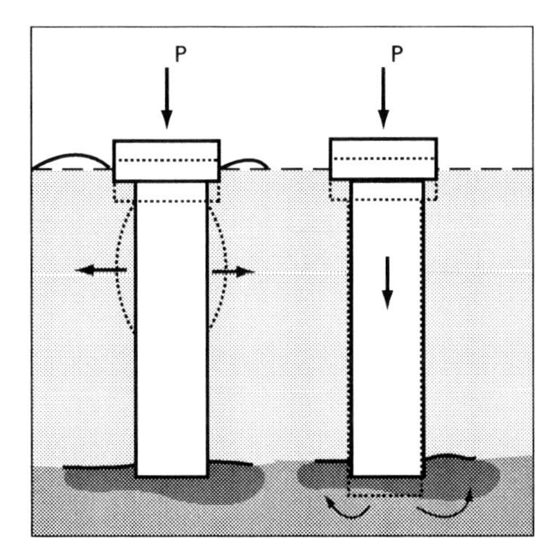

Figure 2.11 Failure mechanism of vibro replacement stone columns under vertical load (acc. to Brauns, 1978).

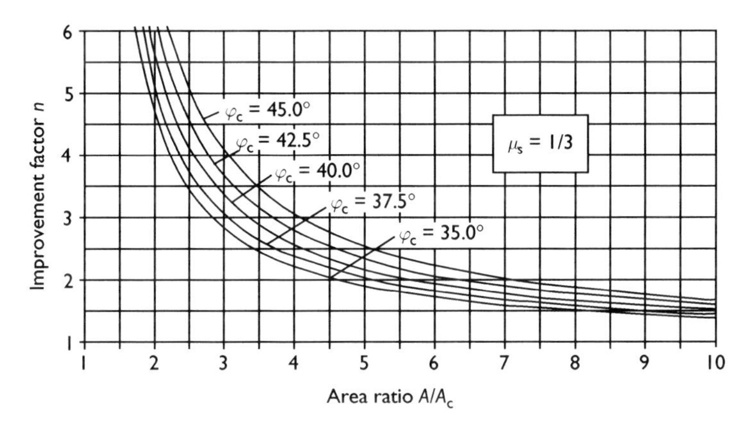

Figure 2.12 Design diagram for improving the ground by vibro replacement stone columns (acc. to Priebe, 1995).

layout for a given number of stone columns beneath a footing. However, in practical applications it is sufficient to determine the grid size required for the calculation by dividing the footing area by the number of columns. The main chart of use in the evaluation of load tests is shown in Figure 2.14.

The application is relatively simple as the relevant settlement ratio depends on the number and diameter of the stone columns together with the treatment depth considered.

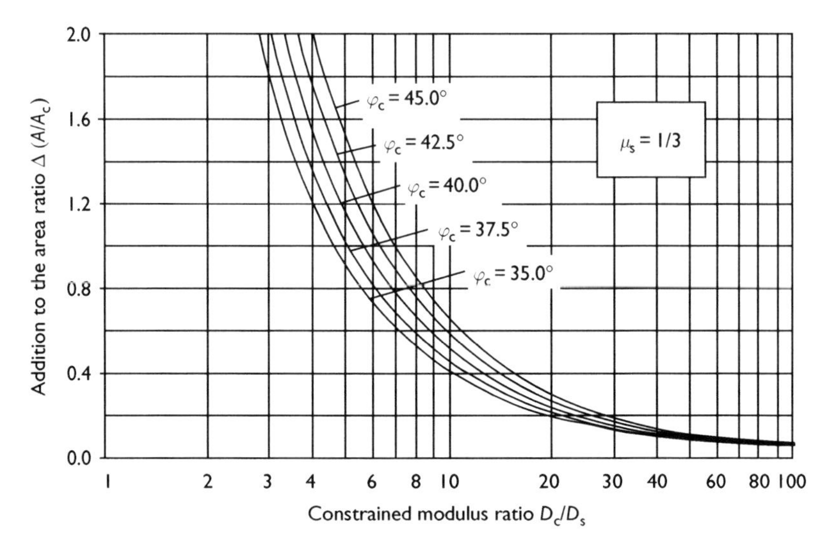

Figure 2.13 Area ratio addition (acc. to Priebe, 1995).

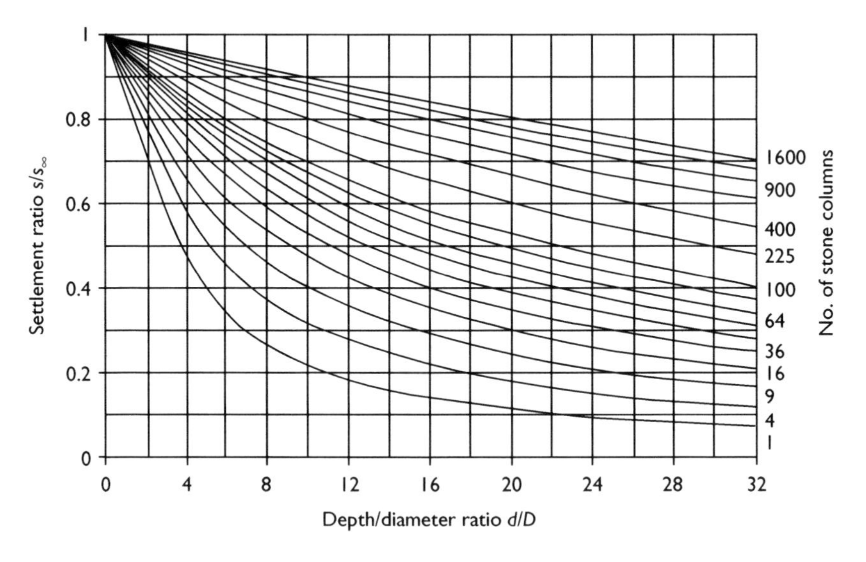

Figure 2.14 Settlement evaluation for isolated footings (after Priebe, 1995).

The United States has seen wider use of Goughnour and Bayuk's (1979) iteration method, even though it is generally considered much more complex. A great number of these calculations are derived from empirical or semi-empirical equations involving simplifying assumptions that do not effectively address the deformation behaviour's intricacies. There is currently a lack of an acceptable design method which can adequately account for all mechanisms that take part in the load transfer process, and also is simple enough for practical use. Therefore, it is best, before making final decisions for the execution of designs for sizeable ground improvement projects, to install test columns and use the achievable column diameters with the load test results to ensure an effective outcome (Chambosse and Kirsch, 1995).

When determining the stress/deformation behaviour in the service load range, simulation calculations, such as the finite element method (FEM) often used in construction, are known to be highly effective. As for vibro replacement stone columns for ground improvement, Schweiger has proposed a method that utilises a homogenised model, the so-called 'ground/column matrix' (Schweiger, 1990). Wehr (1999) has produced noteworthy results regarding the simulation of the failure mechanisms of stone columns, by use of his calculations for single columns and column groups. Brauns' proposed failure modes (1980) and Hu's (1995) model tests (Figures 2.11 and 2.10, respectively) have since been recalculated, resulting in the confirmation that shear zones dictate the settlement behaviour of columns. Currently, numerical analysis by means of FEM has gained acceptance as a valuable tool in stone column ground improvement, when large projects are in the design phase, or current concepts need optimisation.

The phenomenon of liquefaction of granular deposits during an earthquake has been well documented and increasingly studied. Engineering opinion has agreed that the role of liquefaction can be minimised by densifying the soils beyond their liquefaction potential for the site-specific design earthquake. A second method of minimising the role of liquefaction requires the provision of drainage paths, thus allowing rapid dissipation of pore pressure induced by an earthquake. The influence of the drainage capabilities of stone columns have been studied by Baez (1995) concluding that the Seed and Booker (1976) model is useful if allowable maximum pore pressure ratios are maintained below 0.6. Further investigations of *in-situ* stone column compositions indicated that in sands the columns generally have a proportion of 80/20 (gravel to sand) due to the installation process.

A combination of vibro replacement and vibro compaction, where dense permeable stone columns are constructed and the density of the surrounding granular soil is increased, provides an excellent solution to liquefaction problems. Since its first application at Santa Barbara, California in 1974 (Engelhardt and Golding, 1975), it has been used many times. Perhaps the most significant are the documentations of the performance of the Santa Barbara project (Mitchell and Huber, 1986) following a seismic event which induced ground accelerations equal to the design earthquake and the study of 15 sites in the San Francisco area (Mitchell and Wentz, 1991) following the Loma Prieta earthquake. In the latter study, the sites treated by vibro techniques and the buildings founded on them were shown to have suffered no damage during the Loma Prieta earthquake.

The acceleration rates which affect the soil in the immediate vicinity of the depth vibrator greatly surpass those experienced in seismic events. On one project, peak ground accelerations of 1.7 g were detected 0.9 m from the centre of the stone column (Baez and Martin, 1992). As acceleration increases, the soil's shear strength is reduced. Minimum shear strength is attained at approximately 1.5 g (Rodgers, 1979), at which point the soil behaves like a fluid. In saturated sand, complete liquefaction is possible in the event that the increase in pore water pressure generated by the vibrations surpasses the decrease in pore pressure which is naturally caused by filtration/dissipation (Greenwood and Kirsch, 1983).

As long as the treatment medium consists of uniform coarse-grained sands and gravels with a minimum relative density of 80 per cent, the following are attainable: acceptable load-bearing capacities, marginal settlement risk and assurance against liquefaction induced by seismic events (Smoltczyk and Hilmer, 1994). As the percentage of fines increases, higher densities become increasingly difficult to achieve. Therefore, when working in uniform fine-grained or silty sands, it is beneficial to install stone columns that enhance the drainage capacity of the silty sand-like sand drains in clay. Cohesive soils appear to be more resistant to liquefaction than clean sands, but liquefaction is possible as well under seismic action of relatively long duration and high intensity. Many detailed site examples are given by Perlea (2000).

It is not possible to estimate, by statistical analyses, the extent to which the risk of liquefaction is reduced by vibro replacement. The key question is which part of the forces exerted by an earthquake are borne by the columns without any damages. The simple procedure for the design of vibro replacement by Priebe (1995) was modified to account for short-term seismic events (Priebe, 1998). In this case it is more realistic to consider deformations of the soil with the volume remaining constant, that is, to calculate with a Poisson's ratio of 0.5 which also simplifies the formulae. In the procedure above mentioned the improvement factor n_0, which is the basic value of improvement by vibro replacement, is determined initially using

$$n_0 = 1 + \frac{A_c}{A}\left[\frac{1}{K_{ac}(1 - A_c/A)} - 1\right] \qquad K_{ac} = \tan^2\left(\frac{45^\circ - \varphi_c}{2}\right)$$

where A = attributable area within the compaction grid, A_c = cross section of stone columns, φ_c = friction angle of column material. The reciprocal value of this improvement factor is merely the ratio between the remaining stress on the soil between the columns p_s and the total overburden pressure p taken as being uniformly distributed without soil improvement and, as such, can be used as a reduction factor $\alpha = 1/n_0$. On the understanding that the loads taken by the columns from both the structure and the soil do not contribute to liquefaction, it is proposed to use this factor to reduce the seismic stress ratio created by an earthquake and hence evaluate the remaining liquefaction potential according to Seed *et al.* (1983).

A similar approach was proposed by Baez (1995) substituting the above K_{ac} by a ratio between the shear modulus of the soil and the stone column.

It is important to mention that excess pore water pressures play an important role in reducing the effective stresses but are neglected in the conventional design above mentioned. A novel liquefaction approach including pore water pressures was applied by Cudmani *et al.* (2003) to two sites, one of them being Treasure Island influenced by the 1989 Loma Prieta earthquake. Liquefaction was predicted in a concentrated zone comprising both the bottom of a fine sand top layer and an underlying upper part of a silty sand layer. Mitchell and Wentz (1991) reported on the medical building in Treasure Island where the upper fine sand layer was improved with stone columns to a depth of 6.5 m leaving the lower layer unimproved. This resulted in no liquefaction of the improved soil block, but in liquefaction of the silty sand layer below 6.5 m, which was proved by the observation that the bottom 2.5 m of the 6.5 m deep elevator shafts drilled prior to the earthquake were filled with silty sand. Furthermore sand boils were clearly visible outside the improved area.

Another aspect is the design of the extent of soil improvement against liquefaction. An overview is given in Japanese Geotechnical Society (JGS), 1998. Basically there are two questions to be answered about the necessary width and depth of the soil improvement outside the loaded area.

Pore water pressures are transmitted from the liquefied area into the improved area of the ground. It is recommended (JGS, 1998) to improve a lateral area corresponding to an angle of 30° against the vertical axis starting from the edge of the foundation (point A in Figure 2.15). This shall be executed down to a non-liquefiable layer. The area ACD in the figure exhibited particular unstable behaviour during model tests and hence this part should be treated as liquefied in the soil improvement design.

Design guidelines for oil tanks in Japan (JGS, 1998) recommend to improve an area adjacent to the footing corresponding $^2/_3$ of the soil improvement depth (Figure 2.16). Recent research on sand drains for liquefaction remediation yields the lateral extent to be taken as the liquefiable depth (Brennan and Madabhushi, 2002).

In many design codes and standards a maximum treatment depth between 15 and 20 m is given based on experience. A special design chart is available for light-weight and small-scale structures to improve a limited depth leaving a liquefiable soil layer below (JGS, 1998).

The time-dependent behaviour of sand or gravel drains may be analysed using charts proposed by Balaam and Booker (1981). This is an extension of the Barron solution for excess pore water pressure, using the approximate

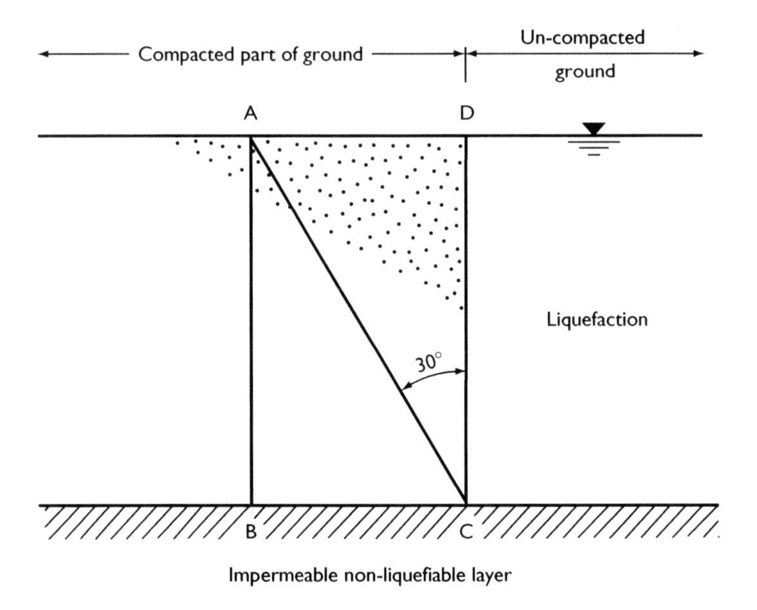

Figure 2.15 Stabilised area (ABCD) adjacent to foundation (JGS, 1998).

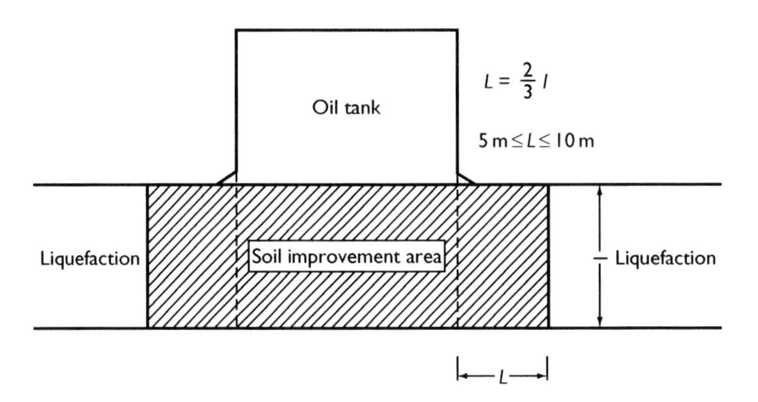

Figure 2.16 Soil improvement area for oil tanks (JGS, 1998).

diffusion theory for consolidation based on Biot's equation of consolidation. The rates of consolidation are presented in charts depending on the diameter ratio of the unit cell and the column d_e/d, the stiffness ratios of the column and the soil under drained conditions E_1/E_2 and a Poisson's ratio of 0.3 which is assumed to be equal for the soil and column.

The design method of Vibro Replacement by Priebe (1995) was extended by Raithel and Kempfert (2000) to account for tensile hoop forces in the geotextile. The hoop force is transformed into a horizontal stress which supports the column additionally to the soil.

Vibro mortar columns (VMC) and vibro concrete columns (VCC) are ideal for weak alluvial soils such as peats and soft clays overlying competent founding strata such as sand, gravels and soft rock. Working loads of up to 650 kN can be achieved in appropriate soils. The 'bulb end' and frictional components of the VCCs enable high safe working loads to be developed at shallower depths than alternative piling systems and thus generally provide a more economical solution.

The Priebe model to design Vibro Replacement was extended to allow also for stiff columns: if the load is higher than the inner strength of the columns, the conventional Vibro Replacement design by Priebe (1995) is executed. But if the column load is lower than the inner strength of the columns, the calculation is modified (Priebe, 2003).

At first the settlement of the soil below the bottom of the VCCs is determined using the stress which corresponds to that of a shallow foundation in a homogeneous half space. This formulation is not on the safe side as the load distribution is smaller than in homogenous soil due to stiffer vibro columns. In a second step the settlement is determined from the difference to the increased stress below the bottom of the columns. This yields the punching effect of the column toe into the soil below. Because there is the difference of the averaged stress, which has been assumed to be quite small before, here a certain compensation is given.

The value determined as column punching has to be added to the settlement of the soil below the columns. A similar model has been developed by Tomlinson (1980) for piled raft foundations.

Once completed, the columns exhibit stiffness that is 10–20 times greater than the adjacent soil. Construction of a supplementary layer of compacted material over the column heads is often performed in order to focus the surface load on the columns. The surface load is focused by means of an arching effect that occurs as this layer thickens. An alternative method involves using a horizontal geotextile. Suspended between the column heads, it prevents the columns from puncturing an attenuated load distribution layer (Kempfert, 1995; Sondermann and Jebe, 1996; Topolnicki, 1996).

2.5 Applications and limitations

Vibro compaction is used to increase the bearing capacity of foundations and to reduce their settlements. Another application is the densification of sand for liquefaction mitigation. In order to reduce the amount of water which has to be pumped during groundwater lowering, sand can be compacted which reduces the permeability. This solution is also possible for dams.

Vibro compaction is limited by the fines and carbonate content (see Section 2.4). Furthermore a certain distance should be kept to existing buildings in order to limit settlements of new buildings. In case of historical buildings vibrations should be monitored. Depths down to 65 m have been improved so far by vibro compaction.

Various ways of creating vibro stone columns have been developed in order to enhance load-bearing capacities of weak soils and limit settlement. For the support of individual or strip foundations, small groups of columns are employed. Large column grids are placed beneath rigid foundation slabs or load configurations that exhibit flexibility, as is the case with storage tanks and embankments. Due to inherent higher shear resistance, vibro replacement columns are a good choice for the enhancement of slope stability. When drainage takes precedence over bearing capacity, vibro sand columns can be employed to function as drains. This drainage type of sand column is constructed simply by lifting the vibrator without compaction, leaving the sand in a state ranging from loose to medium dense.

Vibro stone columns are not suitable in liquid soils with a very low undrained cohesion, because the lateral support is too small. However, vibro stone columns have been installed successfully in soil with 5 kPa $< c_u <$ 15 kPa (see Section 2.7). In case of very hard and/or cemented layers (i.e. caprock) or very well-compacted surface layers pre-boring may be necessary to assist the penetration of the vibrator. Concerning the distance to buildings the same applies as for vibro compaction. Depths down to 25 m have been improved so far by vibro replacement.

2.6 Monitoring and testing

Part of the state of the art is to monitor and record in great detail the operating parameters of any deep vibro work. Details are given in the European standards 'Ground treatment by deep vibration' (European Standard WG12, 2003).

Vibro compaction is monitored online by using devices which record, as a function of time, penetration depth, energy consumption of the motor and, if necessary, pressure and quantity of the flushing media used. If the vibrator frequency can be adjusted during the compaction process, this parameter is also recorded.

For the vibro replacement method all of the essential parameters of the production process (depth, vibrator energy, feed, contact pressure and stone/concrete consumption) are recorded continuously as a function of time, providing the user with visible and controllable data for producing a continuous stone column. Such instrumentation is available for leader-mounted, bottom-feed vibrator systems and has been used in Europe since the 1980s (Slocombe and Moseley, 1991). A typical printout for stone column construction is given in Figure 2.17. Additional to the online control, the final site records include the position and elevation of columns,

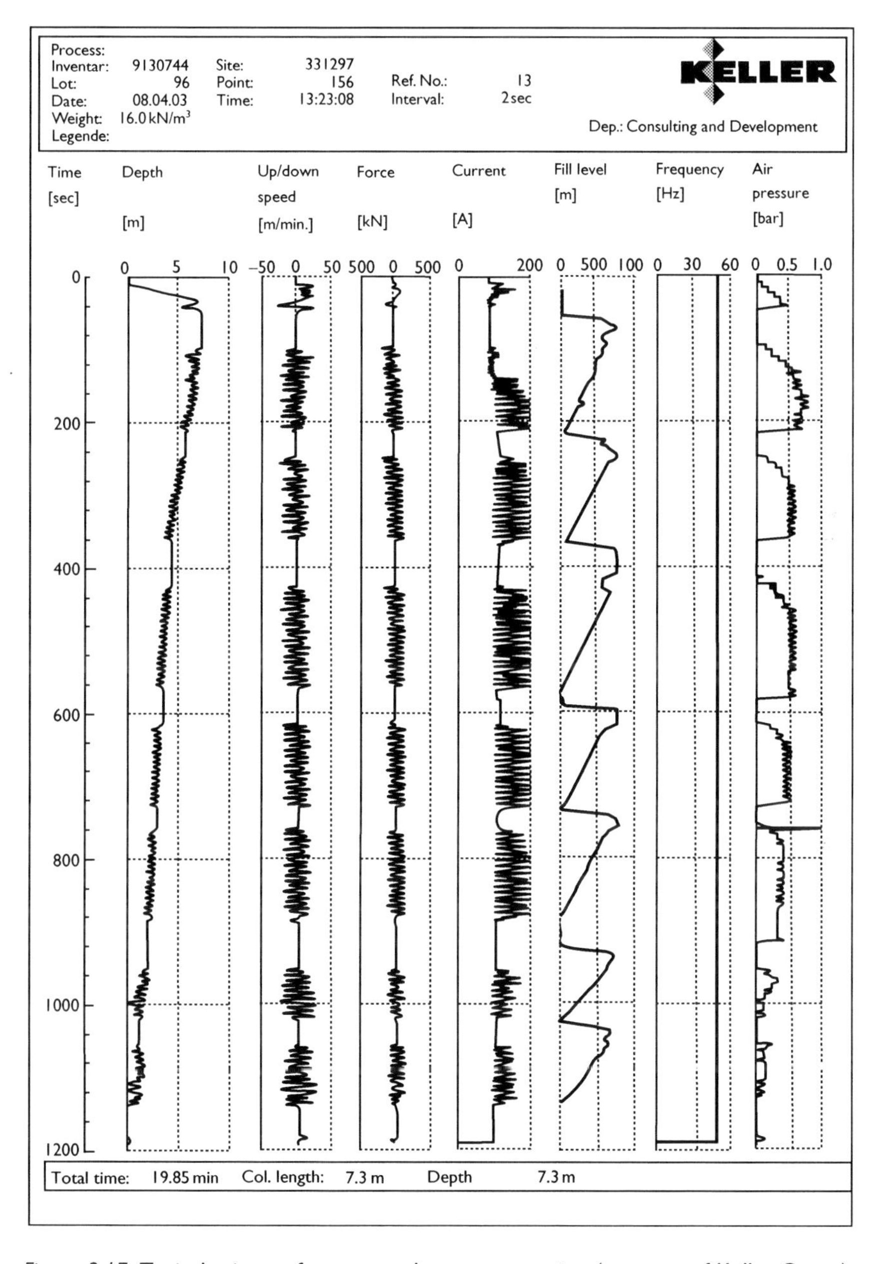

Figure 2.17 Typical printout for stone column construction (courtesy of Keller Group).

the source, type and quality of imported material and, if necessary, environmental factors (noise, vibrations, etc.).

Cone penetration tests (CPT) and standard penetration tests (SPT) are commonly used to verify the success of vibro compaction, with CPT being the better of the two. To compare the initial and final compressibility of the soil, pre- and post-test undergo comparison. When evaluating post-deep compaction-work test results, the ageing effect must be taken into account. This ageing effect on strength goes on for up to several weeks after the column has been installed. Many projects have demonstrated that the strength of compacted sands have the potential, over several weeks, to increase anywhere from 50 to 100 per cent. This substantial increase is attributed to pore water pressure reduction, sometimes in combination with the re-establishment of physical and chemical bonding forces to the column's grain structure (Mitchell *et al.*, 1984; Massarsch, 1991; Schmertmann, 1991). Taking this strength increase into account, it is best to wait at least for 1 week after compaction work before conducting formal compaction tests. The technical literature record contains much information regarding reports on compaction tests and monitoring (Covil *et al.*, 1997; Slocombe *et al.*, 2000).

The performance of vibro stone columns is monitored only for large projects using large-plate load tests which should be carried out by loading a rigid plate or cast *in-situ* concrete pad big enough to span one or more columns and the intervening ground. Zone load tests should be carried out by loading a large area of treated ground, usually by constructing and loading a full-size foundation or placing earth fill to simulate widespread loads.

2.7 Case histories

2.7.1 Vibro compaction in Singapore (2002)

Extensive ground improvement using vibro compaction of reclaimed sand fill was carried out below a future twin crude oil pipeline on Jurong island in Singapore (Wehr and Raju, 2002). Compaction was executed to depths ranging between 20 and 30 m. Onshore compaction was done underneath the future pipeline and offshore compaction on the seaside slope at the (VLCC) very large crude carrier jetty. In total, over 1.9 million cubic meters were compacted over a 3 month period.

The foundation design for the pipeline called for the densification of a 20 m wide strip of the loose reclaimed sand underneath the pipeline to a relative density (RD) of 70 per cent. (see Figure 2.18). Prior to compaction, pre-CPT tests were carried out. Pre-compaction cone resistances varied between 5 and 8 MPa. Friction ratios were generally about 0.5 per cent.

Based on the results of a field trial with varying compaction point spacings, an equilateral triangular grid with 3.5 m spacing was chosen for

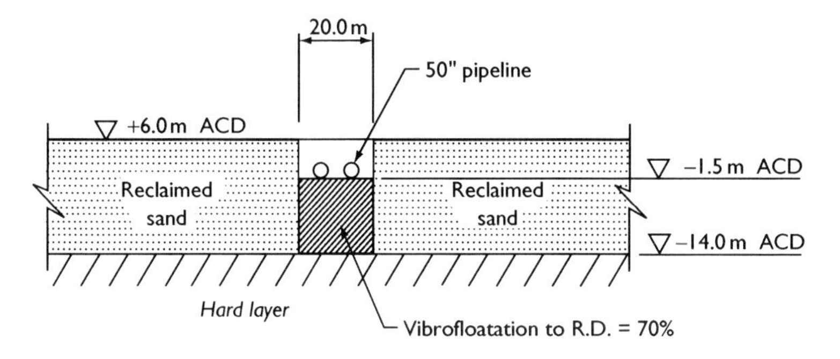

Figure 2.18 Cross-section showing pipeline and treatment area.

compaction to 70 per cent relative density underneath the pipeline. The settlement of the sand surface was approximately 1.5 m. Following compaction a post-CPT test was carried out in the centroid of the triangle formed by the compaction points. Figure 2.19 shows a typical post-compaction CPT result. The required tip resistance corresponding to 70 per cent RD based on a correlation by Schmertmann is also shown.

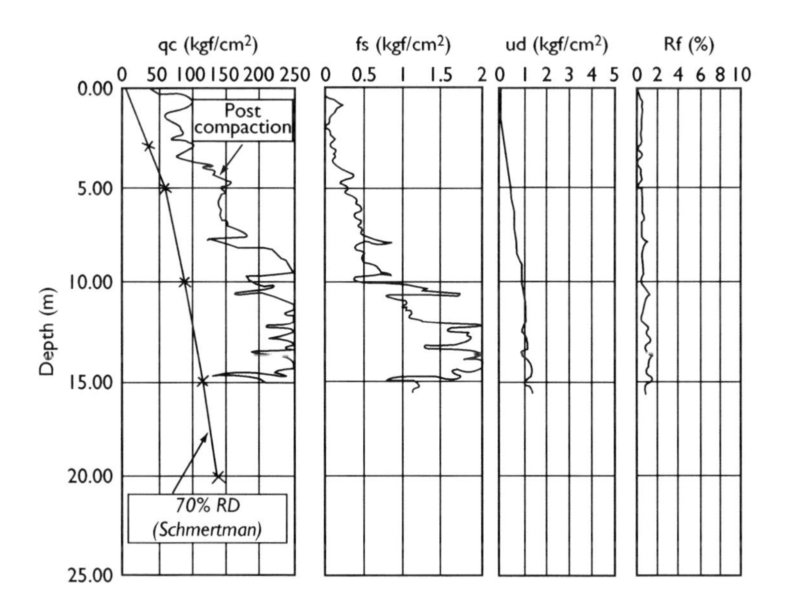

Figure 2.19 Typical post-compaction CPT result including compaction requirement corresponding to 70 per cent relative density.

At the sand bund, compaction was carried out to 70 per cent RD in a 20 m wide strip below the pipelines over the full depth of the sand fill as the pipeline was now located at the top of the sand fill (see Figure 2.20). The average compaction depth was 27 m below the ground. Adjacent to this strip the 1:3 inclined slopes had to be densified to the Jurong town council (JTC) specification which corresponds to a tip resistance of 4, 6, 8 and 10 MPa at a depth of 0, 2, 8 and 24 m below ground level.

A triangular grid with 3.5 m spacing was chosen below the pipelines to achieve the 70 per cent RD requirement. A triangular grid with 4.0 m spacing was chosen on the sea side slope area to achieve the JTC specification. Compaction was carried out using barges (see Figure 2.21). The positioning of the barge-based vibrator units was realised with a GPS-system. Post-CPT tests were carried out on land and in sea. A typical post-CPT result is presented in Figure 2.22, showing also the required cone resistance by the JTC specifications.

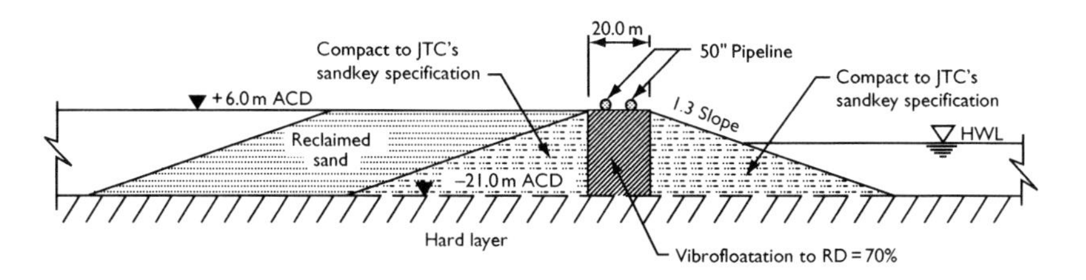

Figure 2.20 Cross section of sand bund showing treatment area.

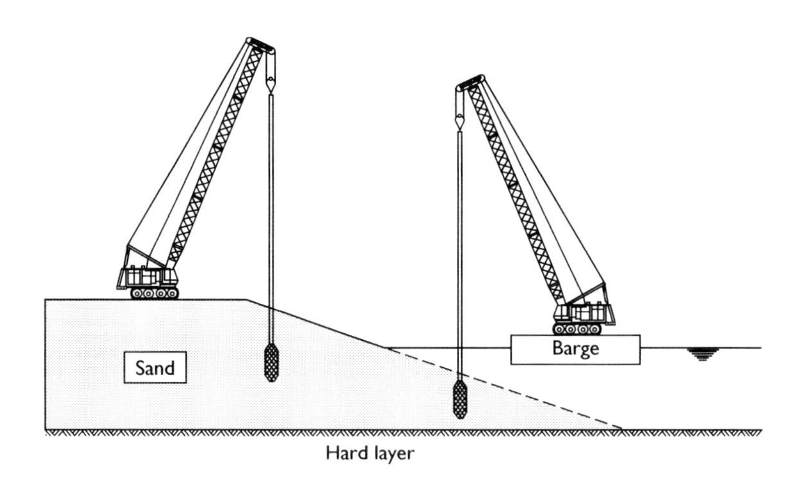

Figure 2.21 Schematic showing on- and offshore compaction rigs.

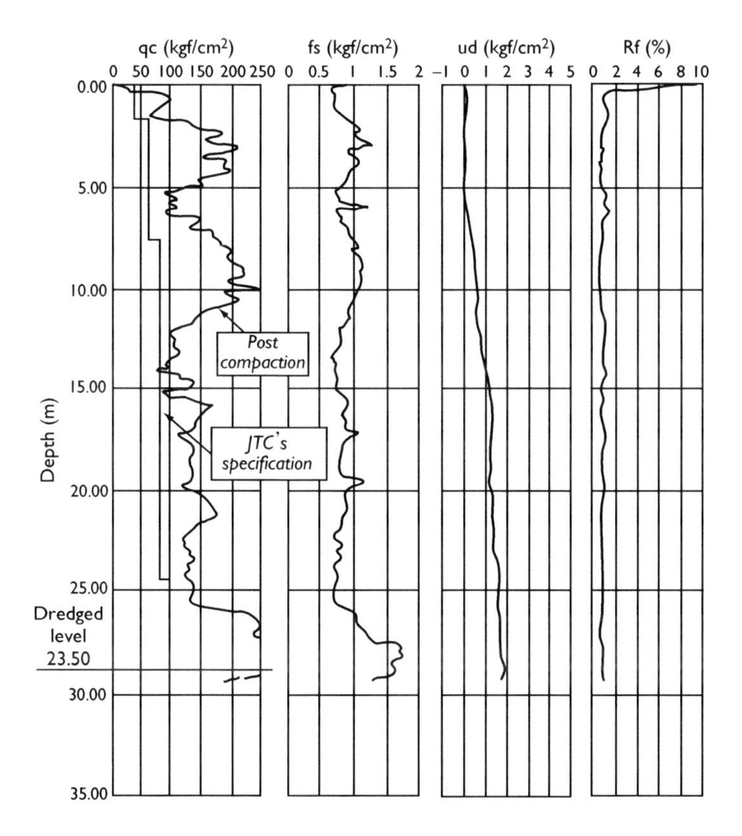

Figure 2.22 Post-CPT with JTC specifications.

2.7.2 *Vibro stone columns underneath embankments in Malaysia (2001)*

The construction of highway embankments and bridge abutments of six projects at Putrajaya, Malaysia required the construction of earth embankments with heights ranging between 16 and 20 m (Raju, 2002). The presence of soft silts and clays with undrained shear strength values between 6 and 20 kPa and to depths ranging between 8 and 12 m posed problems with regard to slope stability and excessive settlements.

The original proposal required the excavation and disposal of the soft soils. *In-situ* treatment using vibro replacement was chosen as an environmentally friendly and economical solution. The treatment was designed to meet strict tolerances with regard to long-term settlements and lateral displacements. Limiting lateral displacements was essential so as to avoid lateral forces on the piles supporting the bridge abutment. Over 420 000 lin m of vibro stone columns were carried out on these projects which demonstrates:

1 The possibility to treat very soft soils: Stone columns are being installed routinely in cohesive soils with shear strengths as low as 6 kPa in Malaysia. Close to the ground level in swampy areas, the shear strength may be as low as 4 kPa. This has not posed a problem since first a sand platform is placed which results in some consolidation under the weight of the platform. The sand platform also assists in providing lateral support to the columns at the top and a safe working ground for the equipment.
2 The ability of the vibro stone columns to carry very high loads: A 20 m high embankment on stone columns installed on a 2 m × 2 m grid implies a load per 'column cell' (defined as the stone column surrounded by the attributable soil in the grid area) of about 1600 kN. This is an exceptionally high load that can exceed the classical bearing capacity of most single columns or driven piles. However a group of stone columns can bear this load when subjected to spread loads. This is because of the very high ductility of the columns. Even for large deformations with axial strains >10 per cent, the bearing capacity of the columns is not compromised.
3 The ability to limit lateral displacements of high embankments: With careful soil investigation, design, column installation and monitoring during embankment construction, it has been possible to ensure stability of up to 20 m high embankments and limit lateral movements to within 10 cm.

2.7.3 Vibro stone columns to prevent liquefaction in the Philippines (2000)

Approximately 550 km in front of the coast of the Philippines, a new gas field will be linked to the existing Shell Malampaya Onshore gas plant. An extension of an existing complex near Batangas, 100 km south of Manila, was necessary.

The soil consists of soft clay in the upper 3 m and a liquefiable loose sand layer 11 m deep, becoming denser with depth. This is underlain by silts and clays down to the base stratum at 25 m.

In order to allow for a bearing pressure of up to 150 kN/m^2 with specified settlements less than 25 mm stone columns were installed with the dry bottom-feed method in a 2 m × 2 m grid. The stone column depth varied between 8 and 18 m. Additionally, an intermediate grid was installed in some areas to limit the settlements in the upper soft clay layer.

According to the recommendations by Priebe (1998), stone columns were installed as well to reduce the liquefaction potential of loose sand, Figure 2.23. The intermediate grid was also necessary in areas with very loose sand to prevent liquefaction due to a maximum ground acceleration of $a = 0.42$ g. To verify the settlement criteria, two-zone load tests with a 4 m × 4 m footing were executed with a load up to 150 kN/m^2. The

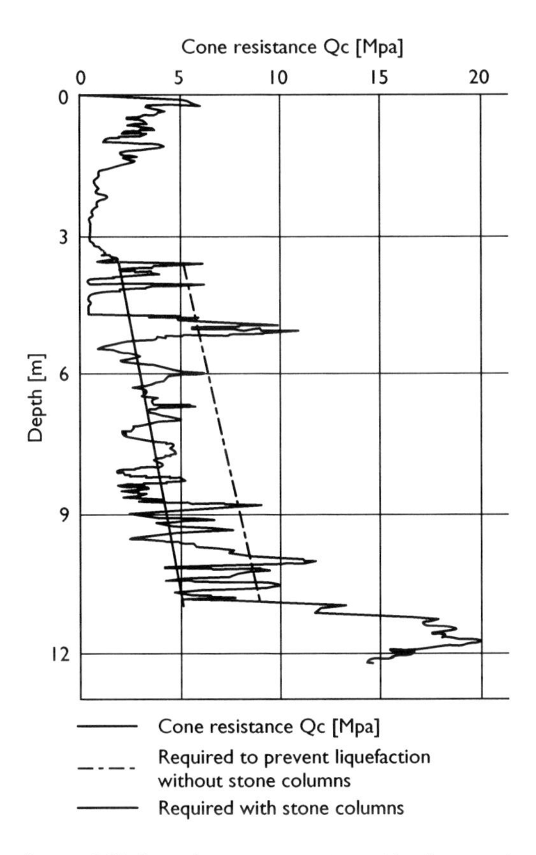

Figure 2.23 Liquefaction mitigation with vibro replacement (by courtesy of Keller Group).

measured settlements were less than 10 mm confirming that the original settlement calculations included a comfortable factor of safety.

2.7.4 Vibro stone columns to prevent liquefaction in the USA (1997)

The geotechnical-earthquake engineering design and construction efforts for the expansion of the facilities at the Albany Airport in New York are presented (Soydemir *et al.*, 1997). The subsurface soil conditions consisted of primarily fine sands with varying amounts of silt, with the silt content reaching to levels above 50 per cent within the top 7.5 m. The groundwater table was at shallow depth within the near-surface deposits.

The relatively loose near-surface deposits were evaluated to be liquefaction-susceptible and subject to significant magnitudes of seismically induced settlements under the prescribed 'design earthquake' (M = 6.0, PGA = 0.15 g, FS = 1.25). Liquefaction envelopes were developed using SPT data (Seed *et al.*, 1983). A ground improvement programme was recommended by the project geotechnical engineer.

Because of the relatively high and variable silt content Keller recommended the utilisation of stone columns. It was believed that stone columns would contribute to seismic resistance by both providing an overall, although somewhat limited, densification (i.e. vibro compaction), and maintaining the excess pore pressure buildup at an acceptable level (pore pressure ratio at or below 60 per cent). The design of the ground improvement scheme was based on these criteria and drainage analyses (LARF computer model; Seed and Booker, 1976). An arrangement of 0.9 m diameter stone columns at a 3 m × 3 m grid was recommended. Subsequently, as a result of the two test areas, the design was modified to a more economical 3.6 m × 3.6 m grid. The modified design for the production stone columns was monitored by compliance testing on a regular basis.

This case study demonstrates that in loose, saturated fine sand with relatively high and variable silt content, stone columns could be effectively designed and constructed to provide resistance to liquefaction and to control seismically induced settlements.

2.7.5 *Vibro concrete columns in Germany (2001)*

The world's largest paper-processing plant was built on vibro concrete columns in Germany. The main building is 240 m long and 60 m wide. All machine foundations including the foundation slab were founded on VCC.

The soil consists of soft silt and clay in the upper 3 m underlain by loose sand and gravel. A pre-densification in a 3.75 m triangular vibro compaction grid down to 11 m was executed in the lower sand/gravel to build a good base for the vibro concrete columns. The CPT cone resistance was increased by a factor of approximately 3 showing a considerable increase even for the large grid chosen.

The vibro concrete columns with a diameter of 0.4 m were installed only in the upper 5 m through the soft layer in a 3–11 m^2 triangular grid to take into account the various loads between 40 and 250 kN/m^2.

With this combination of vibro compaction and vibro concrete columns, the very tight differential settlement criteria of the high-bay warehouse could be met despite of the variable thickness of the soft top layer.

2.8 Conclusions

The vibro systems have proved since over 50 years to offer safe and economic methods of improving weak soils for a wide range of applications. Vibro compaction has been used to densify granular soils to significant depths

(65 m) and the ability of this technique to reduce the risk of liquefaction during an earthquake is well-documented. Vibro replacement is a widely accepted technique for improving cohesive and fine-grained soil to support a wide range of structures. Vibro concrete columns finally are a good alternative to piles.

References

Almeida, M.S.S., Jamiolkowski, M. and Peterson, R.W. (1992) Preliminary result of CPT tests in calcareous Quiou sand, *Proceedings of the International Symposium on Calibration Chamber Testing*, Potsdam, New York, 1991, Elsevier, pp. 41–53.

Arbeitsgruppe Untergrund-Unterbau der Forschungsgesellschaft für das Straßenwesen (AUFS) (1979) Köln, Merkblatt für die Bodenverdichtung im Straßenbau.

Baez, J.I. (1995) A design model for the reduction of soil liquefaction by vibro stone columns, PhD Thesis, University of Southern California, USA.

Baez, J.I. and Martin, G.R. (1992) Liquefaction observations during installation of stone columns using the vibro replacement technique, *Geotechnical news*, pp. 41–44.

Balaam, N.P. and Booker, J.R. (1981) Analysis of rafts and granular piles, *International Journal for Numerical and Analytical Methods in Geomechanics*, Vol. 5, pp. 379–403.

Bell, A.L. (1915) The lateral pressure and resistance of clay and the supporting power of clay foundations 2, *Proceedings of Institution Civil Eng.*

Bellotti, R. and Jamiolkowski, M. (1991) Evaluation of CPT and DMT in crushable and silty sands, Third interim report ENEL C.R.I.S., Milano.

Bergado, D.T. *et al.* (1994) Improvement techniques of soft ground in subsiding and lowland environment. Balkema Rotterdam.

Brauns, J. (1978) Die Anfangstraglast von Schottersäulen in bindigem Untergrund. Bautechnik 8.

Brauns, J. (1980) Untergrundverbesserungen mittels Sandpfählen oder Schottersäulen. TIS 8.

Brennan, A.J. and Madabhushi, S.P.G. (2002) Liquefaction remediation by vertical drain groups, *Proceedings of the International Conference on Physical Modelling in Geotechnics*, St. John's, Newfoundland, July, pp. 533–538.

British research establishment, BRE (2000) Specifiying vibro stone columns.

Chambosse, G. and Kirsch, K. (1995) Beitrag zum Entwicklungsstand der Baugrundverbesserung. Beiträge aus der Geotechnik. München.

Covil, C.S. *et al.* (1997) Case history: Ground treatment of the sand fill at the new airport at Chek Lap Kok, Hong Kong, *Proceedings of the 3rd International Conference on Ground Improvement Geosystems*, London.

Cudmani, R.O. (2001) Statische, alternierende und dynamische Penetration in nichtbindigen Böden, Dissertation, Karlsruhe University.

Cudmani, R.O., Osinov, V.A., Bühler, M.M. and Gudehus, G. (2003) A model for the evaluation of liquefaction susceptibility in layered soils due to earthquakes, *12th Panamerican Conference on SMGE*, Cambridge, USA.

Degen, W. (1997a) Vibroflotation ground improvement, Altendorf, unpublished.

Degen, W. (1997b) 56 m deep vibro compaction at German lignite mining area, *Proceedings of the 3rd International Conference on Ground Improvement Geosystems*, London.

Engelhardt, K. and Golding, H.C. (1975) Field testing to evaluate stone column performance in a seismic area, *Geotechnique*, pp. 61–69.

European Standard (2003) Ground treatment by deep vibration, TC288, WG12, in preparation.

Fellin, W. (2000) Rütteldruckverdichtung als plastodynamisches Problem, *Advances in Geotechnical Engineering and Tunnelling*, Vol. 3.

Foray, P.Y., Nauroy, J.-F. and Colliat, J.L. (1999) Mechanisms governing the behaviour of carbonate sands and influence on the design of deep foundations.

Forschungsgesellschaft für das Straßenwesen (FGFS) (1979) Merkblatt für die Untergrundverbesserung durch Tiefenrüttler.

German Patent: Nr. 22 GO 473.

Goughnour, R.R. and Bayuk, A.A. (1979) Analysis of stone column–soil matrix interaction under vertical load. C.R. Coll. Int. Renforcement des Sols. Paris.

Greenwood, D.A. (1976) Discussion. Ground treatment by deep compaction, *Institution of Civil Engineers*, London.

Greenwood, D.A. and Kirsch, K. (1983) Specialist ground treatment by vibratory and dynamic methods, *Advances in Piling and Ground Treatment for Foundations*, London.

Hu, W. (1995) Physical modelling of group behaviour of stone column foundations, PhD Thesis. University of Glasgow.

Institution of Civil Engineers, ICE (1987) Specification for ground treatment, Thomas Telford.

Japanese Geotechnical Society (JGS) (1998) Remedial measures against soil liquefaction, From investigation and design to implementation, A.A. Balkema.

Jebe, W. and Bartels, K. (1983) Entwicklung der Verdichtungsverfahren mit Tiefenrüttlern von 1976–1982, *European Conference on Soil Mechanics and Foundation Engineering*, Helsinki.

Keller (1993) brochure 13-21D: Sanierung des Hochwasserdammes der Leitha bei Rohrau Pachfurt in Niederösterreich.

Keller (1997) brochure: Rütteldruckverdichtung, Hamburg Elbtunnel, Vierte Röhre.

Keller (2002) brochure 11-31E: Offshore vibro compaction for breakwater construction, Seabird project Karwar, India.

Kempfert, H.-G. (1995) Zum Tragverhalten geokunststoffbewehrter Erdbauwerke über pfahlähnlichen Traggliedern, *Informations- und Vortragstagung über Kunststoffe in der Geotechnik*, TU München.

Kirsch, K. (1979) Erfahrungen mit der Baugrundverbesserung durch Tiefenrüttler. Geotechnik 1.

Kirsch, K. (1993) Baugrundverbesserung mit Tiefenrüttlern. 40 Jahre Spezialtiefbau: 1953–1993. Technische und rechtliche Entwicklungen. Düsseldorf.

Kirsch, K. and Chambosse, G. (1981) Deep vibratory compaction provides foundations for two major overseas projects, *Ground Engineering*, Vol. 14, No. 8.

Kirsch, K. and Sondermann, W. (2003) Ground Improvement. In: Smoltczyk, V. (ed.), *Geotechnical Engineering Handbook*, Vol. 2, Ernst & Sohn.

Massarsch, K.R. (1991) Deep soil compaction using vibratory probes in deep foundation improvement, STP 1089, ASTM.

Massarsch, K.R. (1994) Design aspects of deep vibratory compaction, *Proceedings Seminar on Ground Improvement Methods*, *Hong Kong Inst. Civ. Eng.*

Mitchell, J.K. and Huber, T.R. (1986) Stone columns foundation for a wastewater treatment plant – a case history, *Geotechnical Engineering*, Vol. 14.
Mitchell, J.K. and Wentz, J.R. (1991) Performance of improved ground during Loma Prieta earthquake. Report No. UCB/EERC-91/12, Earthquake Engineering Research Center, University of California, Berkeley.
Mitchell, J.K. *et al.* (1984) Time dependent strength gain in freshly deposited or densified sand, *Journal of Geotechnical Engineering*, Vol. 110, No. 11.
Moseley, M.P. and Priebe, H.J. (1993) Vibro techniques. In: M.P. Moseley (ed.), *Ground Improvement*, Blackie Academic and Professional.
Perlea, V. (2000) Liquefaction of cohesive soils, Soil Dynamics and Liquefaction 2000, ASCE Geotechnical Special Publication No. 107, 58–76.
Priebe, H. (1998) Vibro replacement to prevent earthquake induced liquefaction, *Ground Engineering*, September.
Priebe, H. (2003) Zur Bemessung von Rüttelstopfverdichtungen, Bautechnik.
Priebe, H.J. (1995) The design of vibro replacement, *Ground engineering*, December.
Raithel, M. and Kempfert, H.-G. (2000) Calculation models for dam foundations with geotextile coated sand columns, International conference, Geoeng. 2000.
Raju, V.R. (2002) Vibro replacement for high earth embankments and bridge abutment slopes in Putrajaya, Malaysia; International Conference on Ground Improvement Techniques, Malaysia, pp. 607–614.
Raju, V.R. and Hoffmann, G. (1996) Treatment of tin mine tailings in Kuala Lumpur using vibro replacement, *Proceedings of 12th SEAGC.*
Rodgers, A.A. (1979) Vibrocompaction of cohesionless soils. Cementation Research Limited. Internal report.
Schmertmann, J.H. (1991) The mechanical aging of sand, *Journal of Geotechnical Engineering*, Vol. 117, No. 9.
Schneider, H. (1938) Das Rütteldruckverfahren und seine Anwendungen im Erd- und Betonbau. Beton und Eisen 37, H. 1.
Schüßler, M. (2002) Anwendung neuer, innnovativer Gründungslösungen- ist das Risiko für den Auftraggeber überschaubar?, Baugrundtagung, pp. 339–345.
Schweiger, H.F. (1990) Finite Element Berechnung von Rüttelstopfverdichtungen. 5. Christian Veder Kolloquium. Graz.
Seed, H.B. and Booker, J.R. (1976) Stabilisation of potentially liquefiable sand deposits using gravel drain systems, Report No. EERC76-10, U.C.Berkeley.
Seed, H.B., Idriss, I.M. and Arango, I. (1983) Evaluation of liquefaction potential using field performance data, ASCE, *Journal of Geotechnical Engineering*, Vol. 109.
Sidak, N. and Strauch, G. (2003) Herstellung geotextilummantelter Kiestragsäulen mit Keller-Tiefenrüttlern, Osterreichischer Geotechniktag, 415–434.
Slocombe, B.C. and Moseley, M.P. (1991) The testing and instrumentation of stone columns, ASTM STP 1089.
Slocombe, B.C. *et al.* (2000) The in-situ densification of granular infill within two cofferdams for seismic resistance. Workshop on compaction of soils, granulates and powders. Innsbruck.
Smoltczyk, U. and Hilmer, K. (1994) Baugrundverbesserung. Grundbautaschenbuch, 5. Auflage, Teil 2.
Sondermann, W. and Jebe, W. (1996) Methoden zur Baugrundverbesserung für den Neu- und Ausbau von Bahnstrecken auf Hochgeschwindigkeitslinien. Baugrundtagung Berlin.

Soydemir, C., Swekowsky, F., Baez, J.I. and Mooney, J. (1997) Ground improvement at Albany Airport, Ground improvement, Ground reinforcement, Ground Treatment Developments 1987–1997. In: V.R. Schaefer (ed.), Geotechnical Special Publication No. 69, ASCE, Logan, UT.

Soyez, B. (1987) Bemessung von Stopfverdichtungen. BMT. April.

Tomlinson, M.J. (1980) Foundation design and construction, Pitman publishing.

Topolnicki, M. (1996) Case history of a geogrid-reinforced embankment supported on Vibro Concrete Columns. Euro Geo 1. Maastricht.

US Department of Transportation (USDT) (1983) Design an construction of stone columns, Vol. 1.

Van Impe, W.F. and Madhav, M.R. (1992) Analysis and settlement of dilating stone column reinforced soil. ÖIAZ 3.

Vesic, A.S. (1965) Ultimate loads and settlements of deep foundations in sand, *Proceedings of the Symposium on bearing capacity and settlement of foundations in sand*, Duke University, Durham, pp. 53–68.

Wehr, J. and Raju, V.R. (2002) On- and offshore vibro compaction for a crude oil pipeline on Jurong Island, Singapore; International Conference on Ground Improvement Techniques, Malaysia, pp. 731–736.

Wehr, W. (1999) Schottersäulen–das Verhalten von einzelnen Säulen und Säulengruppen, Geotechnik 22.

Chapter 3

Dynamic compaction

B.C. Slocombe

3.1 Introduction

Dynamic compaction strengthens weak soils by *controlled* high-energy tamping. The reaction of soils during dynamic compaction treatment varies with soil type and energy input. A comprehensive understanding of soil behaviour, combined with experience of the technique, is vital to successful improvement of the ground. Given this understanding, dynamic compaction is capable of achieving significant improvement to substantial depth, often with considerable economy when compared to other geotechnical solutions.

3.2 History

The principle of dropping heavy weights on the ground surface to improve soils at depth has attracted many claims for its earliest use. Early Chinese drawings suggested the technique could be several centuries old (Menard and Broise, 1976). Kerisel (1985) reports that the Romans used it for construction, and Lundwall (1968) reports that an old war cannon was used to compact ground in 1871. In the twentieth century compaction has been provided to an airport in China and a port area in Dublin during the 1940s, and to an oil tank in South Africa in 1955. However, the advent of large crawler cranes has led to the current high-energy tamping levels first being performed on a regular basis in France in 1970 and subsequently in Britain and North America in 1973 and 1975 respectively.

3.3 Plant and equipment

At first sight, the physical performance of dynamic compaction would appear to be simple, i.e. a crane of sufficient capacity to drop a suitable size of weight in virtual free fall from a certain drop height. Most contracts are performed with standard crawler cranes, albeit slightly modified for safety reasons, with a single lifting rope attached to the top of the weight (Figure 3.1). Details such as crane counterbalance weights, jib flexure,

Figure 3.1 Typical crawler crane and equipment.

torque convertors, line pull, drum size, type and diameter of ropes, clutch, brakes, as well as many other factors and methods of working have been subjected to rigorous analysis by the major specialist organisations to improve reliability and productivity. The operation must be performed safely and as a result the Health and Safety Executive in Britain requires that a crane should operate at not more than 80 per cent of its safe working load. Some cranes are better suited than others to the rigours of this type of work, even though on paper they appear to be of similar capacity.

Recent crane developments allow automation of the whole work cycle. This is controlled by a data processing unit which plots for each compaction point its location, number, weight size, drop height, number of blows and measurement of imprint achieved. A particular feature of one European crane is the free fall winch which adjusts the rope length automatically after each blow. Some cranes include the ability for synchronous operation of two winches to lift larger weights than the conventional crane rating.

The majority of British and American contracts utilise weights within the range of 6–20 tonnes dropped from heights of up to 20 m. The majority of UK work is now performed using 8 tonne weights dropped from heights of up to 15 m. Standard crawler cranes have also been used in America for weights of up to 33 tonnes and 30 m height. Specialist lifting frames with quick release mechanisms have been utilised to drop weights of up to 50 tonnes and Menard built equipment to drop 170 tonnes from 22 m height at Nice Airport. In America, and increasingly in Britain, the system is known as dynamic deep compaction.

Weights are typically constructed using toughened steel plate, box-steel and concrete, or suitably reinforced mass concrete where durability is the prime requirement. The effect of different sizes and shapes of the weight has also been extensively researched. Exceptionally, narrow weights have been used to specifically drive material down to depth to form columns in peaty or Sabkha soils. Treatment has also been performed below water using barge-mounted cranes and more streamlined weights with holes cut out to reduce water resistance and increase impact velocity on the seabed.

A form of dynamic compaction, the rapid impact compacter, uses a modified piling hammer to transmit energy through a steel-tamping foot that remains in contact with the ground. This equipment uses higher numbers of lower-energy impacts to achieve treatment depths of about 3.0 m. It is however less successful at treating the mixed soils generally encountered in the UK.

3.4 Terminology

The world-wide use of dynamic compaction has resulted in a large number of important terms, some of which can have different meanings to different nationalities or could be confused with other geotechnical descriptions. The following terms have been adopted in Britain:

(a)	Effective depth of influence	The maximum depth at which significant improvement is measurable
(b)	Zone of major improvement	Typically $^1/_2$ to $^2/_3$ of effective depth
(c)	Drop energy	Energy per blow, i.e. mass multiplied by drop height (tonne metres)
(d)	Tamping pass	The performance of each grid pattern over the whole treatment area
(e)	Total energy	Summation of the energy of each tamping pass, i.e. number of drops multiplied by drop energy divided by respective grid areas (normally expressed in tonne metre/metre^2)
(f)	Recovery	The period of time allowed between tamping passes to permit the excess pore pressures to dissipate to a low enough level for the next pass
(g)	Induced settlement	The average reduction in general site level as a result of the treatment
(h)	Threshold energy	Energy input beyond which no further improvement can practically be achieved
(i)	Overtamping	A condition in which the threshold energy has been exceeded, sometimes deliberately, causing remoulding and dilation of the soil
(j)	Shape test	Detailed measurement of imprint volume and surrounding heave or drawdown effect which permits comparison of overall volumetric change with increasing energy input
(k)	Imprint	The crater formed by the weight at a tamping location

3.5 How dynamic compaction works

There is a fundamental difference between the response of granular and cohesive soils when subjected to the high-energy impacts of the process. It is normal to visualise treatment as a series of heavy-tamping passes with different combinations of energy levels designed to achieve improvement to specific layers within the depth to be treated. The most common approach is to consider the ground in three layers. The first tamping pass is aimed at treating the deepest layer by adopting a relatively wide grid pattern and a suitable number of drops from the full-height capability of the crane. The middle layer is then treated by an intermediate grid, often the mid-point of the first pass or half the initial grid, with a lesser number of drops and reduced drop height. The surface layers then receive a continual tamp of a small number of drops from low height on a continuous pattern.

It is sometimes feasible to combine, and sometimes necessary to subdivide, the basic tamping passes for the reasons outlined below.

The performance of increasing correctly controlled total energy input will normally lead to better engineering performance of the treated ground. However, analysis of over 100 contracts that involved *in-situ* and large loading tests has shown that this is not a linear relationship and that the post-treatment parameters are heavily dependent upon the characteristics of the soil. There is however a broad trend for similar total energy inputs, whether per m^2 of area or m^3 of treatment depth, to provide better performance to granular than mixed than cohesive than refuse-contaminated soils.

3.5.1 *Granular soils*

In dry granular materials, i.e. sand, gravel, ash, brick, rock, slag, etc., it is very easy to understand how tamping improves engineering properties. Physical displacement of particles and, to a lesser extent, low-frequency excitation will reduce void ratio and increase relative density to provide improved load-bearing and enhanced settlement characteristics. A feature that often develops when providing treatment to coarse-fill materials is the formation of a hard 'plug' that inhibits penetration of stress impulses to the deeper layers but is very useful in providing superior settlement performance beneath isolated foundation bases.

When granular materials extend below the water table, a high proportion of the dynamic impulse is transferred to the pore water which, after a suitable number of surface impacts, eventually rises in pressure to a sufficient level to induce liquefaction. This is the theory first proposed by Menard and is a phenomenon very similar to that occurring during earthquakes. Clearly the existing density and grading of the soils will be major factors in the speed at which this liquefied state will be achieved. Low-frequency vibrations caused by further stress impulses will then reorganise the particles into a denser state. This is comparable to the response of sands from the vibro compaction technique for which D'Appolonia (1953) suggested that a vibrational acceleration in excess of 0.5 g was necessary to achieve such a densification effect.

Dissipation of the pore water pressures, in conjunction with the effective surcharge of the liquefied layer by the soils above, results in a further increase in relative density over a relatively short period of time. This can vary from 1 to 2 days for well-graded sand and gravel, to 1 to 2 weeks for sandy silts. The testing programme should therefore recognise the time-dependent response for soils that are normally considered to be free-draining. Longer-term improvement, possibly as a result of chemical bonding or high-residual lateral stresses within the soil matrix, has been reported by Mitchell and Solymar (1984).

There is, however, another school of thought in which the aim is to avoid the liquefied state. While it is recognised that liquefaction cannot be avoided in deep loose sandy deposits with high water table, as often encountered in parts of North America, the Middle and Far East, such conditions are rare in Britain. The treatment is therefore designed to provide compaction by displacement without dilation or high excess pore pressures by using a smaller number of drops from a lower drop height. This method requires substantially lower energy input than the liquefaction approach, with consequent economies. Laboratory and *in-situ* tests have consistently shown that in order to achieve maximum density, the lowest number of stress impulses to attain the required energy input will provide the optimum result. Saturated granular soils will normally require higher treatment energy overall, in a larger number of tamping passes, than if the soils were essentially dry.

Figure 3.2 illustrates the typical volumetric response for granular soils and Figure 3.3 illustrates electric cone results for a site of clean sand with a water table at about 2.5 m depth treated by 15 tonne equipment. This also illustrates improvement with time since the second tamping pass was only capable of treating upto about 4 m depth.

When the individual sand particles are weak, such as the calcareous sands of the Middle East, 'sugar' sands of North America or the Thanet Sands of Britain, crushing tends to occur during the treatment. A similar response affects ash, clinker and weak aerated slags. When these soils are dry, the effect of such particle breakdown is not particularly significant. However, below the water table the higher proportion of fines developing with increasing energy input results in a rapid change from a granular to a pseudo-cohesive soil response.

The existence of very dense layers within the ground can cause anomolous results. Where, for example, cemented layers occur within natural sands, these tend to absorb the energy impulse and arch over the underlying stratum. A similar phenomenon can occur with vibro treatment where the

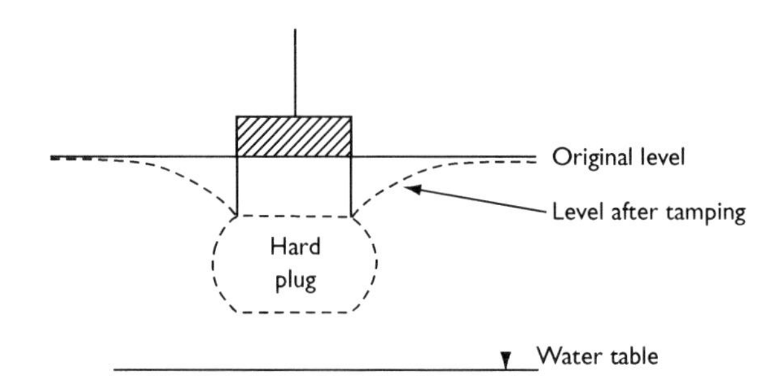

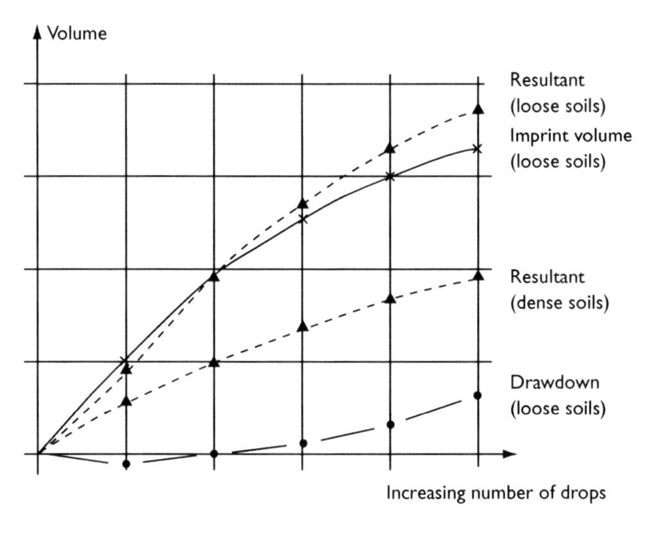

Figure 3.2 Volumetric response – granular soils.

cemented zones do not collapse around the vibrator to permit densification to occur. In these situations, where they occur at shallow depth, the dynamic compaction will break up the cemented layer. However, at greater depth, the energy levels required to break the stratum may be beyond the capabilities of the equipment on-site. The presence of such layers is often not adequately revealed by normal site investigation.

In summary, excellent engineering performance can easily be achieved in dry granular soils. However, care must be exercised for the treatment of soils with significant silt content, particularly below the water table.

3.5.2 Cohesive soils

The response of clays is more complex than that of granular soils. There is again the distinction between above and below the water table.

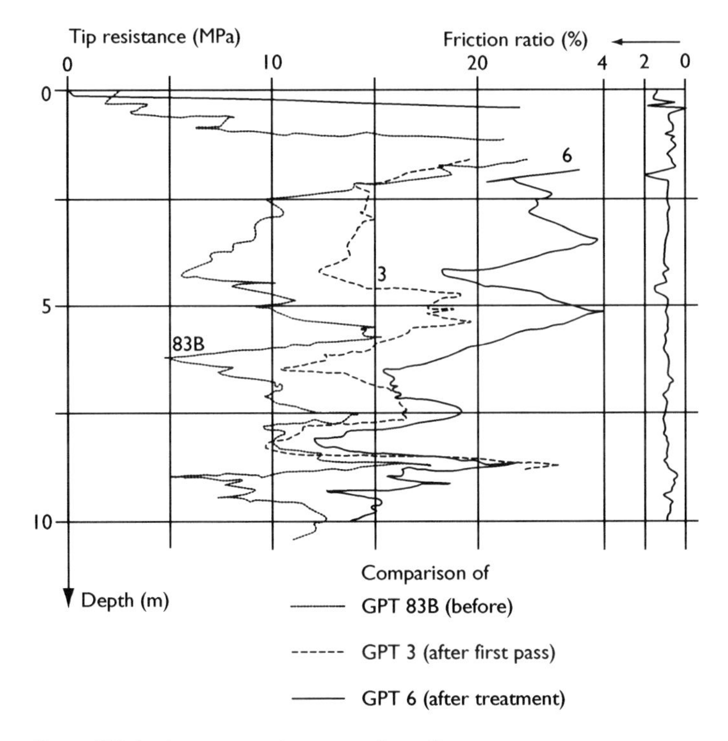

Figure 3.3 In-situ test results – granular soils.

With conventional consolidation theory, imposition of a general static loading will expel water to induce consolidation and increase strength. The rate at which this occurs is dependent upon the imposed load, coefficient of consolidation and length of drainage path. In contrast, dynamic compaction applies a virtually instantaneous surcharge that is transferred to the pore water on a localised basis. This creates zones of positive water-pressure gradient which induce water to drain rapidly from the soil matrix. This effect is further accelerated by the formation of additional drainage paths by shear and hydraulic fracture. Consolidation therefore occurs much more rapidly than would be the case with static loading. Dynamic compaction literally squeezes water out of the soil to effectively pre-load the ground. A typical volumetric response is illustrated in Figure 3.4.

Where the soils occur above the water table, the clays tend to be of relatively low moisture content and even a small reduction can result in significant improvement in bearing capacity. The drainage path is also relatively short. As such, treatment is relatively straightforward and rapid.

Where the clays occur below the water table, a much larger reduction in moisture content is generally required in the presence of a smaller available

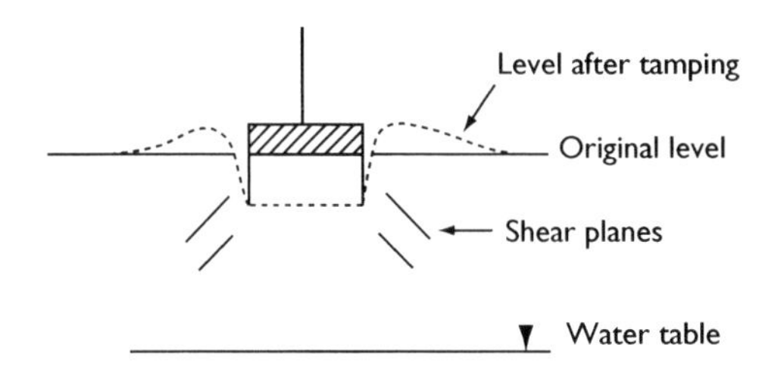

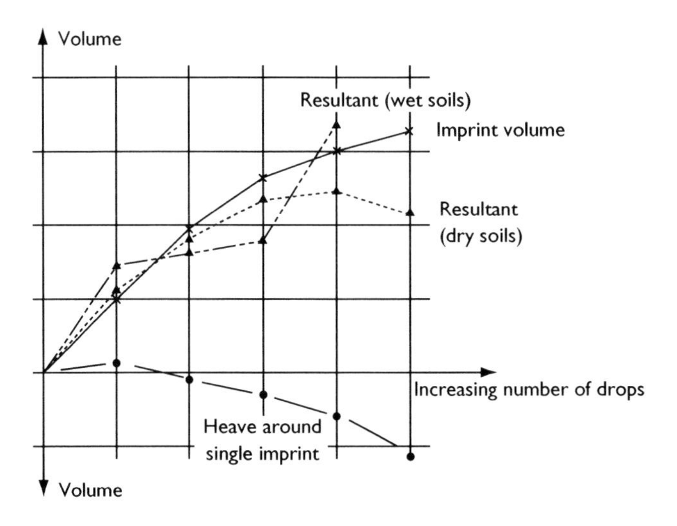

Figure 3.4 Volumetric response – cohesive soils.

pore-pressure gradient and a longer drainage path. These conditions result in the threshold energy being achieved much more rapidly and in many tamping passes of low-energy input requiring greatly extended contract periods in comparison to normal productivity. Only nominal degrees of improvement have been achieved in thick layers of relatively weak saturated alluvial clays and silts, even with additional measures such as drainage trenches filled with sand or wick drains.

Where such layers are relatively thin and require treatment, a better speed of response is recorded due to the shorter drainage path. In some instances coarse granular material is driven into these materials to provide better grading that is more suited to treatment, or to displace from specific locations beneath part of a building area. It is now more common in the UK to adopt vibro stone columns in such soils to more critical locations, e.g. more heavily loaded foundations, and then perform dynamic compaction to preload the ground with the benefit of the stiffer columns also acting as drains to control excess pore water pressures.

For predominantly clay-type fill materials above the water table, the clay lumps can be considered as large weak particles of almost granular response. However, the major improvement is achieved by collapsing voids to provide a more intact structure. Clearly the strength of the lumps and sensitivity of the clay is of paramount importance in such soils. Differing degrees of weathering can also give rise to markedly variant responses on a site, and experienced observation is required to define such locations. Mudstone and shale fragments can break down to a material of clayey response, particularly when heavy rainfall occurs.

For clay-type fills below the water table, the voided structure allows higher mobility of water causing lower excess pore pressures and shorter recovery periods in comparison to natural clays. The constituents would be of higher moisture content but again improvement would be achieved mainly by collapse of voids. Monitoring of excess pore water pressures by means of piezometers is clearly useful but problematic above the water table.

It cannot be emphasised too strongly that the treatment of clays requires very close, experienced control on-site. During treatment, after a small number of drops, heave starts to develop around the edges of each imprint. If tamping continues, the heave can build up to such an extent that it can exceed the volume of the imprint. Clearly this is the precise opposite of what is desired. Also, additive heave can occur by the performance of the adjacent tamping position at too narrow a grid dimension.

Particular care has to be exercised in the timing of successive tamping passes to permit adequate recovery of pore pressures to avoid excessive remoulding of the soils. Such approach can however be relatively slow and, in view of the emphasis placed these days on productivity, the vibro stone column in advance of dynamic compaction method described above is sometimes adopted.

If excessive heave around an individual imprint does start to occur, it is essential that the tamping at that position be stopped. This may only extend over a relatively confined area with better ground elsewhere. In soft areas it is better that twice the number of lighter-energy input tamping passes be performed in a 'softly softly' approach.

Similar considerations apply when attempting to provide treatment to a significant depth where the surface layers are clayey. The strength of the surface soils can reduce in the short term and time has to be spent in improving a disturbed matrix to reconstitute its original, let alone desired, properties. This is particularly difficult where thick crusts to, say, 2–3 m depth of stiff to very stiff clays overlay a granular deposit requiring treatment. In this situation even higher than normal energies are required to attain the deeper layers giving rise to even greater potential for virtually destroying the surface soils.

The treatment of clayey soils will nearly always require a larger number of tamping passes when compared to a similar profile of predominantly

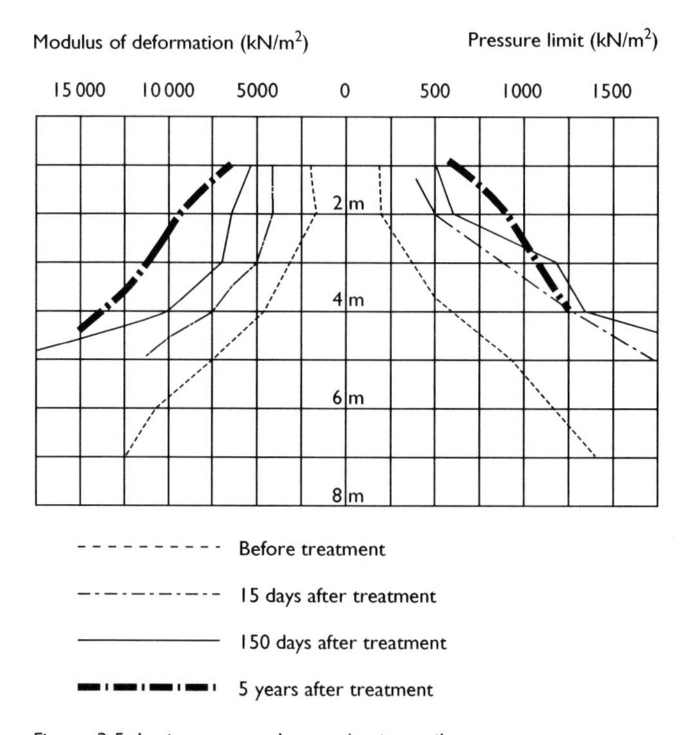

Figure 3.5 In-situ test results – cohesive soils.

granular constituents. Efficient treatment is achieved by attempting to provide as much improvement as quickly as possible while recognising that the response of the soils will dictate the speed of the treatment operations.

Clay soils will continue to improve for a significant period after treatment as reported by West (1976). Figure 3.5 illustrates further measurements on this site taken five years after treatment.

In summary, dry cohesive fills respond well to dynamic compaction. Care must be exercised in the treatment of weak natural clayey soils or clay fills below the water table. The prior performance of vibro stone columns to both stiffen the ground and enhance drainage has proved to successfully combine with dynamic compaction to clayey soils.

3.6 Depth of treatment

Menard originally proposed that the effective depth of treatment was related to the metric energy input expression of $(WH)^{0.5}$ where W is the weight in tonnes and H the drop height in metres. This was modified by a factor of 0.5 by Leonards *et al.* (1980) for relatively coarse, predominantly granular soils, and factors of 0.375–0.7 by Mitchell and Katti (1981) for two soil types. The most exhaustive analysis yet published has been provided by Mayne *et al.* (1984). The author suggests that the range of treatment depths varies with initial strength, soil type and energy input as illustrated in Figure 3.6.

There are many factors affecting this dimension, not least of which are the type and competence of the surface layers, position of the water table and number of drops at each location. Assessment of *in-situ* results to determine such depths also tends to be subjective and will be affected by the recovery period after treatment. As noted in the previous section, a solid 'plug' of very dense material can form beneath the impact locations to inhibit the improvement to depth. Weak surface soils and a high water table can also limit the physical performance of a sufficient number of stress impulses thereby inducing only a minor improvement to the basal layers. However, knowledge of the depth of any stress impulse is a vital factor in both the planning of the treatment operations and the potential for transmission of vibrations as discussed in the next section.

Kinetic energy at the point of impact is clearly a major factor in the depth of treatment and increasing the drop height will increase velocity. In Britain high-speed photography has shown typical impact speeds of about 35 and 50 mph for 8 and 15 tonne equipment to achieve effective depths of treatment of about 6 and 8 m, respectively.

The shape of improvement in the ground tends to be similar to the Boussinesq distribution of stresses for a circular foundation. Modification

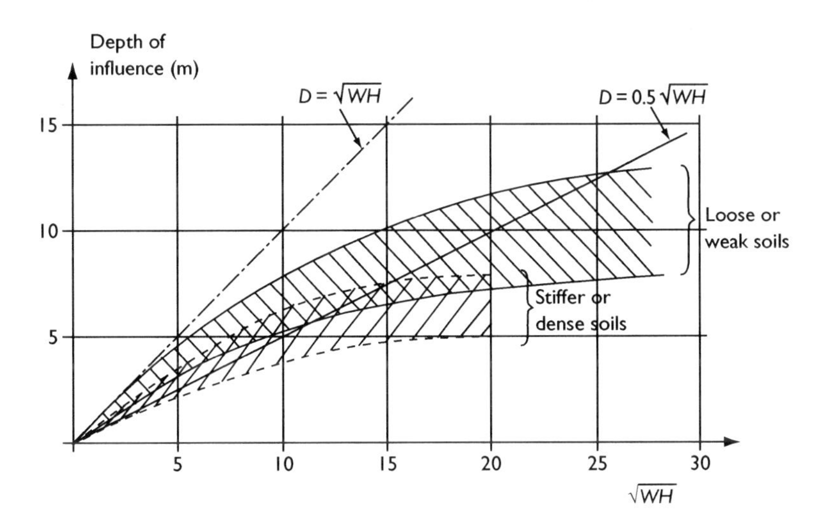

Figure 3.6 Depth of treatment.

of energy levels for each tamping pass can be used to custom-design the treatment scheme to the specific soil profile and engineering requirements. In contrast, the shape of vibro improvement tends to increase with depth. In earthquake areas the required density of soil from the Seed and Idriss (1971) analysis is often better provided at depth by the vibro technique which has the added advantage of forming stone columns to act as drains in the finer soils. Stone columns can also be combined to reinforce weak cohesive soils at a depth that would be difficult to treat when using dynamic compaction for surface fill layers (Slocombe, 1989).

As noted earlier, high-impact energies can weaken the surface layers and the aim is therefore to combine effects to achieve improvement throughout the whole of the desired treatment zone. For example, on a project in Saudi Arabia, drop height, number of drops and the treatment grid were adjusted to provide treatment to three distinct sand layers requiring improvement in a single tamping pass (Dobson and Slocombe, 1982). Clearly, if all structures are founded at depth there is no need for the final tamping pass for treatment to the surface layers, provided the grid of the earlier tamping passes produces overlapping effects at the founding level.

3.7 Environmental considerations

Dynamic compaction utilises large, highly visible equipment. The process creates noise and vibration, both of which must be considered in Britain under the Control of Pollution Act, 1974. The standards listed in the reference section provide further details (BS 5228, 1992; BS 7385, 1990; BRE Digest 403, 1995).

Air-borne noise levels are generated by a number of causes. Of these the point of impact is by far the highest noise level at typically 110–120 dB at source. However, its duration only occupies about 0.5 per cent of the lifting cycle. The considerably lower noise values during lifting and idling when combined with the impact noise using the L_{Aeq} calculation method will normally meet most environmental limitations at distances of greater than 50 m from the treatment operations. Lower than normal noise limits can be achieved by working within a specified zone for only a certain number of hours during the working day. Large-plate glass windows can sometimes act as diaphragms to change the noise characteristics inside a property. Echoes, wind direction and angle of crane exhaust are all factors that should also be considered.

By far the most important consideration, however, is ground vibration. In addition to the magnitude of the vibration, the typical frequency of about 5–15 Hz is potentially damaging to structures and services, and particularly noticeable to human beings. It is suggested that there are three vibration levels that will influence the design of the treatment scheme. Guide values of resultant peak particle velocity at foundation level for buildings in good condition are:

Structural damage	40 mm/s
Minor architectural damage	10 mm/s
Annoyance to occupants	2.5 mm/s

Lower values must be adopted for buildings in poor condition or environmentally sensitive situations such as schools, hospitals and computer installations. Certain major computer companies recognise the importance of the vibration frequency by requiring more onerous limits for frequencies below 14 Hz than above. It should be noted that some amplification can occur as the vibration rises up certain types of structure and that say 1.0 mm/s at ground level could be 2.5 mm/s at the third floor. Services and utilities must be considered on an individual basis depending upon their age, condition and importance with values of 15–20 mm/s normally being considered acceptable, except for higher-pressure gas mains.

The prediction of the level of vibration transmitted through the ground is an imprecise science because of the variable nature of the characteristics of soils. Field measurements of vibrations at ground level have revealed a number of trends which are illustrated in Figure 3.7. The upper dynamic

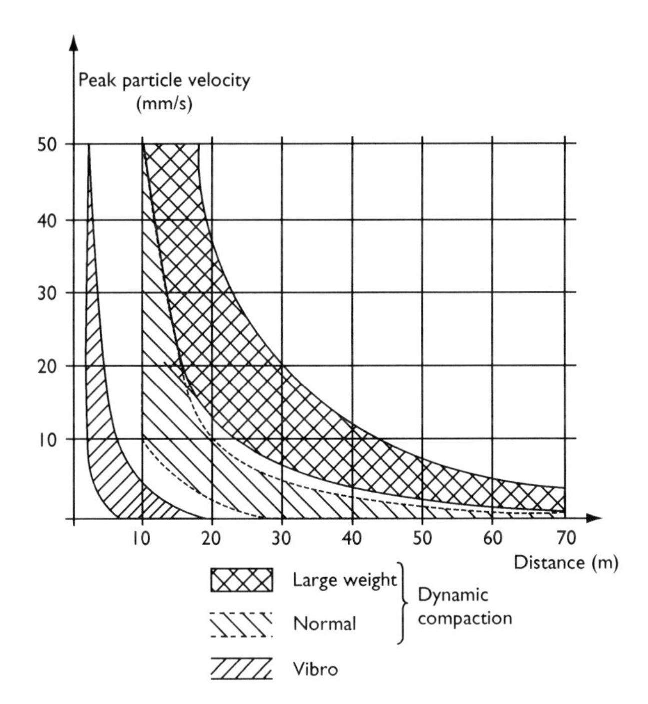

Figure 3.7 Vibrations.

compaction limits tend to occur in the presence of granular or refuse-type soils and the lower limit in cohesive strata. A high water table will also tend towards the higher limit. The upper vibro limit is for vibrators operating at a frequency of 30 Hz and the lower for 50 Hz.

Careful assessment is required where the soil being treated is directly underlain by relatively dense sand, gravel or rock which will tend to transmit vibrations to larger than normal distances with comparatively little attenuation. Pre-existing dense surface or buried layers can have a similar effect of causing the transmission of higher than anticipated vibration levels. The physical performance of the treatment work improves interparticle contact of the soils and as such, vibration levels can sometimes increase towards the end of the treatment operations even though the final impact energy levels are substantially lower than those performed for the initial tamping passes.

When vibrations become a problem there are three main methods of reducing their effect. The first is to simply reduce the height of drop and compensate by increasing the number of drops per imprint. This reduces both the impact energy and penetration of the stress impulse that may have attained an underlying dense stratum. The second method of reduction is to utilise a smaller weight and the third is to excavate a cut-off trench to sufficient depth to intercept the surface wave.

Human beings are particularly sensitive in detecting vibrations and have a psychological reaction in believing that damage is caused even though the values are far below the well-established damage threshold levels. A thorough public relations exercise can sometimes help to overcome concern amongst local residents. Dilapidation surveys prior to the commencement of treatment are often advisable. People are often not aware that vibrations caused by passing lorries or slamming doors can exceed the levels of minor architectural damage. Similarly, very cold or hot weather and snow loading can lead to structural movements that are sometimes incorrectly attributed to vibrational causes.

3.8 Practical aspects

There are a number of practical factors that must be taken into account when performing dynamic compaction contracts. The large crawler crane must be safely supported by a free-draining working surface, the thickness of which will depend upon the type of ground being treated. If the surface layer, say 1.0 m thick, is basically granular, no imported working carpet is generally required. However, during wet weather, sand-sized materials have been seen to eject through the air up to 60 m from the point of impact, and coarser material is essential to overcome this problem if work is carried out near to roads or property. Alternatively, the programme should contain sufficient flexibility to permit treatment to be performed within, say, 50 m of such features only when the surface conditions permit its safe operation.

Where cohesive surface conditions exist, a free-draining granular working carpet is normally required. The thickness can be as little as 150 mm for light-energy treatment in reasonably competent soils up to 1.5 m when treating refuse fills. When aiming for substantial depth of treatment, thick working carpets of 1.0 m or more have been found to inhibit the stress impulse. A more efficient operation, which also provides greater control of backfill quantities, is to start with only 0.5 m thickness and to backfill imprints directly, thus preserving the working carpet for successive tamping passes.

Safe working is a prime consideration. If more than one rig unit is to be used they should be separated by at least 30 m. Similarly, subsequent operations by the main contractor may have to be delayed until the treatment operations are sufficiently remote.

Safety screens have been utilised to attempt to limit the effects of flying debris. However, these must be continually moved to be close to the point of impact, otherwise the materials would simply fly over the top of the screen. Heavy duty membranes have been found to be well-suited to this purpose whereas polythene sheeting tends to tear, and wood hoarding can easily be broken by the velocity of impact of the ejected particles. The continual movement of these screens tends to adversely affect productivity and the former approach of 'dry' weather working has been found to be more efficient.

Winter working will place more onerous requirements on the adequacy of the working surface. The general rule is to increase the depth of the granular working carpet by 25 per cent in comparison to summer thickness. When working in arid climates there is often no need for any working surface, even for clayey soils.

3.9 Induced settlement

The induced settlement is dependent on the total energy input and the manner in which it is applied. Initial shape tests are performed when the soils are loosest. As such, simple extrapolation of these results will overestimate the amount of induced settlement. Mayne *et al.* (1984), as part of his survey of 124 different sites, reported that the magnitude of induced settlement depended on the applied total energy input, also stating that the thickness of the layer was probably an important factor, for six soil types. This analysis does not, however, take into account either the initial softness/density of the soils or the proportion of total energy applied by the high-velocity initial passes or low-velocity final tamping pass that numerically is very significant in determining total energy input. Also, the application of

Table 3.1 Approximate induced settlement as % of treatment depth

Soil type	*% depth*
Natural clays	1–3
Clay fills	3–5
Natural sands	3–10
Granular fills	5–15
Refuse and peat	7–20

too high an energy in clayey soils will result in less than optimum induced settlement occurring in practice.

A convenient simple approach is to adopt approximate percentages of the target treatment depth for 8 tonne (50–100 Tm/m^2) and 15 tonne equipments (100–200 Tm/m^2), the total energies with the 15 tonne energy applying to greater depth of treatment (Table 3.1).

In attempting to induce higher proportions, the increase in energy input will not be linear, e.g. to increase from 10 to 15 per cent induced volume in refuse would require 200–225 per cent of the normal energy because during the treatment the material becomes progressively stronger and there is less and less potential void reduction available. Care has to be exercised to avoid overtamping in these situations since refuse tips tend to be capped by clay soils.

Loose materials will obviously settle more than denser soils. As noted earlier, ash and certain types of slag also tend to break down during treatment to produce induced movements towards the higher value for granular fills given above.

3.10 Additional comments

The shortage of suitable land in Britain has led to the development of many old landfill sites where refuse or garbage is present in significant quantity. There are three main eras of such materials where each have different suitability for the support of structures. The early domestic refuse was typically burnt to ash and cinders before collection and is therefore suitable for treatment for the support of buildings.

Around the 1940s and 1950s the materials tended to be not as well-burnt and contain minor proportions of matter still liable to long-term decay. These materials are suitable for treatment to buildings that are tolerant to some differential movements such as lightweight industrial units. The more recent refuse is not burnt, but is collected in polythene bags. The degradable proportions are high and as such there is potentially a high risk for excessive movements occurring as a result of long-term decay for most structures. However, road embankments, parking areas, services and utilities can all benefit from treatment to these soils. Dynamic compaction has also been performed to reduce ground levels to avoid removal of contaminated fill to specialist tip to permit development at the desired site level.

The response of refuse is normally a combination of both granular and cohesive soils, particularly as there is normally a clay capping present. The aim is to collapse the major voids that have already formed within the surface layers due to the presence of paint drums, barrels, carbodies, etc., and densify the remaining reasonably inert constituents into such a dense matrix that in the event of further long-term decay, any relaxation of the ground would tend to be distributed and not lead to major deformity at ground level. Approximately 50–100 per cent higher total energy levels in comparison to inert soils are normally applied in these situations. Alternatively, by utilising detailed surface observations, increasing energy can be applied to attain a state in which no further practical induced settlement is achieved.

However, long-term decay of the remaining constituents may continue. Measurements of various projects treated up to 20 years ago have revealed total settlements of formation which are typically about one-third of those measured in untreated soils with the additional benefit that no maintenance has yet been required. The α-values for creep settlements with time have been measured to be comparable to those where conventional static surcharge has been applied.

Pulverised fuel ash comprises relatively fine single-sized particles that exhibit pozzolanic properties when properly compacted by conventional methods. When tipped into water, the materials tend to flocculate with little self-compaction occurring with time. Dynamic compaction has been attempted on a number of settlement lagoons with little success. However, dry PFA is considered suitable for treatment, albeit with a number of controls.

Peaty soils can be treated in many different ways, depending on the required end result. High energy can be applied to physically displace the material wholescale from beneath the line of a major road. Discrete columns of sand- to cobble-sized fill can be driven into the peat in a manner similar to stone column theory or normal treatment methods applied to simply squeeze out some of the water and pre-load the ground. The basic fact that must be considered throughout is that peat tends to be a very weak material of high moisture content. As such, the pre-loading method will take time to perform.

Dynamic compaction has been performed to collapse shallow solution cavities. It is important that the extent of these voids be accurately determined prior to treatment so that the crane is positioned sufficiently remote when the cavity caves in to avoid falling into the void. Considerable care has to be exercised during the dropping of the weight to avoid the rope pulling off the crane drum or the weight punching into the void and becoming trapped beneath the surface. Long narrow weights are better suited to this operation.

Many sites of former heavy industry are now being reclaimed in Britain. These often present the designer with the problems of deep fill and massive obstructions from old basements and foundations. Any technique that has to make a hole in the ground will experience difficulty in gaining adequate penetration, and large excavations often have to be performed to remove the obstructions. However, with dynamic compaction such features can be left in place provided they occur at sufficient depth to avoid excessive differential performance. For most industrial or low-rise housing developments this depth would be 1.0 m below the underside of new foundations or floor slabs. Operations would be planned first for advance earthworks to remove all known features down to a specific level, followed by compaction treatment and normal construction operations. In choosing this excavation level the designer should take into account the type of structure and its tolerance to some degree of differential performance and the fact that the treatment will induce a reduction in site level that is slightly higher than normal as a result of the loose nature of the surface materials after the pre-excavation operations. It is also normal to apply a slightly higher than normal energy input to reduce the differential performance between the areas of massive foundations and abutting weaker soils which, if weak clays, could be significant. Where the backfill comprises large concrete posts, slabs or waste that could arch over a void, care must be taken in the design of the foundations since the treatment may not necessarily cause the cavity to wholly collapse and could weaken the member forming the roof to the void. In these circumstances higher than normal energies are preferred to break down such potential.

Brownfield sites with minor contamination are well-suited to treatment since the technique does not create a bore that could permit the migration of leachate. Care should however be exercised to ensure that the impact energies do not shear any underlying clay layer that may prevent the downward migration of contaminants into an aquifer. Similarly, if there is a high water level that is contaminated, attention should be paid to the possibility of a rise in level during the treatment and its effect on adjacent property or features. In some cases installation of drains lined with HDPE membrane or monitoring may be required.

Land contaminated with chemicals or asbestos is now having to be developed. The major advantage that dynamic compaction has over alternative methods is that it can be controlled to avoid exposure of hazardous material to the atmosphere while still compacting soils at depth.

Sites of former quarries are prime situations for treatment in view of the potential for piles to glance off the buried subvertical face between fill and rock, and the possibility of constructing piles to inadequate depth as a result of false readings from boulders or inaccurate historical information on the depth of the quarry.

An area of increasing importance is the treatment of partially saturated fills and natural soils that are susceptible to collapse compression due to inundation for the first time. Dynamic compaction reduces such potential by generally reducing voids, inhibiting surface water from penetrating the ground and by creating a 'stiff crust' over the effect of deeper movements, should they occur.

3.11 Testing

Many contracts have simply involved the measurement of depth of first-pass imprints and monitoring of site levels. *In-situ* and loading tests are often performed and since the technique provides treatment to large areas very quickly, the speed at which such tests provide the necessary information is important, particularly if testing between tamping passes. It is rare, therefore, to recover samples for laboratory testing. In clayey soils, as with the performance of the treatment, it is essential that sufficient recovery period be allowed to avoid ambiguous results.

It is common for dynamic compaction to be performed for sites underlain by coarse fills or including obstructions that would cause penetration problems for vibro or piling methods. Similar problems could therefore be reasonably anticipated in attempting to perform *in-situ* tests. Air drills have been used on a small number of contracts to pre-drill a test location to below the level of the potential obstruction. However, surface loading tests only are more normally performed in this situation.

Table 3.2 describes the relative merits of various test methods. Additional comments on the advantages and limitations of certain *in-situ* tests are:

1 *Standard penetration test*: This is probably the most useful *in-situ* test as it is applicable to both granular and cohesive soils. However, being of a dynamic nature it is particularly sensitive to the presence of residual pore water pressures, quickly liquefying the stratum being tested, and producing lower than expected results. A sample is normally recovered and the speed of provision of information is adequate for most contracts. The main drawback is that a considerable amount of time and money can be spent on chiselling to penetrate the very dense surface layers normally provided by the treatment.
2 *Pressuremeter*: The dynamic compaction technique has been historically associated with this test. While this method is often used in mainland Europe it is now rarely used in Britain.
3 *Dynamic cone test*: This is relatively cheap and robust but is limited by the inability to determine, without the performance of alternative parallel testing, whether a zone of low blow count is caused by loose zones

or cohesive layers that would be expected to respond differently to the treatment. Since the majority of ground improvement contracts performed in Britain require treatment to variable fill sites this method has been found to be of limited value.

4 *Static cone penetration test*: These tests are ideally suited to the testing of sands because they illustrate the soil type by means of the friction ratio. They are considered less successful in clayey soils since experience has shown that this test is particularly affected by the presence of residual pore water pressures. Being relatively sophisticated it is not recommended that these tests be used in the presence of coarse fill materials.

5 *Dilatometer*: This method would appear to have potential. However, no information has yet been published for its evaluation of the treatment of the soils and fills of Britain.

An example of the results of ten zone loading tests on bases 2.0 m × 2.0 m with maximum applied pressure of 300 kN/m^2 is given in Figure 3.8. These tests were performed on a site to areas comprising either wholly granular or cohesive fill materials. The total energy of about 120 Tm/m^2 was applied to all areas using 15 tonne weights with the initial heavy tamping passes being subdivided in clay areas to avoid adverse response. These tests reveal the superior performance that is achieved when treating granular materials in comparison to clayey soils.

Table 3.2 Suitability for testing dynamic compaction

Test	*Granular*	*Cohesive*	*Comments*
Dynamic cone	**	*	Too insensitive to reveal soil type. Has difficulty penetrating densely compacted ground
Static cone	***	*	Particle size important. Can be affected by lateral earth pressures generated by treatment. Best test for seismic liquefaction evaluation in sands
Boreholes and SPT	***	**	Efficiency of test important. Recovers samples
Small plate	*	*	Poor confinement to zone being tested. Affected by pore water pressures
Large plate	**	*	Better confining action
Skip	**	**	Can maintain for extended period
Zone loading	****	****	Best test for realistic comparison with foundations
Full-scale	*****	*****	Rare

Note
* to ***** – Least to most suitable.

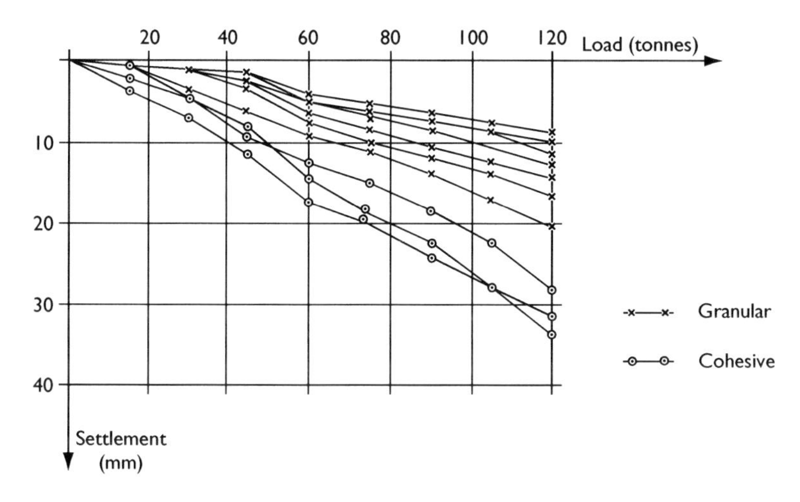

Figure 3.8 Post-treatment zone loading test results.

3.12 Case histories

The treatment scheme to a former steel works site in North Wales required the improvement of about 4–5 m depth of slightly silty fine to medium hydraulic sand fill materials, underlain by competent sands. The ground water was recorded at about 4 m below ground level.

The specification required an area of 67 000 m^2 capable of 100 kN/m^2 bearing pressure anywhere within the treated area and a maximum of 20 mm settlement.

High and intermediate-pressure gas pipelines were present at about 50 and 20 m, respectively, from one site boundary. Vibration monitoring was performed to confirm that peak particle velocities did not exceed agreed limits.

Pre-treatment CPT tests were performed on a 50 m grid to confirm the required energy levels. Post-treatment CPT tests were performed on a 25 m grid to confirm that the desired performance had been achieved with typical equivalent N-values in excess of 25. Short-duration plate loading tests were also performed at a rate of one per 1000 m^2 plus seven zone loading tests on bases of 2.0 m × 2.0 m. These were performed to a maximum applied pressure of 250 kN/m^2 with recorded settlements of 7.5–12.4 mm.

On another site, borehole and trial pit investigations for two large span industrial units with high racking areas had revealed the presence of up to 8 m depth of loose to medium dense dry mixed granular and cohesive fill materials underlain by chalk. The proximity of existing structures and services would not have permitted the use of high-energy dynamic compaction. Vibro stone columns were therefore constructed beneath the proposed foundations, and as a form of investigating the ground, prior to dynamic compaction over the whole building areas. A moveable screen was also provided to avoid potential problems of fly-debris reaching a motorway adjacent to the site.

Treatment was performed to an area of almost 25 000 m^2 for bearing pressures of 125 and 50 kN/m^2 to foundations and floor slabs, respectively, whilst limiting long-term settlements to about 25 mm.

Large-plate tests of 1.5 m × 1.5 m were performed after treatment on the floor slab areas. These were loaded to 150 kN/m^2 or 3 times the design pressure, maintained for at least 12 h. These recorded settlements of 10.9–13.7 mm. Two zone loading tests were also performed on foundation locations treated by both vibro stone columns and dynamic compaction. These 2.0 m × 2.0 m tests recorded settlements of 7.0 and 7.6 mm when loaded to 187.5 kN/m^2 and 12.3 mm when loaded to 250 kN/m^2, clearly illustrating the benefit of the stone columns.

The development of former landfill sites is now a well-established market for dynamic compaction with 20 years experience of long-term settlement performance. The age and type of fill constituents dictate whether certain structures, floor slabs, roads or parking areas can be supported on treated ground or whether other techniques, such as piling, must be adopted. One such site to the south of London was underlain by 8–12 m depth of old refuse that was of a predominantly ashy sandy nature but with local large pockets of metal, paper and cloth.

The proposed structures comprised high warehouses with racking that could not tolerate differential movements. The building frames and floor slabs were therefore supported by driven cast-*in-situ* piles, and a perimeter slurry cut-off wall was installed to prevent the migration of leachate off-site. Dynamic compaction was initially proposed simply to the main spine and service roads plus yards and parking areas to achieve a design CBR value of 5 per cent over a total area of about 20 000 m^2.

However, the site was elevated and surrounded by other industrial units and offices with 80-year-old cottages alongside one boundary. Planning approval had been granted on the basis that the proposed development would not exceed a certain height. This would either restrict the racking heights or require development at a lower level, thus necessitating the removal of substantial volume of spoil. As the fill was contaminated it would have had to be removed to specialist tip at considerable cost.

Advance dynamic compaction trials had shown that it was possible to induce over 1.0 m reduction in ground levels by adopting higher than normal total energy input at about one-sixth the cost of removing excess materials to tip. However, the proximity of surrounding properties would not permit the performance of such high energy over the whole area. Two proposed deep sewers near the cottages were first treated by vibro stone columns. The dynamic compaction impact energies were then progressively reduced towards the site perimeters to avoid the transmission of annoying vibrations being transmitted to the occupants of the surrounding properties.

Levels on the central spine road recorded an average induced settlement of 860 mm whilst movements towards the site perimeters with lesser impact energies were typically 400–600 mm.

Following the German reunification in 1991 uneconomic and environmentally unacceptable brown coal fields had to be closed down. One of these was located to the north-east of the city of Dresden. This decision triggered an enormous rehabilitation and renaturisation programme of derelict industrial facilities and land. Over 150 km of embankments of the lignite mining pits, which were unstable in their composition and prone to sudden collapses with the rising ground water, had to be closed to the public.

In parallel to the monitoring of ground water, extensive ground improvement works were designed and performed to prevent slides and sudden collapses of these embankments causing the mine debris, which consisted of very loose silty sands, to flow like a liquid. One of the design principles to stabilise the shore lines of the pit areas used the concept of 'hidden dams'. This comprised providing improved compacted fill some hundred metres away from the slopes to prevent major failures from extending deep into the hinterland.

Compaction to depths of up to 60 m was and still is being performed using deep vibratory compaction and, to a lesser extent, where environmentally acceptable, by explosive compaction or blasting. Dynamic compaction was performed, sometimes following the deep compaction methods, to prevent surface erosion on gentle slopes and on areas of future roads and other traffic facilities. The treatment was performed to the upper 8 m of soil using semi-automated crawler crane dropping typically 30 tonne weights from heights of up to 25 m with total energies of 240–360 Tm/m^2.

A similar concept was used within the seismically active area of Vancouver, British Columbia. This 29-acre mixed-use development of marina, residential units, arts complex, community centre, school, commercial office towers, retail shops and hotel involved the placement and densification of an off-shore dike and stabilisation improvement of existing shoreline slopes.

The geotechnical engineer, Golder Associates, determined that a 15–21.5 m wide zone of ground improvement would be required to prevent liquefaction and lateral load spreading under the design earthquake peak ground acceleration of 0.31*g*. Dynamic Deep Compaction^{TM} was specified to densify the loose imported graded sand and gravel fill of up to 75 mm maximum size to depths of 10 m from an elevation of +3.0 m. Performance-based densification criteria required standard penetration $N_{1(60)}$ values, depending on silt content, of 18–26 within the 10 m depth of treatment. These were then correlated to the Becker penetration test in view of the

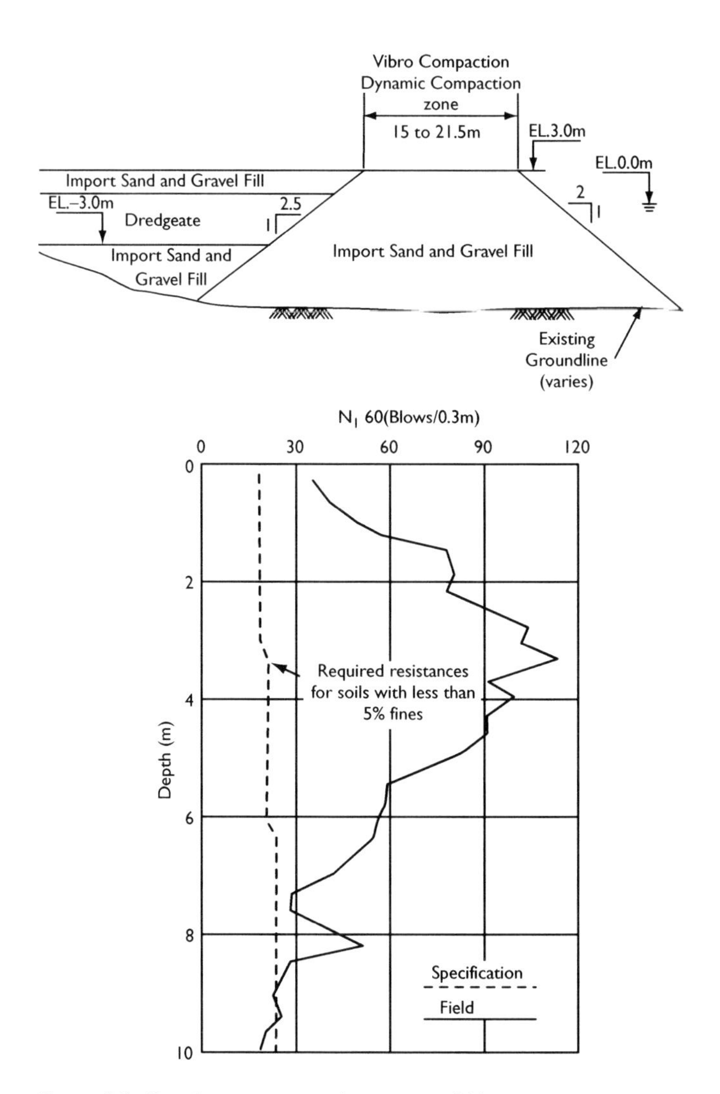

Figure 3.9 Case history – typical section and blow counts.

typical particle size of the fill with soils of greater than 5 per cent fines content being excluded from the criteria.

Vibrations from the dynamic deep compaction were specified to not exceed 12.5 mm/s at any of the adjacent facilities. Initial on-site monitoring revealed that this level would be exceeded at a nearby hotel and vibro compaction was performed for a proportion of the required treatment areas.

The dynamic compaction was performed using a 20 tonne weight dropped in several passes from heights of up to 25 m with SPT values interpreted from the Becker tests generally in excess of 30. A typical test result is illustrated in Figure 3.9. This also illustrates the typical 'shape' of post-treatment results that tend to follow the 'shape' of Boussinesq stress distribution beneath isolated foundations.

3.13 Concluding remarks

The dynamic compaction method is a powerful tool when applied to suitable sites. A large database has been collated over the years to define its limitations and, more importantly, capabilities which can be utilised with confidence.

Acknowledgements

The author extends his appreciation to Keller Ground Engineering, Keller Grundbau and Hayward Baker for the case histories and permission to publish this information.

Bibliography

BRE Digest 403 (1995) *Damage to structures from ground-borne vibration*, Building Research Establishment.

BRE (2003) *Specifying Dynamic Compaction.*

BS 5228 (1992) Part 4 *Noise control on construction and open sites*, British Standards Institution.

BS 7385 (1990) Part 1 *Evaluation and measurement for vibration in buildings*, British Standards Institution.

D'Appolonia, E. (1953) Loose sands – their compaction by vibroflotation, *Symposium on Dynamic Testing of Soils*, American Society of Testing Materials, STP, p. 156.

Dobson, T. and Slocombe, B.C. (1982) Deep densification of granular fills, *2nd Geotechnical Conference*, Las Vegas (April).

Greenwood, D.A. and Kirsch, K. (1983) Specialist ground treatment by vibratory and dynamic methods – state of the art report, *Proceedings, Piling and Ground Treatment for Foundations*, London, pp. 17–45.

Institution of Civil Engineers (1987) *Specification for Ground Treatment*, London, Thomas Telford.

Kerisel, J. (1985) The history of geotechnical engineering up until 1700, *Proceedings, 11th International Conference on Soil Mechanics and Foundation Engineering*, San Francisco (August), pp. 3–93.

Leonards, G.A., Cutter, W.A. and Holtz, R.D. (1980) Dynamic Compaction of granular soils, *Journal of Geotechnical Engineering, ASCE*, Vol. 106 (GT1), pp. 35–44.

Lucas, R.G. (1995) Geotechnical Engineering Circular No. 1: Dynamic Compaction, *US Department of Transportation*, Publication No. FHWA-SA-95-037.

Lundwall, N.B. (1968) The Saint George Temple, *Temples of the Most High*, Bookcraft, Salt Lake City, Chapter 3, p. 78.

Mayne, P.W., Jones, J.S. and Dumas, J.C. (1984) Ground response to dynamic compaction, *Journal of Geotechnical Engineering, ASCE*, Vol. 110(6), pp. 757–774.

Menard, L. and Broise, Y. (1976) Theoretical and practical aspects of dynamic consolidation, *Proceedings, Ground Treatment by Deep Compaction*, Institution of Civil Engineers, London, pp. 3–18.

Mitchell, J.K. and Katti, R.K. (1981) Soil improvement – state of the art report, *Proceedings, 10th International Conference on Soil Mechanics and Foundation Engineering*, Stockholm (June), pp. 509–565.

Mitchell, J.K. and Solymar, Z.V. (1984) Time-dependent strength gain in freshly deposited or densified sand, *Journal of Geotechnical Engineering, ASCE*, Vol. 110(11), pp. 1559–1576.

Seed, H.B. and Idriss, I.M. (1971) Simplified procedure for evaluating soil liquefaction potential, *Journal of Geotechnical Engineering, ASCE*, Vol. 97(SM9), pp. 458–482.

Slocombe, B.C. (1989) Thornton Road, Listerhills, Bradford. *Proceedings, International Conference on Piling and Deep Foundations*, London (May), pp. 131–142.

West, J.M. (1976) The role of ground improvement in foundation engineering, *Proceedings, Ground treatment by Deep Compaction*, Institution of Civil Engineers, London, pp. 71–78.

Chapter 4

Cement grouting

G. Stadler

4.1 Introduction

This exposé on cement grouting is meant to update the general technical knowledge of the reader on grouting using cementicious (particulate) suspensions and to acquaint him with some of the later – and in a way 'demystifying' – experiences and findings related to the different technologies of grouting. It is meant for the learned, semi-experienced who have had some personal contact with grouting projects, or have designed and executed projects.

Grouting generally is used to fill voids in the ground (fissures and porous structures) with the aim to increase resistance against deformation, to supply cohesion, shear-strength and uniaxial compressive strength or finally – and even more frequently – to reduce conductivity and interconnected porosity in an aquifer.

Grouting uses liquids which are injected under pressure into the pores and fissures of the ground (sediments and rock). Liquid grout mixes consist of mortar, particulate suspensions, aqueous solutions and chemical products like polyurethane, acrylate or epoxy. Piston or screw-feed pumps deliver grout through open boreholes into fissures in rock, through lances, perforated pipes and packered or sleeved pipes into sedimentary soils. By displacing gas or groundwater, these fluids fill pores and fissures in the ground and thus – after setting and hardening – attribute new properties to the subsoil. The degree of saturation with – and the properties of – the hardened grout are responsible for the degree of improvement achieved.

Grouting originates from mining and applications in hydro-engineering, and although its history (starting with Berigny in France) now dates back approx. 200 years, these two sectors remain where today's applications prevail. City excavations for high-rise structures and subways (Metro) have been prominently added to these examples. Figure 4.1 (Rodio South Africa) shows that one of the major achievements has been in early days the operation of long-range supply pipe lines for particulate grout in deep level mining. The speedy provision of grout mix to even remote areas of large

Figure 4.1 Deep level mining: ducts for ventilation, compressed air, water and grout (Rodio SA).

mines has more than once helped to save the mine from flooding, gas accident or collapse. Another example shows (Figure 4.2; Insond, Austria) how structural repair of broken concrete in double curvature arch dams has created another modern application of grouting with epoxy resins of high viscosity and strength. The particular feature of this repair was, that cracks

Figure 4.2 200 m high Kölnbrein Dam, Austria (Österreichische Draukraftwerke).

and fissures in concrete and rock had to be bonded under water and still, had to be conditioned so as to transfer even tensile forces of >2 MPa. Lombardi provided the design of this repair work and took the occasion to apply his concept of grouting intensity number (GIN) (see also Section 4.2.5) at this major repair project. More than 130 000 m of mainly small-diameter coredrilling were carried out and 200 tonnes of epoxy resins were injected at pressures of between 60 and 120 bar. Another typical example of applications of grouting may be given with the grouting of horizontal barriers (blankets) in the sands below city excavations in Berlin. Internationally renowned agencies and institutions established headquarters in the revived city centre and around Potsdamer Platz, requiring more than 250 000 m^2 of deep, water-sealing blankets in pervious sands during the 1990s. Figure 4.3 shows respective foundation works for the new Offices of the German President of State. To reduce seepage during excavation of construction pits at gradients of approx. 10, it was necessary to reduce permeabilities to around 1×10^{-7} m/s, which corresponded to seepage values of 1.5 l/s per 1000 m^2. Microfine binders barely met the requirement, and it was mainly silicates and aluminate-hardeners which were used to supply a soft gel in a single-shot treatment campaign (grouting one phase only through one single port outlet), sufficient to achieve the required impermeabilization and to withstand washout for more than 12 months.

Costs are of course at all times a matter of the market, and therefore difficult to generalize. To satisfy commercial requirements however, it is a general rule that grouting is only viable, if the process may be accomplished

Figure 4.3 Single port grout pipes (Berlin, Insond/Züblin).

within an acceptable lapse of time by pumping (in the case of permeation grouting) as speedily as feasible – namely, at pressures below ground fracturing, e.g. where the ground would be detrimentally deformed – fluids into voids.

When considering frequently observed technical and operative 'boundaries' however, like:

- average grouting-rate per pump: 5–20 l/min;
- average man-hour [H] per operative pump-hour [h]: 1.1–3.5 H/h;
- average man-hour per ton of cement of a neat OPC-grout: 5–10 H/ton;
- an average minimum borehole-spacing equal to the thickness of the treatment, or <3 m;
- average percentage of voids in sediments on which to base grout consumption: 28–38 per cent; in rocks however only 0.5–3.5 per cent;
- average metre of borehole per m^3 of soil/rock grouted: 0.25–0.8 m/m^3;
- average depreciation plus interest, cost for maintenance and repair of equipment and machinery, together: 3.6–4.1 per cent per month.

One could arrive at suitable estimates of total time and costs.

Furthermore it could be added here that, on average a minimum of 25 per cent of the operative personnel is required in addition, for supervision and infrastructural services. Mobilization and demobilization costs are to be calculated separately.

4.2 Preparatory works and design

It is important that the designer is aware of both the possibilities and the limitations of grouting. For this purpose a good overview knowledge of available systems and materials is required and a basic engineering understanding of the process. The principles of fluid mechanics might seem to govern grouting in the same way as they do the propagation of fluids in other media like pipes and ducts. The fundamental problem in adapting mathematical solutions from comparable models however is the lack of knowledge of the intricate rheological properties of the grout and of the geometry of the cross-sections in the ground which are exposed to flow. The rheology of the fluid as it can be presented in a rheogram may inform on dynamic, apparent or plastic viscosities and may indicate yield value and thixotropic properties, but it lacks information e.g. on surface tension relative to different kinds of wetted surface – and thus whether *capillary forces* will support or prevent successful penetration of the grout. Furthermore, one has not been able to directly introduce the degree of *dispersion of solids* as a criterion of filtration when confronted with small cross sections of flow. On the other hand, however, it is particularly these small sections, which do create the principle difficulty of allowing classical hydraulics to be applied on the grouting process. Once these minisections of flow take on the same size as the particles of the suspension, *electrostatic forces* seem to take over the role of dominant parameters of the fluid. The interface e.g. between *fluid films adherent* to the mineral surfaces of grains or fissures and the fluid components of the 'stream' of grout passing by (at differing velocities) become important and decisive elements of the process of penetration. This process is not quantifiable for the time being and needs more research.

Semprich and Stadler (2002) do – for the purpose of design and preparatory works – classify grouting:

- according to the type of the ground: grouting in *soil or in rock*;
- according to the aim of treatment: *consolidation (strengthening) or water tightening*;
- according to the period of use: *temporary or permanent*;
- according to the principle of the system: *permeation or displacement* grouting;
- finally, according to the grouting materials applied: *particulate pastes or suspensions*, chemical solutions or chemical products like polyurethane, epoxy, polyamides, etc.

These elements of systematic differentiation provide the user – apart from aspects of classification – with a guide to finding the most suitable grouting system for his project.

If one would opt for entering the aims of a grouting project at one side of such a classification matrix, and the parameters of the ground, rheological properties and composition of the mix at the other side, one could come up with a systems approach in defining the optimal grouting process. Table 4.1 attempts such a framework for respective interdependencies between types of ground, grouting methods and grouting materials – even suggesting suitable commercial structures for schedules of rates and compensation.

And it becomes quite obvious that the designer must have a decent overview-knowledge of what systems and materials are existing on the one hand and what may be achieved with their application in respect of the project targets on the other hand.

Targets for grouting applications are frequently formulated in an overdemanding manner, and it is suggested in this chapter, that specifications should be particularly and carefully reviewed when requirements concerning the targets of grouting are passing beyond the following values:

In alluvial soils

K_f values below 3×10^{-7} [m/s];
global uniaxial strength beyond 2.5 MPa;
deformation modulus higher than 250 MPa.

Table 4.1 Overview of grouting techniques in soil and rock (Stadler)

	Soil							Rock					
type	fine		coarse				type	diffuse fissurization, Kakirites			discrete joints (fracs)		
Grout ways, ports					drillrods & lances			collapsible/ unstable		stable rock			unstable
			perforated pipes				ports	MultiplePackerSleevePipe					
	sleeved manchette pipes								single/ double packer				
	open-ended pipes						system	stage grouting		bottom up grouting			stage
System	displacement (frac, compaction)			penetration (pore grouting, permeation)				frac/(fissure) grouting		permeation grouting			
Grout mixes	Silicate/ Acrylate						Grout mixes	Acrylate/ Epoxy					
		Microfine binder							Microfine binder				
		Bentonite/Cement								Ordinary Portland Cement			
						Mortar							Mortar
virgin Kf	10.E-6	10.E-5	10.E-4	10.E-3	10.E-2	>>10.E-1	virgin Lug.	1	5	10	25	50	>>100
Grouting parameters	Energy and displacement criteria		Grout limited quantities below frac pressure until resurgance or interconnections do occur			Limitation of quantity and pressure	Grouting parameters	Energy- and Saturation criteria			Pressurelimitation/ Energy criteria		
								split-spacing/ from inside outwards/ from outside inwards					
TXT Para	Degree of exploration vers. suggested structure of payable items (Nos. to be identified with TXT)												
2.1.1.1	stratification	stratification	stratification	stratification	stratification	stratification	2.1.2.1	lithological	lithological	lithological	lithological	lithological	lithological
2.1.1.2			porosity	porosity	porosity	porosity	2.1.2.2			frequency	frequency	frequency	frequency
2.1.1.3				density	density	density	2.1.2.3				anisotropy	anisotropy	anisotropy
2.1.1.4					water	water	2.1.2.4					water	water
2.1.1.5						position	2.1.2.5						position
Source/Titel	DCRC	"StilfOs"	ON 2270	DIN 18301 and 18309	Global items	Lump sum	Source/Titel	DCRC	"STILFOS"	ON 2270	DIN 18301 and 18309	Global items	Lump sum
Payable Items	Directly-Attributable-Costs-Reimbursement-Contract; plus Fee for General Expenses	GSE; Rental; Personnel all per time; Production; Materials; Energy	GSE/Mo; /lm Drilling; /Pump hour; /No.Test; Materials	/lm Drilling; /Pump hour; /No.Test; Materials	/lm Drilling; /m³ Ground; /No.Test	Functional; /m³ Ground; /lm Tunnel	Payable Items	Directly-Attributable-Costs-Reimbursement-Contract; plus Fee for General Expenses	GSE; Rental; Personnel all per time; Production; Materials; Energy	GSE/Mo; /lm Drilling; /Pump hour; /No.Test; Materials	/lm Drilling; /Pump hour; /No.Test; Materials	/lm Drilling; /m³ Ground; /No.Test	Functional; /m³ Ground; /lm Tunnel

Terminology:

DCRC	Direct cost reimbursement contract
StilfOs	Stilfontein Gold Mine and Oswaldiberg Tunnel compensation model
ON	Austrian Standard
DIN	German Standard
Nos. in TXT	Meaning of numbers to be looked up in text
virgin Kf	Permeability/ transmissivity before grouting
Kakirites	Decomposed, weathered rock

In rock

Lugeon values below 0.5 [l/min · m];
deformation modulus beyond 1500 MPa;
tensile strength beyond 500 kPa.

In cohesionless granular soils, one would usually obtain – as a permanent result of the treatment – maximum penetration of the porosities by using a combination of stable suspensions (hydraulic binders in water) and solutions (like sodium silicate plus hardener, or similar more modern products) under optimum pressure.

One of the 'classical' ways to provide access for these grouts to the ground (via discrete outlets, or 'ports') consists of the use of so called sleeve(d) pipes (see also Figure 4.1, manchette pipes, tube à manchettes, TAMs) made of drawn or welded mild steel or PVC of 1 in., $1^1/_2$ in. and

Figure 4.4 2 in. PVC manchette pipes, sleeves 50 cm c/c (Insond, Vienna).

2 in. diameter (single-port $^1/_2$ in.-pipes are nowadays also used in uniform sands). The annular space between sleeve pipe and borehole is sealed by a 'plastic' sheathing grout of cement–bentonite, filled in from the bottom of the hole upwards (Because of the low cement content of this annular grout of approx. 175–650 kg PC/m^3, the time to set and harden lasts between 24 hours and 6 days.) The sheath grout is intended to block grout injected later from escaping to surface, instead of penetrating into the ground.

Thus the intended re-use of individual grout ports without having to redrill boreholes is secured. At the same time this system makes it possible to inject different grouts sequentially in the same borehole, not only different mixes at differing ports but also at different times suitable to optimal project specifications. This possibility of a flexible handling of the grouting concept has over decades proven a successful means of adapting originally designed grouting sequences to actual situations encountered during execution of the works. And it is for this flexibility, that this technology clearly dominates alluvial grouting to date. The technique is however not suitable in rock.

Grouting through drillrods or driven *pipes* (*lances*) are systems for grouting applications of lesser requirement and in coarse-grained ground of high porosity. Grout mix in these cases will be placed

- through/via perforated pipes which are driven or inserted in predrilled boreholes;
- via borehole casings during withdrawal from the borehole;
- through the drillbit itself when drilling the hole.

Grouting through rods, casing or pipes may take place only at low pressures, to avoid mix escaping to surface along the unsealed contact between pipe and ground. Following low grouting pressures the reach of penetration and the degree of filling of voids is reduced. The effect of the operation is limited to a partial 'gluing' of grain structures in the ground. Still, these techniques may be used – in their own right – when the requirement is just to provide a decent increase in cohesion of the ground or where open coarse-grained sediments are to be filled.

Fine-grained and cohesive soils are less apt to treatment with particulate grouts or chemicals. This is because the geometry of pores exhibits considerably smaller cross sections for grout to flow and capillary forces in general begin to govern flow regimes, and phenomena originating from surface tensions start to prevail. This is a field which is scarcely researched, however, and the discussion of limiting factors of penetration are concentrating on filter criteria instead. These are based on the relation between characteristic grain sizes of the ground relative to the particle diameters of the solids in the mix.

Schulze (1993) did some research on pore size distribution and penetrability of sediments by relating grain size distribution of OPC and Microfine binders to the sieve analysis of soil samples. Figure 4.5 shows how one might derive a pore diameter distribution from percentage mass distribution (sieve analysis) of the soil. This may be less sophisticated but based on broad experience and hence with similar results are 'groutability ratios' which are much in use in the USA. There the D_{15} (diameter at which 15 per cent of the soil sample is passing) is related to a D_{85} (diameter at which 85 per cent of the grouting material is passing); according to this, groutability is granted at

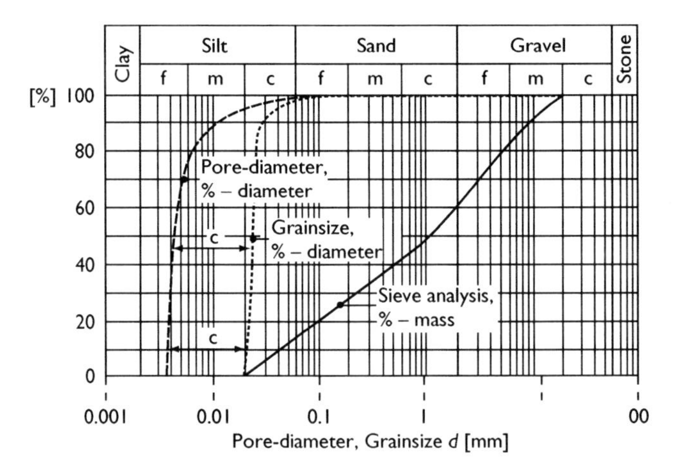

Figure 4.5 Theoretical relation between grain size and pore diameter in granular soils (Schulze).

ratios >24. Practical experience and the fact that for this case the relation between pore diameter and characteristic grain diameter in the suspension amounts to approx. 4, is the reason for these criteria being acceptable generally – even based on common filter criteria only.

According to this, one would not be successful in grouting fine-grained sediments with a silt content >5 per cent when using particulate grout based on PC which at 85 per cent passing contains material of diameter >40 μm. For such cases use of Microfine binders is recommended instead. And if even Microfine suspensions cannot penetrate, the only remaining solution is to use chemical grouts. Frequently sodium silicates combined with organic or (nowadays rather inorganic) hardeners are used, or modern Acrylate grouts (water soluble) or even, hybrid formulations of both. If – with decreasing pore sizes in the ground – the grouting process is slowed down and it becomes a question of time and is therefore getting uneconomical for reasons of high material costs as well, only.

Compaction Grouting (see Chapter 6) and *Frac Grouting* (see Chapter 7) may be resorted to as a means to consolidate or tighten the ground with soil-displacing methods of grouting. Such systems of *Displacement-(Compaction-) and Hydrofrac grouting* do cause – at quite intensive pressures of >40 bar – however, two quite distinct and differing reactions in the ground.

Compaction grouting makes use of the hydraulic properties of pastes (rich in sand) and mortars which are injected to just fill the space below the grout casing during withdrawal in an attempt, to apply its overpressure onto the borehole wall and widen the hole under pressure. This overpressure exerts a consolidating force onto the ground without fracking it. The grout (mortar) develops internal friction during the process and is behaving in the end more like a soil than a fluid. The resulting mortar columns – more or less regular in shape – to some limited extent may even be subjected to vertical loads.

Hydrofrac grouting to the contrary, is rather applying tensile forces to the ground – at the start of the process acting perpendicularly to the orientation of the minor stresses – producing thin sheets of grout parting the ground structure. The orientation in space of these 'sheets' does change during the progressing – stress homogenizing – operation, finally adopting predominantly horizontal orientation in the end. The increase of volume causes consolidation of the ground, densifies the soil structure and reduces the porosity to an extent that consolidating as well as water-tightening effects do occur.

Apart from that, the occasionally intended lifting of the ground below foundation level (and buildings) is caused, and loss of ground – having occurred during tunneling excavations – is compensated.

The grout (usually suspensions), is injected into the ground through sleeve pipes, resulting in the same advantageous multiple use of the grouting ports as is the case with penetration grouting. This makes the application of frac grouting possible in a controlled way. For the specific application of lifting of buildings the sleeve pipes are installed (sub)horizontally. These works require sensitive control of the structurally relevant parts of the building by precision levelling and as such make the process applicable to corrections of level and tilt of slender high rise buildings, bridge pillars and the like.

Grouting in rock – in the majority of cases – aims at tightening against percolation of water. Ground water which migrates and flows under varying gradients in fissures, joints and tectonic discontinuities under dams or in the form of seepage into deep tunnels for instance will be reduced or stopped by grout from such migration.

Figure 4.6 shows that virgin transmissivities before grouting determine which types of grouts may be successfully used to obtain the target result – here below say 0.5 Lugeon. One may see that OPC suspensions are suitable for transmissivities of more than around 25 Lugeon. Below 3 Lugeon on the other hand, only Microfine binders, acrylates or silicates (with organic hardeners) – sometimes even resins – may be efficient.

To arrive at an assessment of likely average fissure widths existing around any borehole Cambefort (1964) has published his view on relations between transmissivities (Lugeon values) frequencies and opening widths. Semprich and Stadler (2002) provide an updated version of Stadler, Howes and Chow (Figure 4.7), including ISRM's view which allows – as an initial approach when grouting in rocks – to estimate likely filtration conditions for particulate grouts, in relation to fissure widths.

For commercial reasons (when considering cost of materials), predominantly suspensions (particulate grouts) are used to tighten and strengthen rocks. Occasionally this aim is, however, indeed even more economically achieved and technically superior by a follow up (or an entire replacing of

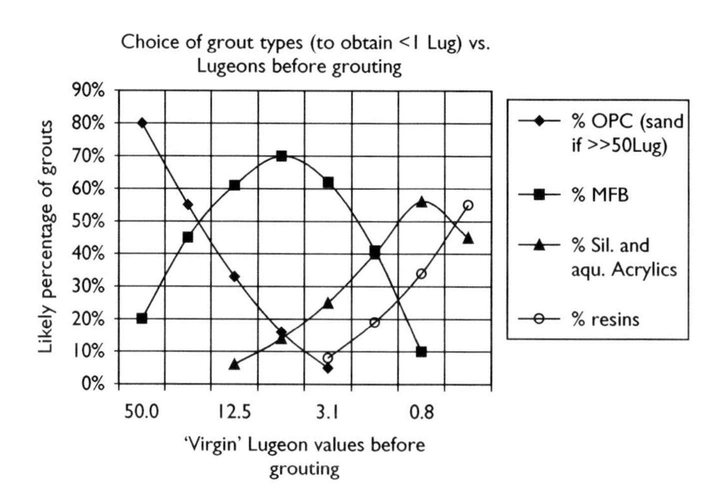

Figure 4.6 Effective grouts for different fracture transmissivities (Stadler).

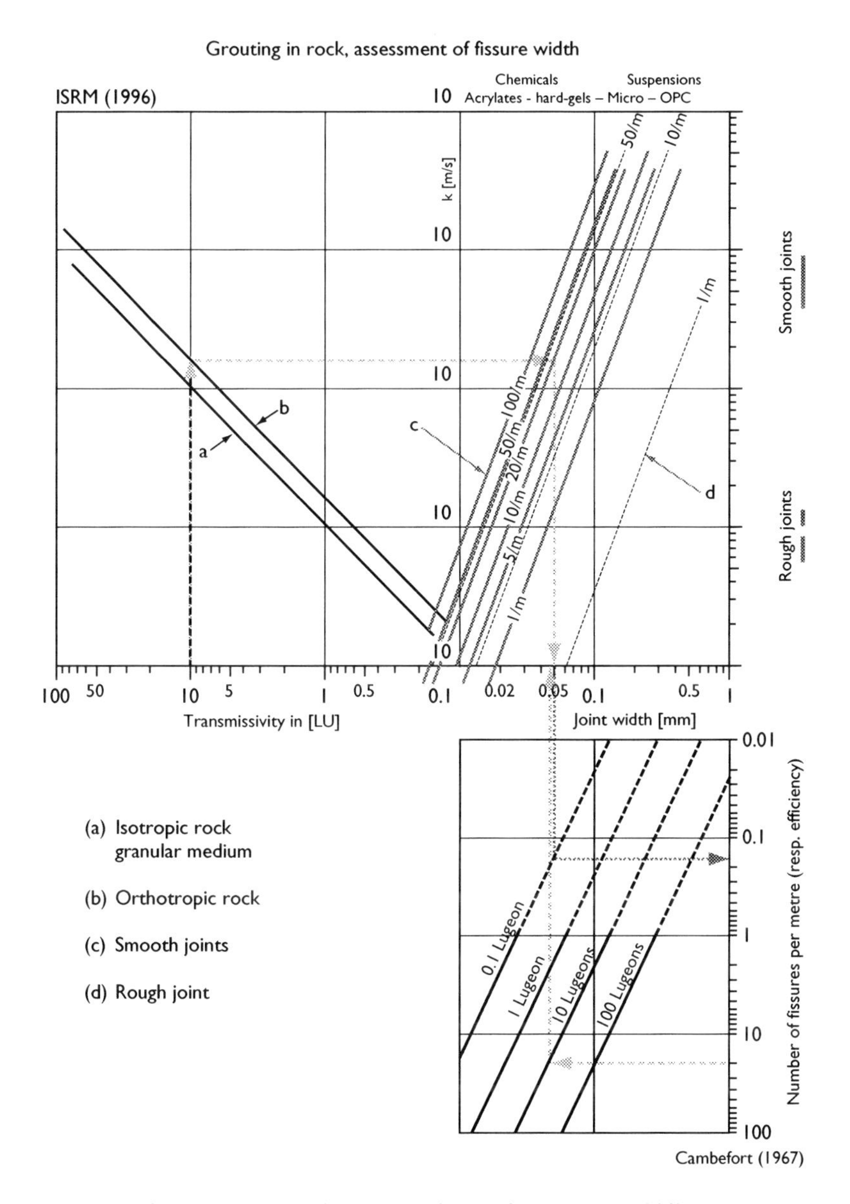

Figure 4.7 Grouters diagram (ISRM, modified by Stadler *et al.*, 1998).

OPC grouts even) with chemical grouts such as acrylates, polyurethanes or polyamides (Karol, 1990).

The filling of all of the accessible interconnected voids/porosities obviously results in a maximum reduction in transmissivity. The better the engineering terms of the grouting process in respect of hydraulics are understood, the better their application in the field and their results. For a more profound understanding of rock grouting it is therefore quite important to know that the process is distinctly different from the one in sediments where grout travels through comparably isotropic, homogeneous porous structures (if we disregard preferential horizontal permeabilities, for the sake of argument).

Discontinuities in rock, are dominantly two-dimensional in shape and moreover, are frequently intersected by other sets of fissures and joints. All of which neither appears plane and parallel nor is the opening width constant. These planes are often in contact with each other at innumerable points and sectors, and are frequently filled partly or entirely by weathering material or tectonic debris.

Consequently, the flow regime in fissures varies from 'channel flow' at low transmissivities (of <5 Lugeon), to concentrical 'planar flow' starting off a singular intersection of a grout-hole with a fissure plane, and finally to 'spherical flow' which activates a multitude of criss-crossing fissurizations – suggesting quasi isotropic conditions. Flow equations are proposed by Hässler and Gaisbauer (Widmann, 1993).

Other accompanying phenomena are a consequence of hydraulic grouting pressures, which are exerting normal forces onto the sidewalls/flanks of these fissures and cause *displacement in the rocks* around the respective zone. Some of the fissures are widened elastically or permanently, others will be compressed and contract, and sometimes closing penetrability. The flow regime is laminar already shortly after grout has entered the fissure. Grouting pressures generally drop exponentially with increasing distance from the hole. Bingham fluids tend to accentuate this pressure drop compared to Newtonian fluids, and it is for this reason that the problem of fissure widening or the danger of frac propagation are linked more to the use of the latter.

In stable rock, boreholes for grouting are produced using roto-percussion drills, flushed with compressed air or water and, with drilling diameters of 36–76 mm. Drilling depths normally are limited to 60 m when using external hammers; deflection increases depending on oriention of fissures, change of hardness of strata (silica content) and downpull on the advancing drillstring; drilling progress decreases with number of rod couplings, insufficient flushing or loss of drilling fluid and condition of the drillbit. After reaching final, drilling depth the hole will be separated by single packers in sections of 1.5 to 6.0 m and grouted from 'bottom up'. Mixes for grouting in rock used in earlier days unstable suspensions (separating >5 per cent free water in 2 h under gravity) relying on the phenomena of pressure-filtration for the success of grouting. Today, stable suspensions are preferred. Grouting to a pre-determined refusal pressure has been abandoned as well, in favour of a dual criterion using the product of quantity and effective grouting pressure (grouting intensity, which in essence corresponds to a limitation of grouting energy per m^3 of treated ground) as specification for a general break-off criterion.

In collapsible rock formations grouting technology differs as does drilling technology. Caused by the tendency of the hole to collapse, boreholes are drilled in a step by step technique. The sequential production of a hole is characterized by the intermittent operation of the drill and the grouting operation, so that the typical sequence of operations consists of: drilling, setting of packer, grouting, leaving the grout set, redrilling of the grouted zone, drilling of fresh rock below grouted zone (as deep as borehole stability permits, or 6 m), setting of packer and repeating in sequence. The packer is here generally set at the collar of the hole.

Sleeve pipes with inflatable jute bag packers (or multiple packer sleeve pipes (MPSP), Rodio) occasionally were successfully used in collapsible rocks where the drilling to final depth was possible, either using casing or drilling muds. This special sleeve pipe consists of a $1\,^1/_2$ in. sleeved steel pipe, manchettes placed at intervals of 50 cm–1 m. After insertion in the hole the external jute packers (mounted at 1.5–6 m spacing) are inflated and thus the borehole is separated in sections in which sleeve pipe grouting may be carried out, comparable to alluvial grouting. In the sections between the jute bags the pipe is not sheath grouted so as to provide better circulation of grout into the hydraulically conductive zones of the rock. Because of the relatively low W/C ratios of the grouts used of below 1.3 (and their rapid hardening), no regrouting is possible if not done into the fresh unset grout.

Sleeve pipe and frac grouting in ‘rock’ will of course only be successful in ground which is decomposed by weathering to a great extent or where tectonics did remould intact rock structures into consistencies comparable to sediments. If in such decomposed rocks groutholes have to be collared below water table, special care has to be applied so as not to allow broken (Kakirites) or otherwise ‘liquifiable’ ground to enter the working area (uncontrollably forced by water under high pressures). Drilling in such cases is performed through stuffing boxes or annular preventers equipped with two independent systems of locking the borehole securely.

When designing the borehole grid, making the choice of a grouting method and specifying grouting parameters, it is essential to properly adapt general ‘rules’ onto local geological geotechnical conditions, considering the quantifiable aims of the treatment, topography and other limiting circumstances of the project.

Whenever possible, the design and specification for the works should be tuned to the specific local project situation by *grouting trials*, and the flexibility to consider amending and adjusting design and operational parameters during execution (based on processing data and improved reconnaissance – in particular relating to the structure and properties of the ground) should be incorporated. The reach of grout and consequently the borehole spacing is generally influenced – inversely proportional – by the *yield value* of the grout mix. It is on the other hand increasing proportionately with the grouting pressures applied. *Viscosity* of the grouting fluid on the other hand is responsible for frictional losses which must – when specifying maximum allowable pressures – be duly considered, as well as pressure losses occurring when grout enters small fissures from a borehole. Thus the *maximum allowable pressure* is defined rather 'by itself' and on the basic of the hydraulic interaction of geometry of voids and rheology of the grout mix, than on the weight of the ground over the point of injection! (Figure 4.8 shows borehole grid for horizontal curtain grouting at JFK Airport, ICOS.)

From the above it is evident that every grouting project needs expert preparatory action and special attention to its design definitions. Particular

Figure 4.8 Chemical grouting of sands (JFK airport, ICOS).

care in this context has to be applied to the elementary question *whether grouting truly is the best technical solution to the problem* (strengthening or tightening of the subsoil), which of the parameters of the ground – as it exists – have to be improved, and from which state (as is) to which desired state these parameters (properties) will have to be changed. Only measurable properties should be addressed. And only parameters which have a defined relation to grouted or ungrouted soils should be selected. Visual inspection of a grouted strata in this respect frequently fails as suitable criterion (e.g. of acceptance).

The *designer has obviously therefore to be an expert.* His realistic judgement of the above questions will highly influence the outcome of the treatment. And regardless of even the best efforts in properly assessing all the technical aspects it remains particularly important to define specifications and contract conditions (particular specification of the works, bill of quantities and schedule of rates) in a likewise suitable manner.

Grouting materials play an important role in as much as (a) their initial choice depends on the design strength and the desired degree of impermeability and durability, and (b) they might during execution have to be adapted to the requirements and conditions of the ground. The resulting flexibility has also to account for environmental aspects which have to be duly considered as well (Stadler, 2001).

Suspensions are characterized by solids, homogeneously dispersed in liquids. OPC and Microfine binders are usually mixed in water. Water to solid ratios generally range between 0.65 (when using OPCs) and 5.0 (MFB). W/C-ratios below 0.5 result in pastes. The addition of sand (0.1–1.0 and 1.0–3.0 mm) at ratios around 1 (OPC):2 (sand) generates mortars. The grain size of solids is limiting penetrability and therefore, finely ground materials do penetrate small voids in the ground more easily. MFBs are exhibiting max. particle diameters of 15–20 μm. OPCs range upto 100 μm. Solutions nowadays are – for environmental reasons – limited to the use of sodium-silicates and inorganic hardeners. These do provide relatively soft gels ($\tau = 5$–$25\,\text{g/cm}^2$ after 7 h) and are designed for temporary performance of water barriers in sands. Organic hardeners – whenever permitted – may add strength to the grouted soils as well.

Chemical grouts relate to the following:

- Polyurethane foams and gels (in the mining industry, in mini tunneling and to block rapid flowing currents underground). For details refer to Karol (1990).
- Acrylates (of MBTechnologies or Sika) seem to be a new 'follow-up' type of chemical grouts subsequent to earlier generations of materials, like acrylamides as were AM9 in the USA, the Japanese Nitto SS and Rocagil BT of Rhone Poulenc, France – to name a few examples. Environmental effects seem to have been overcome by these quasi

non-toxic formulations, still exhibiting the positive properties of penetrability and strength. Hybrid formulations with silicates are on the market as well.

- Epoxies in special compositions have been used recently for structural repairs of concrete and rock (Zeuzier dam in Switzerland, Al Atazar dam in Spain, Kölnbrein dam in Austria). High viscosities of more than 5000 cps, high affinities to wetted surfaces and high strength (tensile strength of up to 80 MPa) make their application somewhat special but highly successful.

Proper knowledge of the subsoil conditions is of prime importance for grouting applications. And it cannot be overemphasised that – grouting being a predominantly hydraulic process – soil-exploration campaigns therefore require reconnaissance primarily of hydraulic properties of the ground. In particular the stratification of sediments is an important feature as far as grouting is concerned. Figure 4.9 shows an example of stratified ground (Vienna, Subway Lot U3/4), where hydraulic properties distinctly follow the sedimentary layers.

Figure 4.9 Typical stratification of alluvial deposits (Vienna Subway, U3/4).

Of soils the following information is required:

- stratification, typical grain-size distributions, conductivity profile, K_H/K_V;
- porosity, saturation, specific surface [m^2/m^3];
- density of packing (CPT, SPT), grain shape, deformation modulus;
- ground water table, gradient, GW-chemistry;
- position of wells, rivers, sewers, gullies, lines and ducts relative to the intended grouting area; building foundations, basements, underground structures and their respective conditions and properties near the grouting field.

Extensive experience indicates that, the more differentiated the schedule of rates for grouting works is structered, the less may be the intensity of exploration – but never (!) dropping below the items mentioned under bulleted items above and below. Particularly when – in rare cases – the efforts of grouting to obtain certain targets have to be included in other rates (say: costs for grouting works to be included in any m^3 of tunnel excavation) then, the standards and the quality of information gained from exploration must be all-encompassing, conclusive and fully comprehensive.

In rocks the following information is required (Figure 4.10):

- lithological stratification, stereoplot of discontinuities, transmissivity-profile;
- frequency of discontinuities, modulus of deformation, porosity;
- anisotropy of transmissivity, RQD, mineralogical composition, weathering;
- ground water table, gradient, sources, barriers and wells, ground water chemistry;
- position and conditions of underground facilities and structures.

Conductivity tests in stratified sediments primarily should not only lead to the commonly tested horizontal capacity of flow in the respective individual strata but also the collection of information on horizontal and – at the same time – vertical permeabilities. This is becoming obvious e.g. from grouting tests in uniform sands (Figure 4.11, Berlin, Germany), where polyurethane foam was injected from the end of a pipe to find out the difference in reach of grout, travelling from this single point of injection in all directions. The results were quasi-elliptical bodies of grouted sand exhibiting a relation of approx. 1:2 between vertical and horizontal dimension. This relation might grow to as much as 10 or 15 if grout is injected in lesser 'classified' sediments, which had been transported only a few kilometres

Figure 4.10 Bedding and jointing in Gneiss (Central Alps, Austria).

Figure 4.11 Test grouting in sand, anisotropic spread of silicate grouts (Berlin).

instead of hundreds of kilometres, as was the case for the sediments of the Brandenburg sands (comparable to the *sables du Beauchamps* in Paris). For the design of the expected spread and vectorial travel of the grout in the ground this information therefore is of major importance and frequently underestimated.

Apart from grain size distributions to be established for each characteristic strata in a sediment, it is also important to know about the conductivity values of the respective strata in the grouted area and in its immediate vicinity, individually and in profile. This may be accomplished by point tests (Lefranc) or by measuring the vertical velocity-profile of the flow in mini-wells (of 80 mm filter diameter and, using Micromoulinet-type vanes) when creating an artificial depression. Only this type of field-testing identifies the true hydraulic information required of the conductivities (caused by sedimentary stratification).

The dimension of the measured unit (permeability coefficient) is [m/s]: A 'velocity', which must however not be misunderstood as the true flow velocity of the fluid in the ground but as a velocity which is related to the ground as a whole. The true velocity may be arrived at, when considering the porosity of the ground only.

Hydraulic testing in rock is standardized in the form of the *Essay Lugeon* (Cambefort, 1964). This test is carried out in open boreholes of 75 mm diameter, in which sections of 1.5–6.0 m length are isolated by single or double packers, and water is infiltrated into the rock under pressure. The absorption in liter per minute and per metre of borehole at 10 bar (over)-pressure is called Lugeon value; in other words, 1 Lugeon represents the inflow of 1 l/min per lin. m at 10 bar pressure into rock. The dimension consequently is [m^2/s], the appropriate term – transmissivity [T].

It is recommended that this test be carried out in a number of pressure steps (say: 0.5–1.5–4.5, –7 and 10 bar on the way up, and 8–4 and 2.5 bar on the way down). The plot of pressure [*p*] against rate [*q*] will indicate by its shape and hysteresis, whether deformation (e.g. dilatancy of fissures) or washout (of infill) is taking place, or whether clogging by suffusion or, filtration of particulate grout occurs. Elastic or permanent deformations of rock, turbulent flow conditions, etc. may be identified as well.

Lugeon testing as such gives valuable information on geotechnical and hydraulic conditions of the underground. There is however, no relation between high Lugeon values and high grout takes or vice versa[1], and there is no sound reason for a relation between height of overburden, type of rock and maximum allowable grouting pressure either.

Attention is drawn here to the fact that

- horizontally bedded (sedimentary) rocks tend to frac and heave at pressures of as low as 3 bar (Ewert, 1985);
- optimum grouting pressures are better established as described below;
- hydraulic testing has to take into account hydrostatic (phreatic) levels of ground water tables and the resulting hydraulic counterheads.[2]

Operative criteria were again recently addressed by Semprich and Stadler (2002) in a comprehensive way. They are of the opinion (in line with requirements published in EN 12715) that the *specifying of grouting parameters* has to be already accomplished in the design itself, even if – despite of all improved theoretical background – this may still only be based on experience and empirical data from past project realisations. And in fact, this is important also, for to be able to adjust during execution against specified pressures, quantities and rates, as they may have been laid down – either per passe[3] or per unit volume of ground.

In general, single parameter criteria are being replaced by *dual criteria strategies*. Whereas formerly the grout take alone (grouting rate and amount at which a passe is able to be injected) determined the next steps (Weaver, 1993), today it is a dual strategy which facilitates a more global assessment of the proceedings.

Pressure and rate (or quantity) together frequently supply break-off criteria which together with movement of ground, interconnections between holes and outbreak of grout to surface jointly or individually provide sufficient data to assess saturation or indicate the necessity to change procedure or materials.

The *maximum allowable grouting pressure* at the collar of the hole frequently will be fixed at around 80 per cent of the prevailing frac pressure. This frac pressure (at which the ground is separating and/or is loosing its cohesive state) may only be established by testing the ground at the individual project site. The test consists of registering grouting pressures measured at different locations (which representatively cover the treatment area), and different depths, while grouting with systematically differing pumping rates. Generally the frac pressure is then manifesting itself, when increased pump rates at a test point result in decreasing grouting pressures. This methodic approach makes the (guess work) specification of maximum pressures on the basis of depth, or in relation to the surcharge weight of the ground, obsolete. And it is only with this system, that optimum grouting rates without causing detrimental deformations (heave on surface, spalling in tunnels) may be established. For common diameters and lengths of grout lines in alluvial grouting (taking into account tiny cross-sections of flow in packers and sleeve pipes, narrow fissures in sheath grout and tiny pore-structures in sands) between 5 and 35 bar do result as a respective grout pressure. *Grouting rates* of particulate grout and chemical solutions vary between 5 and 15 l/min respectively. For fissures in rock, exhibiting <0.15 mm width and, using highly viscous epoxies, these pressures might rise to even >120 bar without causing any damage, the main reason being that, the pressure drop at the entry of the fissure is already consuming most of any destructive energy.

The specification of a maximum *quantity of grout* to be injected per passe or per unit volume of ground is based on the plausible estimate of accessible porosity. Accordingly, for sediments these estimates vary between approx. 25 and 40 per cent. For rock these quantitative limits are specified –

frequently rather for economical than technical reasons – to prevent uncontrolled loss of grout. Porosities in rock generally vary between 0.5 and 5 per cent. Only in karstic limestones and dolomites, or in highly deformable broken rocks are these porosities to be estimated higher.

The grouting rate usually results from the interaction between cross section exposed to flow and rheology of the fluid (grout mix). Common applications of particulate suspensions are operated at rates (as mentioned above) between 3 and 20 l/min. In karstic rock this value might rise to even 100 l/min, or the limiting capacity of the pump. Highly viscous epoxies on the other hand, might have to be grouted into fissures of <0.15 mm at rates of as low as <1 l/min.

The importance of grouting parameters *(which are meant to be adapted, adjusted or are indicative for breaking off the process altogether)* become obvious when considering the fact, that no direct means exist to measure the degree of accomplishment of the process during execution. All relevant specifications therefore are on the one hand to be based on exploratory investigations of the ground and values derived from experience. On the other hand, these specifications must be flexible enough to allow for their adaptation and change if processing data and other observations suggest such adjustments. This recommendation has been expressively taken over as a 'requirement' when drawing up the EuroNorm 12715. In consequence, this same flexibility is recommended to be introduced when formulating the construction contract and the relevant items in the Bill of Quantities. Primarily, these criteria are to suit the necessary assessment of achievements in accomplishing the aims of filling of voids and reducing permeabilities in the ground. For this purpose the *pressure development against time* is playing a key role. For classical penetration grouting this curve generally reflects a steady gain in pressure until optimum saturation is reached. The interpretation of such curves is particularly informative when keeping rate and rheological properties of the mix constant. In sediments the pressure development is to be seen in relation to the designed total quantity; in rock however, the successive drilling of primary and secondary holes makes it possible to investigate the actual reach and effect of the preceding grouting operations under the respectively applied pressures.

A *pressure drop* frequently indicates progressive opening of flow paths – in rocks not so much a consequence of fissure widening but rather the effects of erosion of weathering fills under the gradients and pressures actually prevailing in such fissures; in sediments such digressive pressure development frequently indicates displacement of fines under erosion or suffusion. Sudden pressure drops when grouting alluvial soil indicates the development of (mainly) subhorizontal fracs (fr.: *Claquages*) which may cause unintended heave on the surface or loss of grout.

The observing of the *transitional pressure development* after a borehole cave-in caused by deliberate stoppages of the grout pump is called

TPA-technique (transitional pressure analysis, see glossary EN 12715). This method allows a reasonable estimate of effective pressures in the ground and indicates actual transmissivities of the ground relative to the mix in use. Finally it may indicate (if carried out repeatedly at short intervals of say 15 to 5 min) to assess the degree of saturation achieved. This technology was derived from oil well techniques (Stadler *et al.*, 1989). Figure 4.12 illustrates this technique: pressure development after pump stop is plotted and interpreted from trend lines of each individual interception. The smaller the drop value of the curve the smaller becomes the relative transmissivity. For the first time, this technique has been used in construction practice at a large scale in the 1990s (Stadler, 1992). TPA is mainly applied in rock grouting. The technique is based on information received when the pressure recording after a pump stop is continued. The diagnostic character of this information regarding the grouting process makes it possible to quantitatively discuss the applied grouting parameters against the original design. Together with the limitation of applied grouting energy which corresponds to the *grouting intensity number* (GIN value, which is the product of quantity of mix grouted and grouting pressure per linear metre of hole subjected to grout, [bar · l/m]) this technique provides a simplified engineering approach for assessing the grouting process. It remains important however, that the control of the procedure is not carried out without proper geological–geotechnical assistance, and furthermore that the GIN values calculated

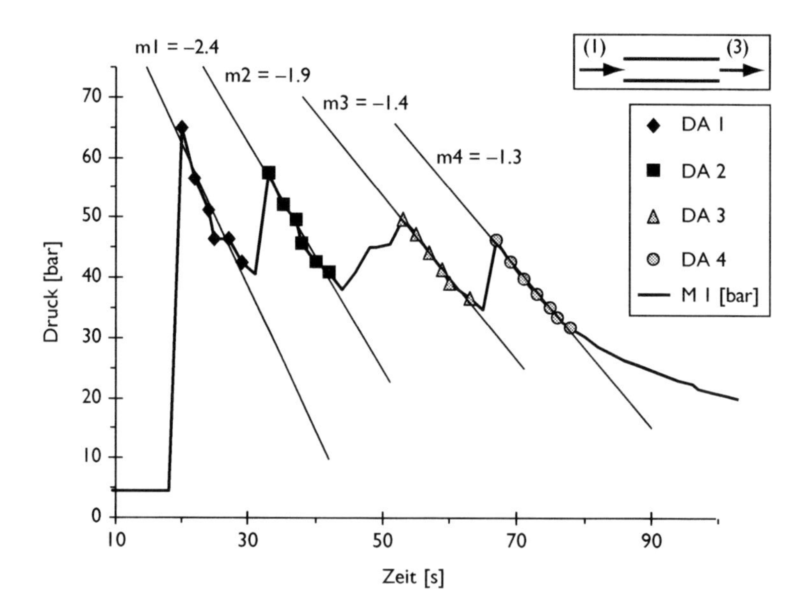

Figure 4.12 Transient pressure development after grout stops (TPA pressure sensitive grouting, Stadler, Zettler).

are using effective pressures only (not operative pressures taken at the pump or the collar of the hole).

Regarding adjustments to the grouting procedure and the recommended steps which lead to the final halt of the grouting operation, Weaver (1991) formulated respective criteria which have been successfully applied in grouting under dams and may be adapted to similar applications. His flow charts supply the respective logic indicating when to change either rate, mix or pressures of grouting. Based on such or comparable considerations (TPA and GIN) it becomes possible to formulate the design of modern grouting practice, particularly for grouting in rock.

The definition of operative parameters for penetration grouting of sediments on the other side, is more dependent on the relation between geometry of pore sizes, composition of particulate grouts and rheology of the mix. Diagnostic interpretation of the process during the grouting operation itself at present remains limited. The interpretation of success or failure of grouting in so distinctively orthotropic situations as is the case with the stratigraphy in most alluvial sediments, will therefore be even more dependent on the relation between a $K_{\text{horizontal}}$ and K_{vertical} than on an – even intricate – observation of the development of the grouting pressures or rates. One of the methods practised is to either observe and interpret rates at constant pressures, or pressures while keeping the rate of grouting at a constant value.

Grouting design in rock tunnelling may be discussed and quoted[4] here as one of the most actual examples concerning technical and commercial requirements: Design of grouting in tunnels in rock means essentially

- the development of orientation and density of drilling patterns;
- the choosing of appropriate grout materials;
- the methods and procedures (operative parameters like pressure, rate, quantity, trigger values for intervention against movements or seepage rates, etc.) to be applied during execution.

These are the variables which can be controlled by engineering and which are varied according to local conditions in the tunnel, with the purpose of achieving a specific result. The respective design efforts, however, are based on exploratory information only, and it is no wonder that the type of result which will be achieved cannot be accurately predicted, the reason being that, nobody can directly observe what truly happens in the ground during grouting. Technical (and commercial) flexibility is required therefore to follow the actual conditions encountered.

Even the evaluation of carefully controlled full-scale tests may be difficult. Such tests are carried out mainly in holes which are drilled vertically, whereas tunnels are obviously driven horizontally. The uncertainty about unforeseen changes in ground conditions from one test location to the next thus is surpassed by the imponderables of anisotropic hydraulic properties and therefore cannot be accurately quantified. However, most of the principles of pre-grouting are backed by seasoned experience and are supported by the results of several thousand tonnes of grout injection in mining and tunnelling, and thus the understanding of the principles are of less guesswork than is sometimes claimed.

The word 'design' however needs to be clarified here. This need arises from the difference to the normal (contractual) understanding of the term when used in structural engineering. Design of a bridge or a high-rise building will include the necessary drawings, materials specifications and calculations to define the dimensions, the geometry, the load-bearing capacity, the foundations and the general layout of the object. The respective analysis has to be based on the given physical surroundings, the owner's requirements, life expectancy of the project and other.

The reality in tunnel grouting however is that it is not possible to design the work with comparable precision in advance, so it is in no way comparable to the 'design' process referred to in the previous paragraph. The design of tunnel grouting operations is based upon the best estimates of the permeability and geometry of fissures in the rock through which the tunnel is to be driven, frequently based on the average values only. The design will usually include calculations of the likely water ingress, drawings showing matters such as the depth, angle and pattern of the intended drilling, execution procedures covering all aspects of the operation and the materials specification, so as to aim at satisfying the required water tightness of the tunnel. There is thus no question of drawings being produced showing what the finished job will look like or to give accurate dimensions for the result.

The pre-investigations for rock tunnel projects will never give sufficient details about the rock material and the hydrogeological situation for the full length of the tunnel, so as to allow a 'bridge design' approach. Furthermore, the calculation methods available are not refined enough to accurately analyse the link between the required result and the necessary steps to produce it. Risks in terms of construction time and financial implications therefore are, in the proper sense of the word, not calculable, and must remain with the owner.

The basic design for the grouting operation as referred to above has to be applied in practice on an empirical, observational basis as described as follows.

Once the 'tightness' requirements are defined, the project data and all available information about rock conditions and hydrogeology can be analysed and compared with the requirements. This often includes indicative calculations of potential ground water ingress under different typical situations. Based on empirical data (previous pre-injection tunnel project experience) a complete pre-grouting method statement can then be compiled.

However, irrespective of how elaborate this method statement (or 'design') is, and whatever tools and calculations are employed to produce it, it will not be more than a prognosis. This prognosis will express how to execute the pre-grouting (under the expected range of ground conditions), to meet the required tightness of the excavated tunnel.

During excavation the resulting tightness in terms of water ingress achieved can be measured quite accurately. This means that it is possible to move to a quantitative comparison between targeted water ingress and the actual result and accurately pinpoint if the situation is satisfactory or not. If the results are satisfactory, the work will continue without changes, and only a continued verification by ingress measurements will be necessary.

If the measured water ingress rate is too high, this information will be used to decide on how to modify the 'design' to ensure satisfactory results compared to the requirements for the remaining tunnel excavation. This may have to be executed in stages, until satisfactory results are achieved. Excavated tunnel sections which do not meet the requirements of the specification will have to be locally post-grouted until the overall result for such sections are acceptable unless it is possible to compromise on the water tightness requirements.

For grouting in TBM-driven tunnels it is vital to focus on the need to efficiently drill ahead of the tunnel face (between 1 and 4 holes generally, for sections of up to $10\,m^2$), to be able to detect water, locate hydraulic or strength anomalies, and to execute the required pre-grouting. This may seem like an obvious thing and unnecessary to state, but unfortunately, experience shows through numerous examples that this basic fact had not been handled properly in all instances. The best grouting materials and injection techniques are of no use if the necessary precautions for these trial-holes are not made.

Since the TBM occupies a major part of the excavated space close to the tunnel face, there is a slim chance of improvizing efficient drilling installations after the TBM has started excavation. The only realistic way is to custom design and install such equipment on the TBM in advance of equipment commissioning in the tunnel. Once such equipment has been mounted for collaring of holes in free positions, a short distance behind the face, not only probe drilling and drilling for pre-grouting becomes a viable option, also coring ahead is possible. The risk of having to cross shear zones causing TBM advance rates of less than 5 per cent of the normal is thus substantially reduced.

If one would try to generalize systematic requirements of drillholes and grout mixes to obtain results (as they are frequently contained in actual design specifications) of say 5×10^{-8} m/s (or 0.5 Lugeon, if one accepts this simplified transformation of units), the following types and amount of drilling and grouting may be required (Figure 4.13).

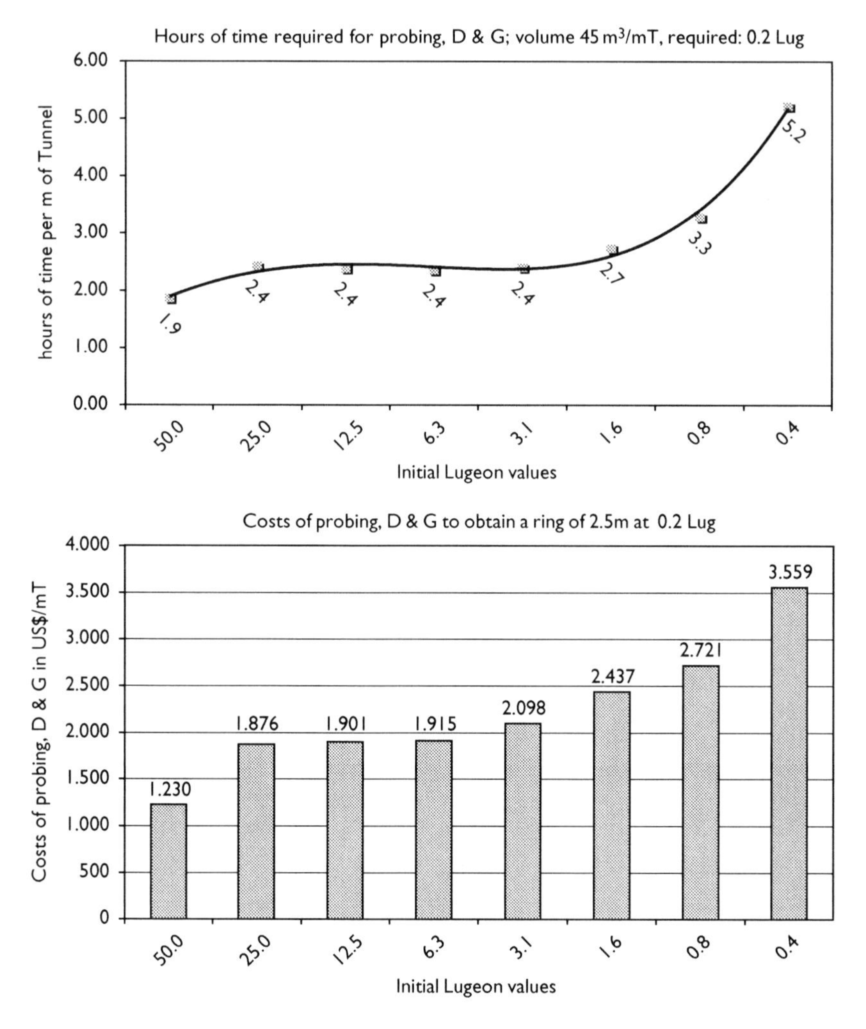

Figure 4.13 Time and costs required for probing and grouting ahead of a 3 m diameter TBM tunnel in rock (exemplary, Stadler).

4.3 Execution of works

In general, grouting works should always be carried out by trained and skilled personnel under competent and experienced supervision.

Drilling should basically make use of systems which least disturb the access for the grout into subsoil porosities. This means that the drilling process should not reduce or clog permeabilities around the hole, as this is an area where frequently success or failure of a grouting treatment is decisively influenced (oil well technologists comparably talk of the 'skin effect' around a hole).

In this way *percussive systems* generally tend to compact alluvial soils and thus, principally reduce porosity. Percussive systems in rocks tend to shatter intersections with fissures, and promote intrusion of cuttings which – in consequence – some times lead to clogging.

Still however, roto-percussive systems are favoured – mainly for economical reasons – and make use of external or down the hole hammers, with or without casing, in rock as in alluvium.[5] Figure 4.14 shows details of such a hydraulic powerhead (hammer 2200 blows/min, torque 600 mkg, pulldown 6 tons) mounted on a diesel-hydraulic crawler drill weighing 6 to even 12 tonnes. This rig can handle rods of $1^1/_4$ in. diameter, and casings of up to 133 mm diameter.

Rotary drilling in loose ground may be a very efficient alternative too. Drilling tools mounted on $2^1/_2$ in. or $3^1/_2$ in. rods are fish-tail bits or tricones, and drilling mud is used to stabilize borehole walls and provide for a suitable flushing media. Of course these muds are to be chosen carefully so as not to pre-maturely clog permeabilities around the hole.

Figure 4.14 Telescopic drilling of cased 4 in. drill holes (Swissboring).

Direction and inclination of holes must follow the intentions of the design. Two per cent deviation is normally accepted down to 20 m depth. But in all cases the actual deflection of holes must be such that the aims of grouting are not upset. It has to be accounted for that horizontal, and holes drilled by percussion tend to deflect more than others. Holes in rock should be flushed before grouting. Flushing with the aim to wash out fines or clayey materials from the ground has limited effect and should in any case not be carried out at length.

Lances and prefixed piping is used where drilling may not be required or special systems do allow for mounting grout ducts in advance.

Packering is the second most important part of the entire process!

What is to be understood under packering or the setting of packers with the intention of providing grout just to that particular point in the ground, or section in rock?

The reason, why this aspect is so prominently discussed here is, that the most effective way to fully provide grout into all underground voids would of course be to address individually each and every fissure, and each and every individual stratum of sediment on its own. Each of the porosities' hydraulic properties could then be matched by the application of a rheologically corresponding mix, applied at optimum pressures and supplied at optimum pumping rates.

However, this is neither technically feasible nor economically viable (nearest to these above requirements would get – the admittedly slow and expensive, but efficient system of – 'top down' treatment, see 'pumping' below for details), and therefore, an 'averaging' process is chosen as an economical compromise, having the grouting ports installed at predetermined intervals (tube á manchettes, see Section 4.2) – regardless of the details of sedimentary stratification, or – as is the case in stable rock – separating the individual borehole sections by packers at regular, uniform distances of say 1–6 m.

Only when therefore, during the production of grout holes – or other 'access ways' and ports for the grout – the individual position in space, and hydraulic capacity of the porous and permeable features are recognized in some discerning way, only then a system may be chosen, aiming at the maximum (engineering) success.

All different designs of grout-pipes and packers should therefore be considered at the time when deciding the drilling method!

- *Manchette pipes* see Figures 4.3, 4.4 and 4.8; their undisputed advantage being the re-use of the individual ports when grouting successive phases of differing grouts.
- *Single port* outlet (see Figure 4.3) mounted as a non-return valve at the bottom end of a $^1/_2''$ pipe. This grout pipe may also be installed in bundles of several individual supply lines, connecting to ports at different

elevations in the same hole, the advantage being, that no maneuvering of packers is required when grouting at different depths. The pumps are connected at the collar of the hole only.

- *Multiple packer sleeve pipes* consist of a combination of manchette pipes activated between jute bags inflated by cement grout. Thus, even collapsible ground may be systematically treated in well-defined sections.
- *Open ended* or perforated lances driven into the ground by hammer or hydraulics. These grouting devices provide access for grout in situations of lesser requirement, or loose ground exhibiting high conductivities.
- *Single or double packers* are used when grouting in rock or, the latter, when grout is pumped into TAMs. Single packers set at the collar of a hole in rock are frequently screw type expandable rubber packers whereas, gas-inflatable single or double packers (between 0.3 and 1.5 m in length) may be lowered into holes as deep as 50 m. At greater depth the risk of packers getting stuck and lost, or being ruptured at crevices or cut at sharp edges of rock increases.
- *Self inflating* rubber packers are using the backpressure of the grout (being pumped through a nozzle in the packer or breaking through a metal-membrane of defined bursting pressure) to inflate the sealing element. A non-return valve prevents the packer from deflating when grouting is interrupted. The packer may however not be retrieved.
- *Prefixed TAM-piping* (PVC, $^1/_2$″ dia.) is used under dams for contact grouting or, in the contact between rock and concrete of hydropower waterways (pressure tunnels) for the purpose of pre-compression grouting. Figure 4.15 shows the instrumentation in such tunnels. Convergences of upto 2/1000 are achieved by pumping cement grout quasi concentrically behind the lining.

Mixing of suspensions sounds like a trivial task; however, it is an art if performed well to pre-determined requirements. The emphasis lies in the introduction of enough shear forces into the mix as to tear apart all agglomerated solid particles and have them perfectly wetted by the mixing water, and kept apart and afloat during the entire process. Figure 4.16 shows the vortex of the backflow from the mix into the tank of a 'colloidal' centrifugal mixer. True 'colloidal' mixers however had wear plates in use, next to the turbine itself, to reduce particle sizes of cement grout by additional grinding during mixing. This technique has become obsolete to a great extent, as the fineness of grouting cements now reach Blaine values of >4500 cm^2/g (specific surface) as a standard.

Figure 4.17 gives some impression of a *mobile mixing and pumping unit* (Häny, Switzerland); Figure 4.18 provides some detail on containerized versions of mounting, weighing, batching, mixing and storing tanks of cement

Figure 4.15 Pre-compression grouting of waterways for Drakensberg Hydropower Project, Natal, RSA (Rodio/LTA).

mixes (including two pumps and registration units) all within a 20 ft standard container (Insond, Austria).

Stationary plants using silos not only for cements, but also for pre-mixed bentonites (for full hydration) and fine sands (in the case of using mortars and pastes) do provide for sufficient automatic functions so as to limit manpower and increase capacity and accuracy.

Storage tanks should be provided with sufficient capacity (approx. 500 l per pump and per hour) and located between mixer and pump. Paddles are installed to stir grout and prevent solids from settling. Stable mixes are preferred.

Stable mixes means that under gravity no more than 5 per cent free water should appear in a settlement test in 2 h. Pressure filtration according to ASTM should not give more than 100 ml of filtration water. It is of course reality, however, that suspensions never remain at their consistency after mixing. WG 6 of the German Society for Geotechnics and Ground Engineering

Figure 4.16 High-turbulence mixing of water and cement (Häny, Switzerland).

Figure 4.17 Compact mobile grout station (Häny, Switzerland).

recently came up with results on research to this topic. According to these, as much as 30 per cent of volume is lost within 50 cm of travel of grout from the point of injection into the ground. Not only volume is lost by filtration, but also viscosity, yield and setting time are seriously effected by this phenomenon.

Figure 4.18 Containerized grout station including electronic data acquisition (Insond, Austria).

Grout pumps are mainly of the double acting piston or reciprocative plunger type. They are hydraulically driven and regulate any flow rates within the range of the capacity of the pump (usually 3 to 20 l/min). Pressures may range up to 250 bar (for highly viscous epoxies), usually pumps should however be able to handle up to 100 bar at the corresponding minimum rate, i.e. 50 bar at 6 l/min, or 20 l/min at around 15 bar. Other pumps in use (for minor applications and standards) are sometimes of the screwfeed type.

Every pump is connected to a single grouting port (packer position). Manifolds connecting more than one hole to a pump are reducing the quality of the treatment (holes with little or no take tend to 'freeze', while one permeable hole may take the entire capacity of the pump).

Pumps are operated by specialist personnel. Not only the characteristic properties of the mix, the mechanics and the hydraulics of the equipment will be familiar to him, but also the sounds and reaction of parameters during this interactive travel of the fluid pushed into the piping and into ground by the pistons of his machinery. His observations must be given room in reporting, and must be questioned every shift.

Table 4.2 Grouting strategies according to EN 12715 (CEN)

	Rock				*Soil*		
	Stable	*Collapsible*			*Drillrod*	*TAM*	*Lance, casing*
	Open borehole	*Open borehole*	*TAM*	*Drillrod*			
Single phase	x			x	x		x
Multiple phase			x			x	
Bottom up	x		x	x	x	x	x
Top down		x	x			x	x

Pumps are connected to recording systems which the operator must be instructed about under the guidelines of proper quality assurance measures.

In Table 4.2 some of the more frequent strategies of grouting are presented, as these are published in European standards.

In stable rock it is common to drill the grout hole to the designed/required depth and to start grouting in passes from bottom to top. Length of the pass may be variable. Single or double packer (the latter sealing the section on top and at the bottom) may be used. The use of single packer might result in re-activating flow of grout in the preceding pass below. A new hole has to be drilled if the same grouting area is wanted to be taken up for a second time.

Open boreholes in collapsible rock are either treated top down (stage grouting, Table 4.1), through TAMs or drillrods, or with multiple packer sleeve pipes (page 147). ‘Top down’ stage grouting means, that in a first step the hole is drilled to a depth, where the borehole walls still remain reasonably stable (but less than 6 m to assure decent spread of grout penetration), a single packer is set at the collar of the hole and grout pumped into this first section (see also para 2.4.9). TAMs will only work, if rocks permit sufficient deformation for the sleeves to open; therefore MPS pipes are used, where the section between the jute packers remains unsheathed by sealing grout, and thus open for the proper cement grout to spread and flow into fissures.

In soils the procedures as indicated in Table 4.2 refer to descriptions of cohesionless soils (page 124).

The principle of filling larger voids first and take up porosities of smaller dimension with a more penetrable grout in a subsequent phase is found in the technique of ‘multiple phase’ grouting:

- Mortars and pastes (low W/C ratio, use of OPC only, addition of stone dust or sand) are applied during the first phase of cement grouting works (if permeabilities do permit).
- Bentonite or polymer-stabilized OPC grouts at higher W/C ratios may follow.

- The remaining finer void structures are grouted using microfine binders (MFB). W/C ratios tend to be higher (due to the macro-surface of the binder per unit weight of solid); stability and pressure filtration have to carefully be monitored.
- Instead of MFBs, acrylates and similar may be used alternatively.

Another important consideration concerning grouting strategies is to be given to the grouting 'from inside out' or 'from outside in'. These two options have to be in mind:

1 not to trap water in the pores or fissures between already grouted areas;
2 or (in the second case) to prevent grout from escaping into areas where the treatment is not foreseen, or to reduce grout being lost outside the treatment zone.

A proper reporting system is recommended and indeed required to keep track of the operations, and take adequate and timely decisions on the changes to the procedures during the process.

Electronic data acquisition therefore is the standard today for the reporting of grouting parameters such as rate, quantity and pressures. It is important in this connection to make up one's mind concerning the intervals at which the relevant signals are to be timed. One can imagine that it makes a difference, whether for example the actual pressure at the pump is registered each minute, or whether remotely (at the collar of the hole) the pressure signal is picked up at intervals of milliseconds, averaged over say, half a minute, transferred to the grout station and registered there as the representative grouting pressure. Under these aspects it may even be of advantage to programme the recording system for the actual requirements of a project.

By comparison it may be found occasionally appropriate enough, if grouting rates are calculated from the – electronically registered – consumption time per batches (of say 25 l), instead of mounting flow metres which record an inductive signal derived from flow velocities through an electronic gauge.

Online transfer of these data may occasionally be arranged even for remote control and interpretation. Storage and handover to the engineer on discs for documentation purpose, is standard. Interference in the daily routines of a grouting operation from remote interpretation of data is however, not recommended.

Reporting on drilling operations is to be drawn from international standards like EN 12715 or local routines.

Reporting on the grout mix and its consistency is a prime quality assurance act!

- Density is to be checked for correct content of solids (Areometer).
- Viscosity (at least Marsh flow cone time in seconds, if not shearometer).

- Yield (fluid cohesion; Kasumeter, 1996).
- Setting time (not at ambient but at ground temperature)

  – of dehydrated grout (remainder of an ASTM pressure filter metre test);
  – of regular 250 ccm samples (including 28 day uniaxial strength);
  – of a film (adhesive layer shed over a suitable base).

- Dispersion test (drop of mix squeezed between two 10×10 cm glass plates and viewed under scaled magnifying-glass: 50x, or microscope, against light).

4.4 Monitoring, controls and acceptance tests

During the execution of grouting, controls of the efficiency of the treatment are required and at the end, acceptance tests must form part of the design and must be part also of the relative agreement in the construction contract. This requires a decent effort as to the specifying of measurable target values.

Grouting data: The acquisition of grouting data makes an interpretation of the grouting process possible; conclusions may be drawn from these data sets as to the success of the treatment.

1 Pressure/rate development
2 Hydraulic fracs/movement of ground
3 Interconnecting holes/*resurgences* (spurt of grout to the surface) and *renards* ('fox-holes' with comparable effect to above) of grout.

Permeability tests before and after grouting to assess the degree of saturation achieved by grouting.

- The higher the virgin permeabilities of the ground the greater the chance of a considerable improvement.
- In soils this means that from an initial level of a K_f-value of say $\geq 10^{-2}$ m/s (coarse gravel) it might be relatively easy to obtain $<5 \times 10^{-7}$ m/s using a 2-phase-treatment with bentonite/OPC and silicate grouts, whereas starting from 10^{-5} m/s (silty sand) the same target may not be achieved by comparable means at all.
- In rocks a similar statement may be made for >50 Lugeon virgin transmissivity and 0.5 Lugeon to be obtained after grouting with OPC and MFB, whereas starting from a diffuse transmissivity of <5 Lugeon chemical grouts may be required from the start to obtain a comparable result, and certainly more drilling per m^3 of ground will be necessary on top of that.
- Wherever possible, upstream/downstream piezometers or seepage rates should form the acceptance criteria for a successful grouting scheme, having to achieve an impermeabilization effect.

Upstream/downstream piezometers for the testing of the eventual horizontal/vertical separation of aquifers resulting from a successful grouting operation.

Reduction of take of subsequent passes.

Continuous penetration tests (CPT) to judge on the improved cohesion, compressive strength and deformation resistance.

Drilling energy measured when drilling (roller or fish-tail bit, borehole supported by drilling mud) testholes before and after grouting.

Plateload testing either down the hole or in open pits to document the improvement in deformation characteristics.

Core sampling for laboratory testing; this is recommended more in rocks than in soils. The system is inevitably failing in soils if – even using diamond core drills – uniaxial strength of the cored material is dropping below 5 MPa. No intact sampling is possible which would satisfy laboratory requirements.

Open pit inspection sometimes is sufficient to ascertain an improved cohesion of grains or the visible presence of grout in the ground.

4.5 Resources and equipment

Drill rigs for the production of groutholes are of diesel hydraulic design with different mast configurations and kinematics. The power heads might provide (hollow stem if rods are passing through the power head) spring loaded or hydraulic jaws, or swivel-type connections (rods only connected below drill head) between driven rotating parts of the head and the drillrods. The length of free travel (of the head on the mast) is essential for productivity; for every breaking and connecting manoeuvre of rod couplings does of course reduce production time for drilling. In Europe Casagrande, Hütte, Wirth and Klemm have produced respective machinery, which has satisfactorily performed over decades. Core drills (or non-coring systems with full face bits) may alternatively be used to great advantage in rock. Light weight design and small dimensions (with electro-hydraulic power packs of 30 kW capacity) facilitate their use in dams and tunnels (typically the rigs of the Diamec generation of Atlas Copco, Figure 4.19).

Mixers and pumps are electrically driven and (as far as pumps are concerned) have secondary hydraulical systems installed.

Batch mixers need to complete the entire mixing cycle (typically 3 min for dosing, filling, mixing and transfer) before the mix can be used. Continuous (srewfeed) mixers are less accurate and bulky, but do provide an uninterrupted flow of primary mix into a secondary agitator tank, where the mix may be refined (tuning of final W/C ratio).

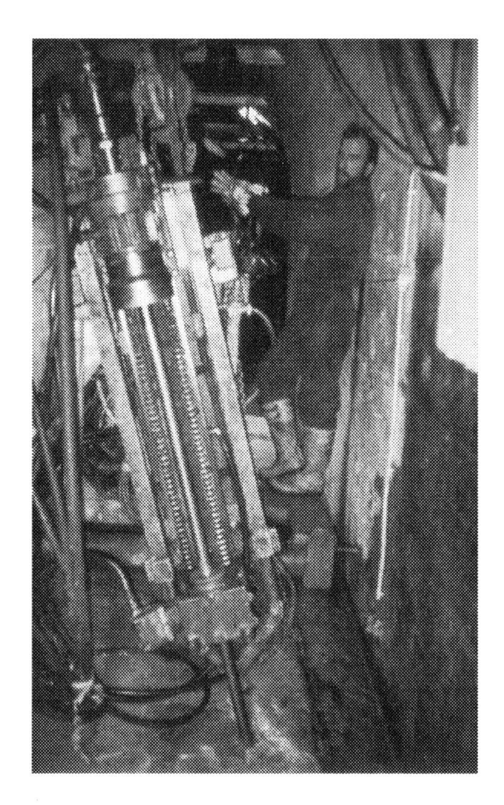

Figure 4.19 Slim hole coring with Diamec 250 for grout holes in Kölnbrein Dam, Austria (Insond).

Mixing may be categorized into:

- Mixing by agitation (paddle mixers, unstable mixes, batch mixing).
- Mixing by generating high shear forces. High speed or turbo mixers have therefore either close tolerances between the impeller and the casing of the mixer-pump, or recessed vortex impellers. The latter do generate less wear and are less inclined to clog or block because of allowing larger particles to pass. Conventional centrifugal pumps might serve a similar purpose but obviously, less effectively.
- Mixing (because of imparting high energy) should be limited to 30–60 s. Overlong mixing heats the grout and triggers the hardening process of hydration at too early a stage. The mix should rather be kept (if at all) in agitating tanks (1.5 times to twice the volume of the mixing tank) where low energy paddles keep the grout in motion and prevent particles from sedimenting.

Eccentric screw pumps and twin double acting piston (or plunger) pumps are used for transporting and forcing grout into the ground. Many experts are two minded about the requirements on the evenness of the flow, though a slightly pulsating regime finds many supporters. But what is even more essential is the possibility to regulate flow and pressure of the pump in a way, which makes either a constant energy concept possible (see page 150) or a constant rate or constant pressure scheme. Some recommendations in this context may be:

- the maximum grouting pressure and flow rate should be adjustable over the whole range;
- the controls should be situated at the drive end not at the grout end; pressure limiters with ON/OFF function are not suitable/recommended;
- pumps with a direct drive need a bypass system which is prone to early wear;
- abrasive grouts should be handled by low wear plunger pumps;
- valves and areas for grout to pass through must be large enough to pump viscous and sandy grouts;
- ease of cleaning and maintenance is of great importance (downtime!).

4.6 Technical summary

Literature has seen over the last decades a large number of successful grouting applications with reports about their execution and performance, including many hydro- and irrigation dams in the register of International Commission on Large Dams (ICOLD). Several conclusions may be drawn from this experience, and combined with the recommendations of this report.

In order for cement grouts to be successfully injected it is necessary for the cement *particles to remain in suspension* during injection. When injected under pressure the mix will lose water into the surrounding ground (pressure filtration), and this loss of water will cause a thickening and reduction of volume of the mix. The generation of internal friction, increased viscosity and yield of the grout will finally block any further flow of the grout into the ground.

For effective cementitious grouting, the *effective grouting pressure* should be sufficiently high to overcome substantial pressure losses when entering fine voids, and to enlarge fissures elastically and to facilitate the entry of the grout particles. The ability of particulate grouts to *penetrate into fine aperture* voids of the ground is not only controlled by the ability of the particles themselves (size and dimension), but also by the degree of elastic deformations of the ground that occur during the grouting process in its immediate vicinity (fostered by pulsating flow?).

Simple theoretical considerations and elementary experimental evidence show that, as soon as *internal friction in a particulate mix* occurs, grouting is no longer possible. If the cement grains are not transported freely by the fluid but come into contact, friction between the particles will develop and effectively stop grouting. This phenomenon is particularly important (see 'pressure filtration' above).

The penetrability of a cement-based grout into fissures depends on two main factors: the *grain size of the cement used and the rheologic properties* of the suspension. However, merely studying the size of a single dry grain is misleading: single dry grains have a tendency to grow in size during hydration and agglomerate, producing 'flocs' larger than the single dry particle. Therefore to improve the penetrability of a particulate grout, it is necessary to both keep the grain size low and reduce/prevent the tendency for single grains to flocculate in the mix (plate glass test, page 153).

Cement as a material requires a *water cement ratio of about 0.38* (by weight) to gain hydration. However in this form it would be an extremely stiff paste, hence for injection purposes additional water is added to the mix, to enable effective transporting of the cement grains (W/C of 0.65–3.0). The addition of water has the combined effects of reducing viscosity, yield and strength, and increasing the bleed (pressure filtration), shrinkage and setting time. The higher the water–cement ratio, the weaker the grout, the greater the shrinkage and the longer the setting time that will result.

The question of *setting time* is important for the management of the process against time, and the choice of a correct treatment system altogether. Cements are manufactured so that they have a setting time for industrial applications of about 4–5 h. If we greatly dilute cements the setting time is first delayed (10–16 h may result for water cement ratios of 2:1 and 3:1 respectively) and then accelerated again during filtration. The addition of clays, bentonites or accelerator admixtures reduce setting times (simultaneously increasing the viscosity of the mix). So it is quite clear that the rheological behaviour of the suspension follows delicate relationships which not only have to be monitored and engineered once at the beginning of the grouting operation, but continuously.

To conclude, the essential ingredients for a successful grouting project are

- to go about any grouting project as open (educated) and engineering minded as possible;
- to perform under continuous questioning/reaffirming of the geotechnical model of the ground (possibly in partnership with the engineering geologist);
- under permanent perception of the phenomena observed and interpretations derived from these (possibly cross-checking them with an experienced grouting foreman).

Notes

1 And this for obvious reasons; because the same transmissivity may be caused by few but wide fissures as it may be caused by many tiny fissures; see also Cambefort and Figure 4.7.
2 This is particularly important when hydraulic exploration tests are carried out say from surface (e.g. from 110 m above tunnel alignment), testing taking place at 100 m below GW, whereas construction excavation later does create drained conditions under which the actual stress level changes significantly – and so do fissure openings and transmissivities.
3 'Passes' are characterized as individual sections of a hole which are separated from the remaining borehole length by (rubber)seals (or similar). Each pass is grouted individually.
4 This section relates also to the opinion of Mr Knut Garshol of MBT, provided to me in personal communications.
5 Compared to percussive drilling methods the costs for rotational systems (coring or destructive full face borings) tend to be more expensive at a ratio of 2 (up to even 10) in rocks.

References/Literature

Akjinrogunde, A. (1999) *Propagation of cement grout in rock discontinuities under injection conditions*, Institute for Geotechnics, University of Stuttgart, Vol. 46.
Baker, W. Hayward (ed.) (1982) Grouting in Geotechnical Engineering, *Conference Proceedings*, New Orleans.
Cambefort, I. (1964) Injections des sols.
EN 12715 (2000) Execution of special geotechnical work – *Grouting*, July.
Ewert, F.-K. (1985) *Rock Grouting with Emphasis on Dam Sites*, Springer Verlag Berlin–Heidelberg.
Houlsby, D. (1989) Cement Grouting, Dep. of Water Affairs, Sydney, Australia.
Karol, Reuben, H. (1990) *Chemical grouting*, 2nd edn, Dekker.
Kasumeter, International Society for Rock Mechanics, Widmann, R. (1996) Commission on Rock Grouting, *Int. J. Rock Mech. Min. Sci & Geomech. Abstr.*, Vol. 33, No. 8, 803–847.
Lombardi (1989) The role of cohesion in grouting. ICOLD Proceedings, Lausanne.
Nonveiller, E. (1989) *Grouting, Theory und Practice*, Elsevier, Amsterdam.
Schulze, B. (1993) Neuere Untersuchungen über die Injizierbarkeit von Feinstbindemittel-Suspensionen. Berichte der Int. Konf. betr. Injektionen in Fels- und Beton, A.A. Balkema Rotterdam, 107–116.
Semprich, S. and Stadler, G. (2002) Grouting. In: U. Smoltczyk (ed.), *Geotechnical Engineering, Geotechnical Handbook*, Ernst & Sohn, Berlin.
Shroff, A.V. and Shah, D.L. (1993) *Grouting Technology in Tunnelling and Dam Construction*, A.A. Balkema, Rotterdam.
Stadler, G. (1992) Transient Pressure Analysis of RODUR Epoxy Grouting at Kölnbrein Dam, Austria, Diss. Thesis, MUL, Leoben.
Stadler, G. (2001) Seminar on Waterproofing of Tunnels, CUC-Seminars, F. Amberg, Sargans, Switzerland.
Stadler, G. (2001) *Permeation Grouting*, ASCE Seminar Publications, New York.

Stadler, G. (2002) 'Was hat die internationale Normung der Injektionstechnik gebracht', Injektionen in Boden und Fels, Ch. Veder Colloquium, Graz University of Technology, pp. 1–20, April.

Stadler, G. *et al.* (1989) Technical University, Vienna, 'Pressure Sensitive Grouting'.

Stadler, G., Howes and Chow (1998) '100 years of Engineering Geology', Symposium, Tu-Wien.

Verfel, J. (1989) *Rock Grouting and Diaphragm Wall Construction*, Elsevier Amsterdam.

Weaver, K. (1991) *Dam Foundation Grouting*, American Society of Civil Engineers, New York.

Weaver, K. (1993) Selecting of grout mixes – Some examples from US practice, *Proceedings of Grouting in Rock and Concrete*, A.A. Balkema Rotterdam, 211–218.

Widmann, R. (ed.) (1993) *Grouting in Rock and Concrete*, A.A. Balkema, Rotterdam.

Chapter 5

Jet grouting

R. Essler and M. Shibazaki

5.1 Introduction

Of all forms of ground improvement systems, jet grouting must be regarded as one of the most versatile. With this technique it is possible to strengthen, cut-off groundwater and provide structural rigidity with a single application. It can also be regarded as one of the most technically demanding of ground improvement systems requiring equally both technical excellence in design and construction because failure of either component will result in failure of the product.

Figure 5.1 shows the principle method of application whereby either high-pressure water or grout is used to physically disrupt the ground, in the process modifying it and thereby improving it. In normal operation the drill string is advanced to the required depth and then the high-pressure water or grout is introduced while withdrawing the rods.

As discussed below in Section 5.2, jet grouting has a considerable history of development from its initial use to current practice. In the field of jet grouting, the most notable advancements have been in Japan where the technique has been refined to its present day capability by careful attention to detail in all aspects of the system. Through the years, careful research and execution has resulted in increasing column diameter and range of applicable soils. This development is set out also in Section 5.2.

Within this chapter it is hoped that the practicing engineer will understand how jet grouting came into existence, the technical complexity and design requirements needed for a successful application and indeed the range of applications that jet grouting can be put to.

Reference should also be made to the contents of the chapter on jet grouting in the first edition of Ground Improvement for further reading and in particular details of column layouts typically adopted which is not repeated here.

5.2 History

The scouring power of water has probably been employed as a soil excavation method since early times, especially in the mining industry where use of it is documented in the Middle Age.

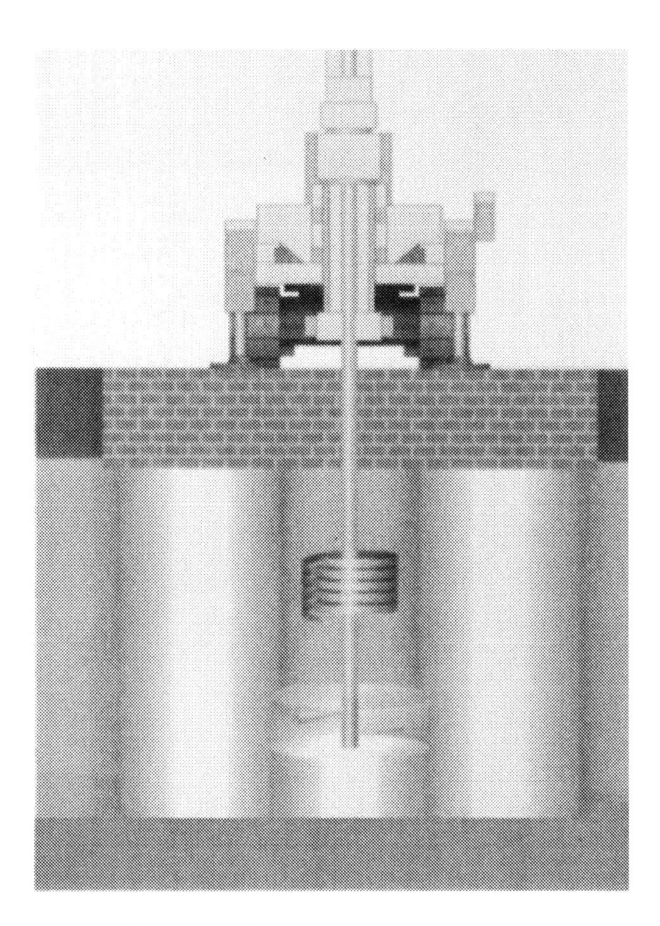

Figure 5.1 Jet grout column construction.

The earliest patent regarding jet grouting was applied for in England in the 1950s; however, the real practical development of jet grouting took place for the first time in Japan. This technology was initially aimed at improving the effectiveness of water tightness, in chemical grouting, by eroding the untreated or partially treated soil which was then ejected to the surface for disposal being replaced with a cement-based slurry for imperviousness. Subsequently, jet grouting was first applied to create thin cut-off walls as shown in Figure 5.2.

For preventing water ingress, a derivative of panel jet grouting was evolved which sealed the gap between declutched sheet piles, for example. This derivative allowed the formation of part columns (shown in Figure 5.3) by causing a twin-angled jetting motion or a windscreen wiper motion of the monitor during lifting.

Figure 5.2 Exposed jet grout panels.

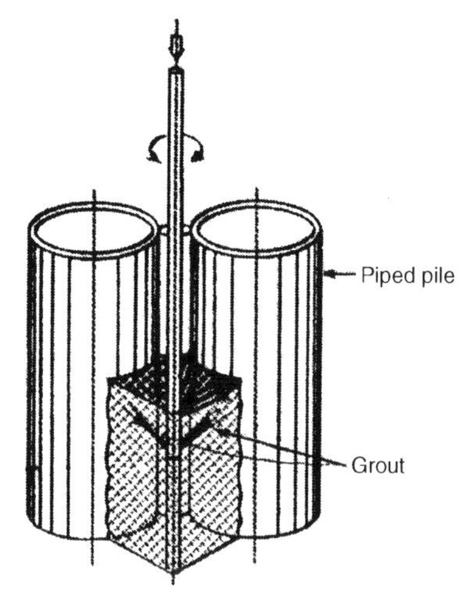

Figure 5.3 Jet grout sealing between piles.

In the early 1970s, rotating jet grouting emerged in Japan because of the fact that panel jet grouting could hardly create satisfactory products with varying thickness and somewhat fragile strength.

In the mid 1970s, jet grouting was exported to Europe and has become popular worldwide since then. According to required geometry, three main variants of jet grouting have emerged in the same period, of which conceptual schematics are illustrated in Figure 5.4.

One of the variants is called the single system (S), which is the simplest form of jet grouting, ejecting a fluid grout to erode and mix with the soil.

Spoil cannot easily travel up on to the surface and heave may consequently occur. When drilling significantly below the ground water level, eroding distance can be considerably shortened on account of the absence of the shrouded air, which increases cutting energy.

The double system (D) adds compressed air which surrounds (shrouds) the grout jet to enhance the erosive effect, especially below the water table. However, the system still retains a defect in that a considerable percentage of the grout may be lost to the surface due to the airlift. Such behaviour may also decrease the quality of improvement.

The third method which is called the triple system (T) utilizes three fluids, namely grout, jetting water and compressed air shrouding the water. This system normally consists of a grouting nozzle approximately half a metre below a water jetting nozzle in order to convey as much excavated soil particles as possible to the surface while limiting the grout ejected. While the double system may produce more spoil than expected based on the eroded volume of soil, the triple system achieves erosion and grout injection independently and can thus be optimized for the required performance. In other words, it is a system superior to the other two systems from the view point of quality control.

In the 1980s, experience and confidence with jet grouting now spanned a very wide range of application.

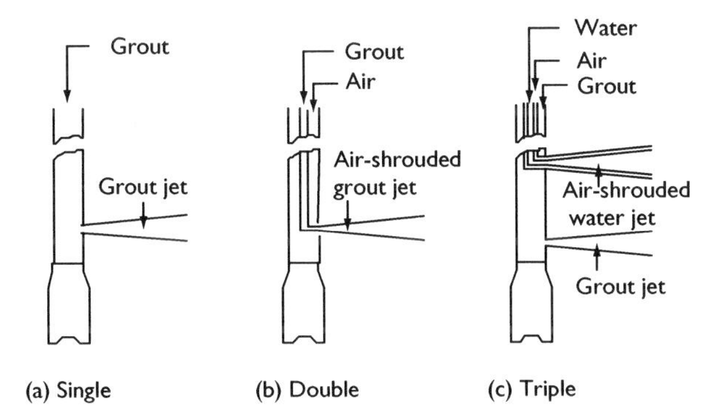

Figure 5.4 Single, double and triple jet grouting.

Since the early 1990s, newer methods of jet grouting capable of a considerably larger columnar improvement have been developed on grounds of cost and programme. This enabled jet grouting to obtain a column with a diameter in excess of 5 m or even 9 m in softer ground (Figures 5.5 and 5.6 show an example of such an oversized body). This method could improve volumes of soil 20 times as large as the previous conventional systems, due to equipment development providing significantly higher flow rates at higher pressures.

Figure 5.5 Early exposed superjet column.

Figure 5.6 Trial superjet columns.

The successful construction of a large column requires the use of focused jets, maintained in pristine condition, otherwise a large proportion of the jetting energy is lost within the system itself. Thus jet grouting emerged capable of spanning a very wide range of applications.

The result of jet grouting can vary according to both equipment and soil types. Given these constraints, many measurements have been taken by varying the values of key parameters as a basis of theoretical solutions; however, even these trials cannot provide exact solutions because of the limited investigation into the soil.

In the late 1980s, a new concept provided an innovative progress for jet grouting systems, namely, dual jets colliding with each other to limit their eroding capability, thus achieving an exact intended diameter regardless of soil type. The arrangement of these jets is shown in Figure 5.7a while an exposed column is shown in Figure 5.7b.

The conceptual comparison of conventional and colliding methods is shown on Figure 5.8, non-colliding jets producing columns of variable diameter in variable ground. Colliding jet grouting has raised the required design quality since its appearance under the name of 'Crossjet grouting'. In the early 1990s, colliding jetting was further evolved to include the deep mixing method to substantially increase the range of application. Conventional *in-situ* soil mixing suffers from a serious drawback of imperfect continuity when executed adjacent to walls; however, attaching an assembly of colliding jetting equipment at the tip of a drilling bit or blade (Figure 5.9) has enabled the construction of optimal interlocking as shown in Figure 5.10.

(a)

(b)

Figure 5.7 (a) Cross jets; (b) Columns produced by cross jetting.

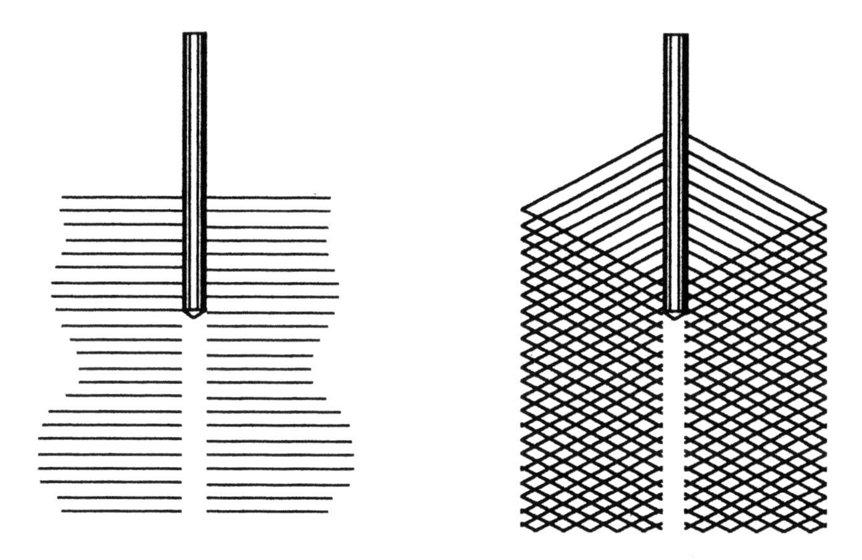

Figure 5.8 Principles of cross jetting.

Furthermore, the enhancement in this *in-situ* mixing system results in more than 4 times the treated volume using the same equipment. This is shown in Figure 5.11, the conceptual schematic of the jet and churning system management (JACSMAN) system.

5.3 Theory of jet grouting

Many factors influence the efficiency and effectiveness of the jet grouting process and require consideration when both designing and constructing jet grout columns.

5.3.1 *Effect of pressure*

When eroding soil with a high-pressure water jet, the eroding distance radically increases as the water jet pressure exceeds the unconfined compressive strength of the soil, as shown in Figure 5.12 (the relationship between eroding distance and jetting pressure). It is possible, with a lower pressure, to erode the same distance over a longer time period, however the high pressure saves time for most practical applications. Typically, water pressures between 30 and 60 MPa for an overburden soil such as silt, sand, etc., and more than 200 MPa for rock formation are employed.

Figure 5.9 JACSMAN tool details.

Figure 5.10 JACSMAN column abutting sheet piling.

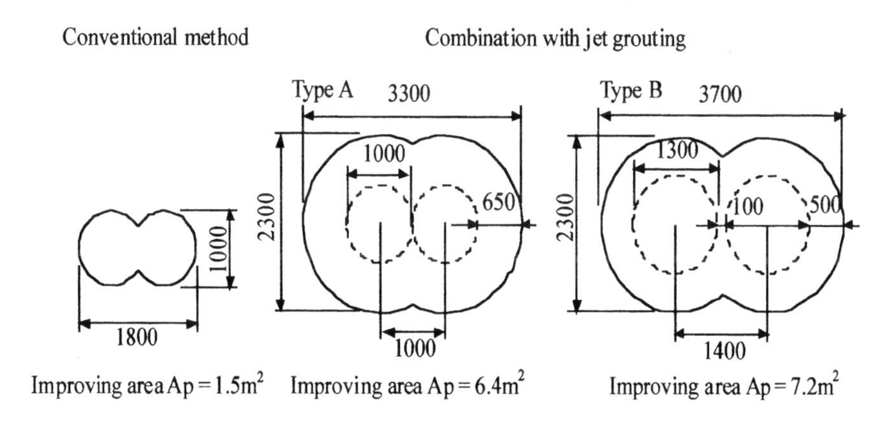

Figure 5.11 JACSMAN system concepts.

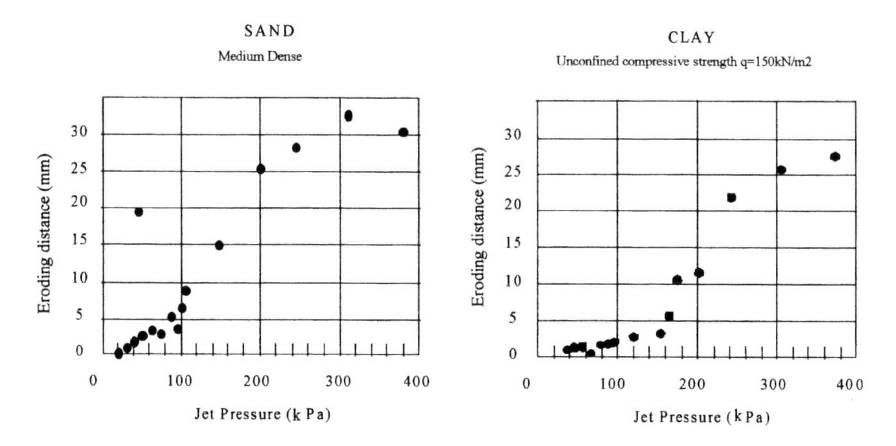

Figure 5.12 Eroding distances vs jet pressure.

5.3.2 *Effect of flow rate*

When pressurized water passes through a circular nozzle, the following equation is obtained from the law of conservation of energy.

$$V_0 = m\sqrt{2gp_0} \tag{5.1}$$

where

p_0: initial pressure from a nozzle, V_0: initial speed from a nozzle, g: acceleration of gravity, and m: nozzle efficiency.

A practical example of a calculation for pressure effect and flow effect is given below. If a water jet is discharged at 40 MPa through a fine nozzle of 2 mm in diameter such that the velocity of shrouded air is 100 m/sec, we can obtain an eroding distance of 1 m (2 m in diameter) at the point of 4 MPa from Figure 5.13 (Dynamic pressure 0.1 times nozzle pressure). This may be regarded as the effective limit of the column for most practical purposes.

Since an excellent nozzle has $m = 0.92$ as an efficiency coefficient, substituting $m = 0.92$ into equation (5.1) to get a flow rate below:

$$\begin{aligned} Q &= VA \\ &= m\sqrt{2gp_0}\frac{\pi}{4}d^2 \qquad (5.2) \\ &= 49\,\text{l/min} \end{aligned}$$

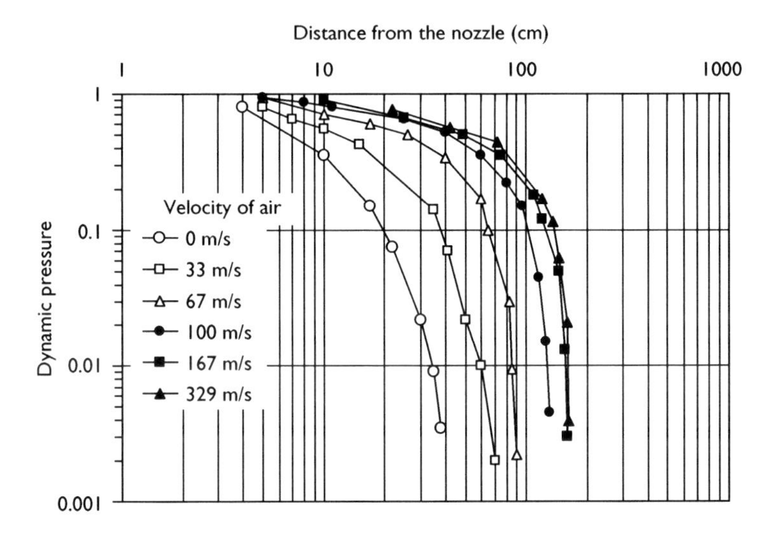

Figure 5.13 Relationships of dynamic water pressure and air flow rates with distance.

If a 5 mm nozzle of the same efficiency is used instead then, in order to achieve the same required column diameter, the flow rate must be altered in accordance with the square of the nozzle diameter:

$$\frac{Q_1}{Q_2} = \left(\frac{d_1}{d_2}\right)^2$$

Hence the required flow rate is 306 l/min.

5.3.3 *Effect of compressed air*

An increased air flow rate with even low pressure can extend the eroding potential considerably as illustrated in Figure 5.13 (the relationship between dynamic water pressures and air flow rates with distance). Jet grouting requires compressed air for successful operation in several respects. It is first indispensable for obtaining maximum eroding energy and then of vital importance for conveying spoil up to the ground surface.

5.3.3.1 Effect of compressed air shrouding

A water jet as a fire extinguisher is totally effective; however, it's effectiveness is significantly decreased in water. Because jet grouting mostly treats

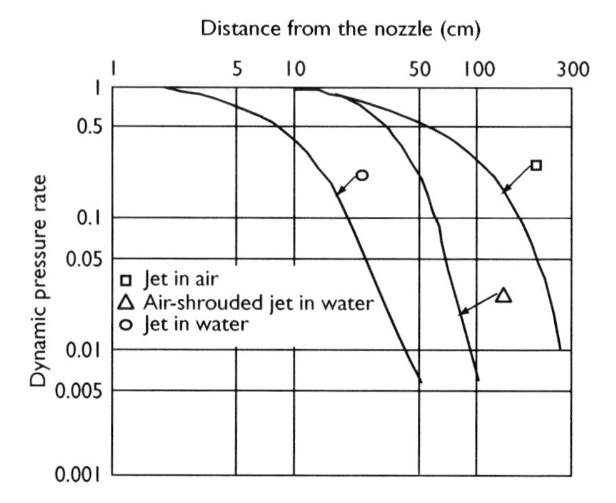

Figure 5.14 Relationships of dynamic pressure rates and distance from nozzle in various media.

the soil beneath the ground water level, a water jet alone cannot cause significant ground improvement. In this respect, compressed air shrouding of liquid jets is a primary technique in eliminating ground water around the jets, thus forming an atmospheric condition.

Figure 5.14 sketches the eroding distance of respective jets in air, in water, and in water with an air shroud.

This chart clearly demonstrates the jetting principle that a liquid jet with a pressure of 40 MPa through a nozzle with a diameter of 2 mm, can attain a distance of 3 m in the air, is reduced to just half a metre in water; however, with the addition of the compressed air around the water jet, it is extended again to $1.1 \sim 1.2$ m.

5.3.3.2 *The velocity and volume of compressed air*

As stated previously, merely the presence of the air shroud does not always prove successful, but it should also maintain a higher velocity than half the sonic velocity to ensure the formation of an atmospheric condition, as is clearly outlined in Figure 5.13. Additionally, an air nozzle has to be ring shaped or annular, surrounding the nozzle which preferably includes a minimum straight length before the air discharge point.

The width of this annulus must be approximately 1 mm thick as standard which should provide sufficient air flow and yet does not allow any foreign particles like sand to flow upstream. Compressed air may be generated by a low-pressure compressor rated at 0.7 MPa for work up to 20 m deep;

however, a high-pressure compressor is required to withstand the ground water pressure for deeper works.

5.3.4 Other effects

The quality of the material and internal finish of the nozzle is of vital importance as well as its dimensions and geometry. Furthermore in reality, care must be taken that even a nozzle perfect before use may be easily damaged owing to impurities in the jetting stream and/or foreign particles in the soil.

In order to account for this, inspection of the condition of nozzles before and after each jet grouting operation has to take place. An optimal inspection technique employs a special measurement system of dynamic testing in association with pressure-sensitive films with a pre-determined range.

If the jet is sound, the pressure-sensitive film reveals an annulus, with the centre destroyed, which is the so-called core of the jet still maintaining sufficient eroding energy to penetrate the film.

For a defective jet, the film reflects a totally coloured spot, with no central penetration as sketched in Figure 5.15 (a turbulent flow).

Apart from dynamic pressure and flow rate there are other parameters which have an influence on the eroding power of a liquid jet. An experimental equation explains these:

$$R = KP^{\alpha}Q^{\beta}N^{\lambda}V_n^{\delta} \tag{5.3}$$

where R = eroding distance, K = factor, P = initial dynamic pressure, Q = flow rate, N = repeating frequency of erosion, and V_n = moving speed of nozzle.

(a) Turbulent flow (b) Focused flow

Figure 5.15 Turbulent and focused flow.

Figure 5.16 provides experimental results for the optimal repeating frequency of the eroding jet, indicating that frequencies in excess of 5 marginally increase the column diameter.

Lifting up the jetting rods in steps provides the necessary rotation using an integral number which is not possible with a steady lift as shown in Figure 5.17 (lifting methods). Each step corresponds to an intended diameter; however, practical experience gives a 5 cm lift for up to 2 m in diameter, and a 10 cm lift for more than 4 m in diameter, as optimal increments.

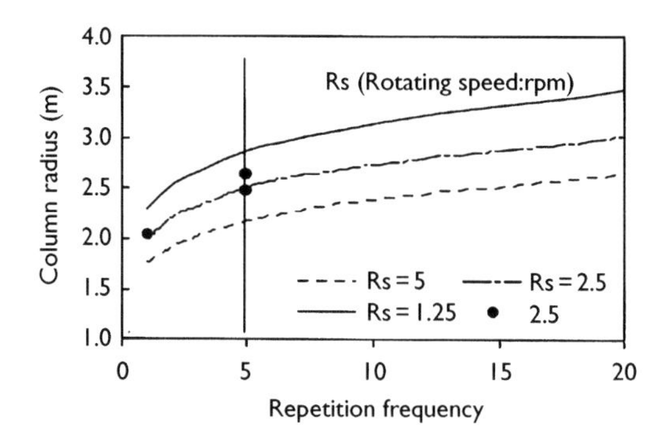

Figure 5.16 Experimental results for optimal repeating frequency of eroding jet.

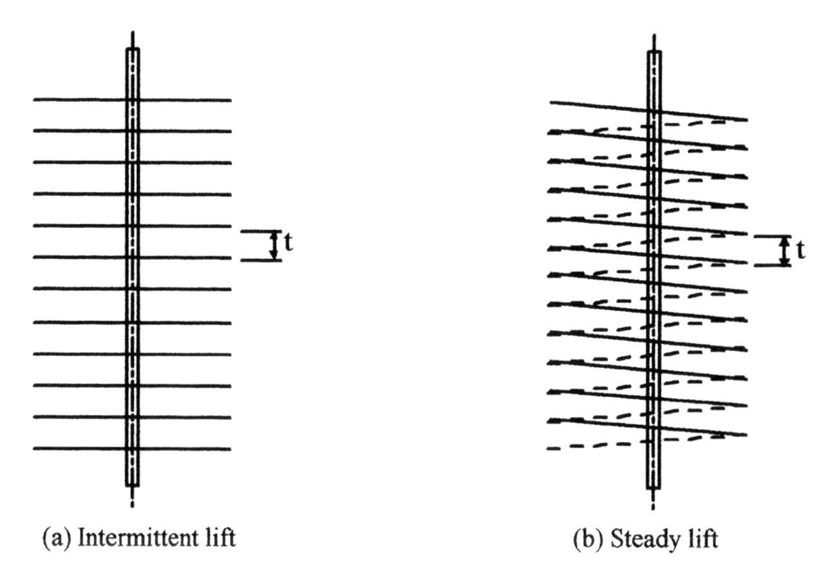

Figure 5.17 Lifting methods.

5.3.5 *Practical considerations*

In order to successfully design a jet grout project, both theoretical and practical considerations need to be taken into account. For a successful project both the columns must be installed correctly and the achieved properties must be in accordance with those values required by the design.

5.3.5.1 *Design parameters for jet grout material*

Strength of treated ground is usually assessed on the basis of unconfined compressive strength tests on samples obtained by coring. The histograms shown in Figure 5.18 demonstrate experiential unconfined compressive strengths in granular and cohesive soils. The Japan Jet Grouting Association has adopted these distribution charts, defining the unconfined compressive strength to be taken for design to be the minimum safe values which range between 1 and 3 per cent from the least values in the whole group.

This definition gives the standard unconfined compressive strengths as follows (where the water/cement ratio of the grout is 1).

qu = 1 MN/m^2 (Unconfined compressive strength in cohesive ground);
qu = 3 MN/m^2 (Unconfined compressive strength in granular ground).

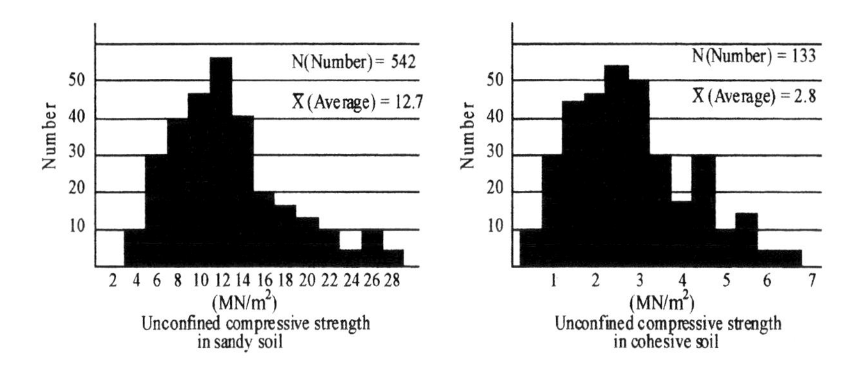

Figure 5.18 Experimental unconfined compressive strengths of treated ground.

Table 5.1 Standard strengths in designing

Soil type	*qu: Unconfined compressive strength* (MN/m^2)	*c: Cohesive strength* (MN/m^2)	*f: Bond strength* (MN/m^2)	σ_t: *Bending tensile strength* (MN/m^2)
Cohesive	1	0.3	0.1	0.2
Granular	3	0.5	0.17	0.33

The design standard strengths of cohesion, bond and tension in bending are then determined with reference to the values as shown in the Table 5.1 above.

5.3.5.2 Drilling tolerances

Drilling tolerances are particularly relevant with jet grouting as overlapping of columns is vitally important.

Inadequate interlocking not only takes place through drilling deviation which increases the offset from a neighboring column with depth, but also through penetrating into a neighboring column that has already set.

The latter problem results in jetting within a set and rigid material consequently leading to unsuccessful works as no column is formed, as diagrammed in Figure 5.19. Inadequate interlocking can only be limited by excellent drilling coupled within hole survey techniques. Because of this, jet grouted holes should be surveyed whenever possible to ensure deviation is within acceptable limits.

In general, drilling tolerances of up to 1 in 100 can be achieved but special consideration for the specific risk have to dictate the definite tolerances for the radius in the depth of interest.

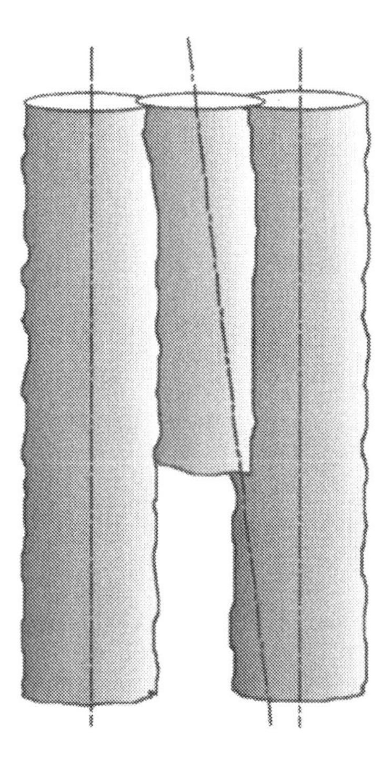

Figure 5.19 Effect of poor drilling tolerance on column construction.

5.3.5.3 Quality control and validation

Section 5.3.5.2 above alludes to problems of deviation but there are a number of other factors that must be considered:

- Column diameter
- Column position
- Column properties.

And it is therefore of importance wherever possible to record and validate the installation of individual jet grout columns. Most specialists have the instrumentation to record the following parameters during installation:

- Depth
- Withdrawal rate
- Air pressure
- Grout or water pressure and volume
- Rotation.

In addition some specialists have developed inclinometers built into the jet grout monitor that measure deviation of the drill string. It is also equally important to carry out quality control testing on the grouts used. This normally includes specific gravity, viscosity and strength by 28 day cube strengths.

The knowledge of all these parameters allows the site engineer to review the column installation and come to a decision as to whether any column is misplaced or incorrectly installed. This is of paramount importance for base slabs or tunnel break-in or break-out where the omission or misplacement of a column can have the most serious effect on performance or safety. A further difficulty is the repair of these jet grout bodies as usually the location of a failure is difficult to locate.

Franz (1972), Fritsch and Kirsch (2002), Kirsch and Sonderman (2002) list standards and publications relating to control and execution of jet grouting.

Validation of jet grouting can be problematic. In order to fully validate a project column diameter, position and strength or permeability must be checked. Techniques typically carried out are as follows:

Column diameter: The most appropriate technique is to construct trial columns and then expose them to measure diameter directly. This is an excellent method but can only be used at shallow depths due to the expense of accessing columns at depth.

Coring of columns can be successful but often suffers poor core recovery leading to difficulty in interpretation of diameter or strength.

Borehole calipers can be lowered and extended to measure the extent of a column prior to initial set but are restricted to shallow depths and are not very successful often remaining behind in the column.

Some geophysics companies are developing non-destructive techniques utilizing 3D borehole radar, as yet still remaining at the research stage but offering a promising solution.

Column position: Column position relates to measurement of drilling tolerance and as discussed above this is either accomplished by built in inclinometers or by survey of the hole prior to jetting.

Column properties: This is most commonly measured using coring techniques although some companies offer sampling within the column prior to initial set. Some forms of non-destructive techniques can be used as discussed above.

5.4 Application of jet grouting

As set out in the introduction, jet grouting is an exceptionally versatile tool when considering ground improvement as part of a project. There are many applications that suit jet grouting but they can be grouped together under the following headings as follows:

- groundwater control
- movement control
- support
- environmental.

Groundwater control applications include:

- preventing flow either through the sides or into the base of an excavation;
- controlling groundwater during tunneling;
- preventing or reducing water seepage through a water retention structure such as a dam or flood defence structure;
- preventing or reducing contamination flow through the ground.

Movement control applications include:

- prevention of ground or structure movement during excavation or tunnelling;
- supporting the face or sides of a tunnel during construction or in the long term;
- increasing the factor of safety of embankments or cuttings;
- providing support to piles or walls to prevent or reduce lateral movement.

Support applications include:

- underpinning buildings during excavation or tunnelling;
- improving the ground to prevent failure through inadequate bearing;
- transfering foundation load through weak material to a competent strata.

Environmental applications include:

- encapsulating contaminants in the ground to reduce or prevent contamination off site or into sensitive water systems;
- providing lateral or vertical barriers to contaminant flow;
- introducing reactive materials into the ground to treat specific contaminants by creating permeable reactive barriers.

The above lists show that jet grouting has a multitude of uses, all of which must be understood, designed and executed accordingly.

Some important main applications are described below in more detail.

5.4.1 Groundwater control

The last two decades have seen an increasing number of large excavations constructed in water-bearing soils. The use of conventional ground water lowering techniques has been reduced as a result of the increasing importance of:

- economic water control
- environmental aspects of the aquifer
- observance of existing water rights
- protection of existing buildings.

Conventional chemical-based injection systems have been almost completely replaced by jet grouting techniques where the use of cement-based grouts reduces alkalinity.

Typical waterproofing elements are vertical and horizontal walls with and without an additional structural function in deep excavations or for dams and dykes, break-in and break-out blocks to assist tunnel-boring machine operations. While with jet grouting columns a permeability of 10^{-9} to 10^{-10} can normally be expected, the permeability of the system as a whole ranges from 10^{-7} to 10^{-8}. As a rule the excavation cannot commence until the allowable flow rate has been achieved and proven by a pumping test. Excess seepage is generally a result of a defect in the wall or slab or at the joints between wall and bottom seal.

The detection and location of leaks is extremely difficult, sometimes even impossible, and full or partial drawdown of the water table and the observation of piezometers or the measurement of the ground temperature during rerise of the water are the most promising methods of leak detection. The necessary remedial works are often time consuming and extremely expensive, clearly therefore the proper design and execution of jet grouting sealing elements is vital to the success of the project. The design requires the definition of sufficient strength, minimum permeability, homogeneity and dimensional accuracy. It is essential to remember that water will not forgive any mistakes.

Defects in jet grout bodies can occur as a result of:

- insufficient overlapping of individual jet grout columns;
- jet shadows caused by natural or man-made obstructions;
- inhomogeneities in the ground (hard layers imbedded in sand, peat layers);
- instability and subsequent collapse of jet grout columns before they set;
- process deficiencies and interruptions, errors.

To mitigate these risks, a thorough quality assurance plan is essential and indeed is state of the art. The plan should include the following elements:

- setting out of the jet grout columns by *x-y* coordinates;
- drilling depth determined by efficient levelling systems;
- definition of drilling and jetting parameters;
- execution of test columns, documentation and evaluation of results;
- definition of the sequence of the works;
- identification of obstacles and countermeasures;
- grout composition and measurement of characteristics by sampling at mixing station and in back flow;
- measurement of drilling accuracy and countermeasures;
- process documentation during execution in real time of
  - speed of insertion and extraction of monitor;
  - pressure and flow rate of grout, water and air;
  - drilling and jetting rotation;
  - data secured on RAM cards.

When looking at the evolution of grouting techniques in contractual terms it is clear how much injection of sediments has departed from rock grouting. It must be remembered that the completed jet grout body is not homogeneous and therefore generally does not exhibit a constant strength or hydraulic characteristic. Design, specifications and quality control must therefore reflect an uneven distribution of strength and permeability due to the variability of the soil under treatment.

Horizontal jet grout barriers in deep excavations should therefore be designed and executed with the following considerations:

- minimum slab thickness not to be less than 1.0 m and to be increased by 0.1 m for every metre in excess of 10 m depth for safe uplift slabs;
- large slab areas to be divided into compartments of 2000 m^2;
- increase of slab thickness in the immediate vicinity of vertical walls;
- avoid different slab elevations in one compartment;
- avoid location of slab within unsuitable soil conditions;
- time schedule to allow for possible remedial work;
- avoid anchored jet grout slabs;
- prepare emergency plan.

Similar recommendations apply for vertical jet grout barriers as structural members:

- applications with water pressures in excess of 5 m require special attention (redundant design, appropriate checking procedures, emergency plan);
- identify soils with erosion potential in case of leakages;
- avoid slender construction elements;
- special care required when ground anchors are necessary.

5.4.2 Underpinning

Underpinning of structures using jet grout normally involves the construction of a body of improved ground beneath the structure such that the structural load is transferred to depth. If the underpinning is carried out next to an excavation then the jet grout body must be designed accordingly and the stability checked for bearing capacity, sliding and overturning. There is sometimes an economic relationship between the creation of a gravity underpin (i.e. a body that is self-supporting and stable) and a propped or strutted body where overturning or sliding is restrained by props or even anchors (as for the case history below).

An example of jet grout underpinning adjacent to an excavation is shown on Figure 5.20. This example is taken from a project in London where a self-supporting underpin was required adjacent to a new basement construction.

The design of a jet grouted underpin is exactly similar as for any gravity structure except that consideration needs to be taken into account that the strength of the jet grout body is usually significantly lower than brick or concrete.

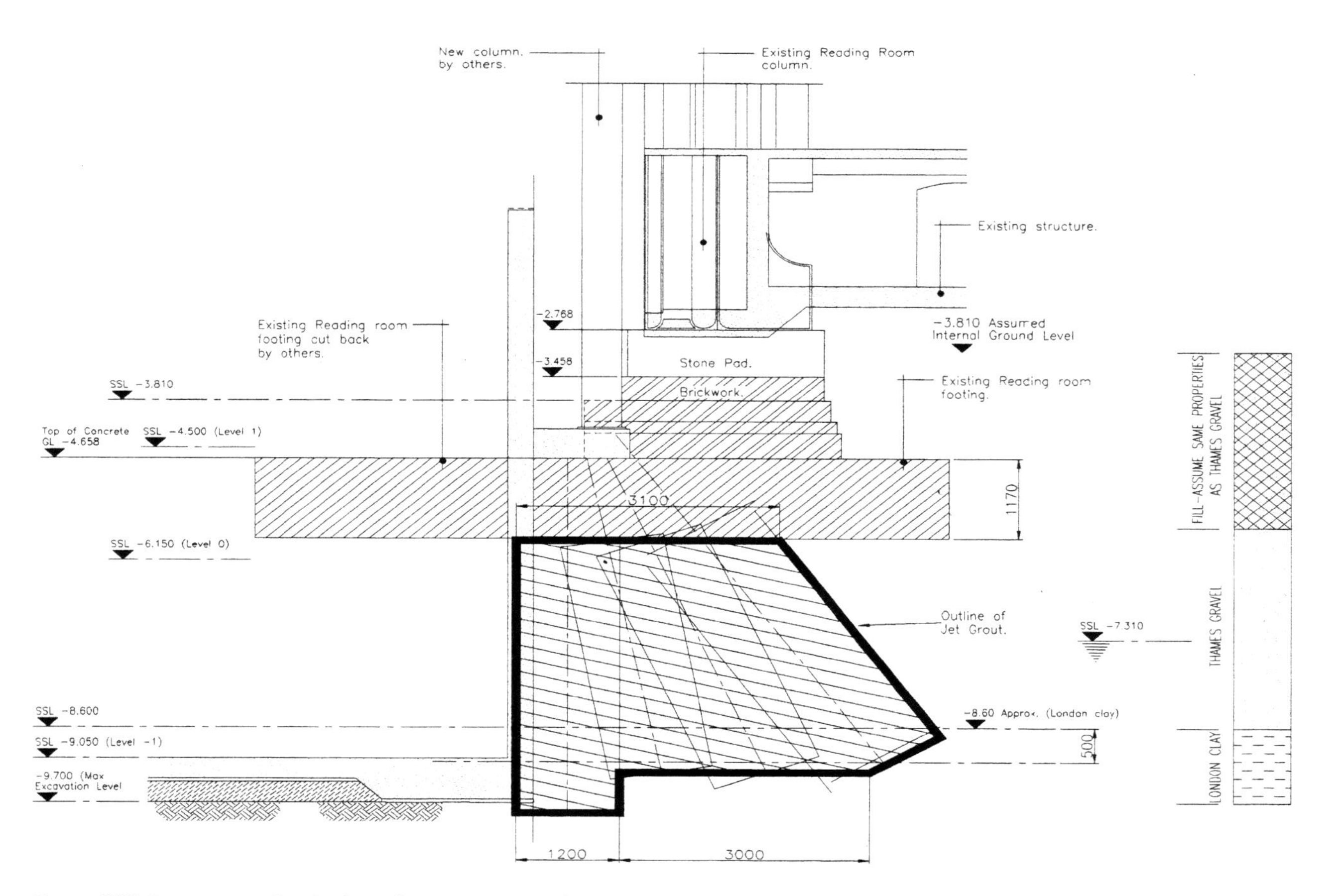

Figure 5.20 Jet grout underpinning adjacent to excavation.

5.4.3 *Tunnels and shafts*

5.4.3.1 *Bottom slab*

Base sealing of the slabs of shafts for tunnelling can be designed for the application of jet grouting to prevent base heave or piping in cohesionless soils saturated with groundwater. As discussed above in Section 5.4.1, these constructions are risky if incorrectly executed and require careful design and application.

The Academia of Japan dictates that normally the thickness of this slab must exceed half of the span between shaft walls. However, thinner slabs are possible by employing circular arc beams on which only compressive stress acts, as illustrated in Figure 5.21. This method of design results in an arch prop, 3 m in thickness even at a position of 40 m below ground level.

A tentative calculation gives a maximum value of 1.1 MN/m^2 and a minimal value of 0.95 MN/m^2 as compressive stress on both sides of the arch. As the average unconfined compressive strength of treated soils by jet grouting commonly exceeds 3 MN/m^2, this gives a high assurance of success.

5.4.3.2 *Subsurface props*

Displacement of walls is always of primary concern in open excavation. Late propping during excavation often causes tilting and/or settlement of

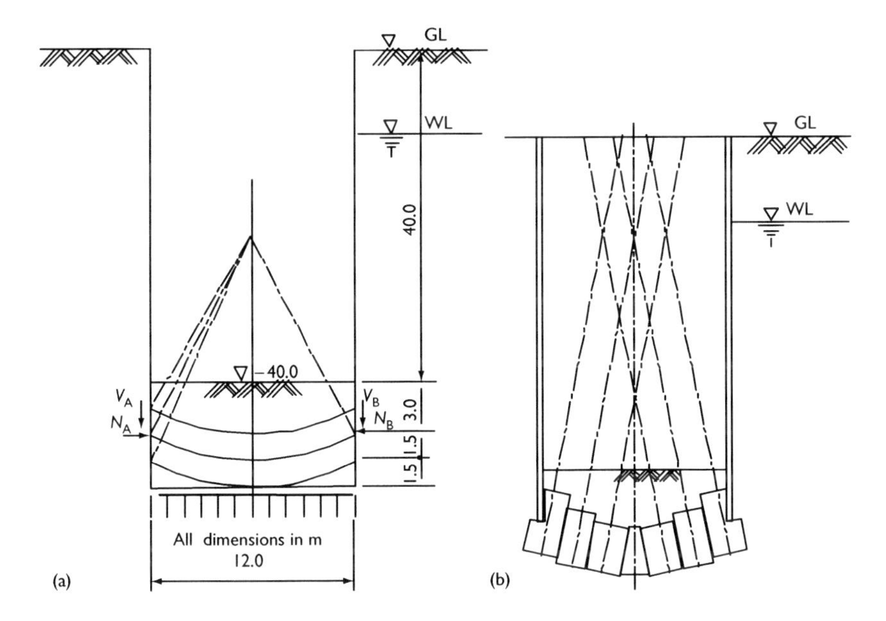

Figure 5.21 Base sealing of a shaft: (a) the conceptual cross-section; (b) layout of jet-grouted arch.

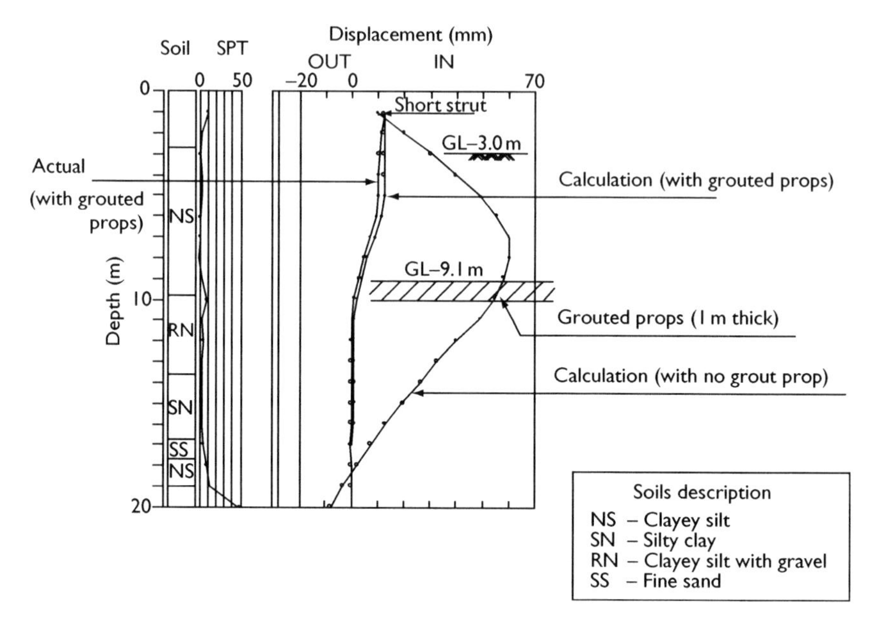

Figure 5.22 Comparison of calculated and actual wall displacements.

not only adjacent buildings but water supply, sewer lines and other underground facilities. Therefore, jet grouting offers the radically different approach of an *in-situ* soil-mix propping prior to excavation.

A practical case history briefly explains the result. The work required an excavation of 10 m depth in a soft clayey layer for basement construction but adjacent houses were so close to the site that they were afraid of being largely undermined due to displacement of walls for shoring, as shown in Figure 5.22.

Consequently, jet grouting-produced props of just 1 m thickness at the bottom of excavation have proved successful together with a row of conventional strutting at ground level.

Adding a row of grouted props enabled the reduction of displacement by approximately 80 per cent as clearly shown in Figure 5.22.

5.4.3.3 Roof barriers

In starting a tunnel-boring machine (TBM) through a wall of a shaft into an alluvial deposit, the soil surrounding the TBM may be lower in strength due to the loosening effect of the construction of the structure. This could trigger collapse or settlement because of extension of this loosening to the ground surface, especially in the case of shallow tunneling. Given such difficulties, jet grouting offers theoretical advantages in designing roof barriers. The design geometry is explained by reference to Figure 5.23 which illustrates how to obtain the zone to be treated (R—a), the property of which is to be reinforced by jet grouting.

A successful design follows an achieved line of shear strength to exceed a failure envelope of Mohr circle of the original ground. Figure 5.23 also shows that the radial and tangential stresses balance each other on the boundary line of the elastic region from the plastic one and consequently derives the equation below.

$$\frac{\partial \sigma_r}{\partial r} = \frac{\sigma_\theta - \sigma_r}{r} \tag{5.4}$$

where σ_r = radial stress, σ_θ = tangential stress and r = variable radius.

Next, since a failure takes place when the failure envelope becomes horizontal and the internal friction angle becomes zero, another equation is derived below:

$$\sigma_\theta - \sigma_r = 2c \tag{5.5}$$

where c = cohesion.

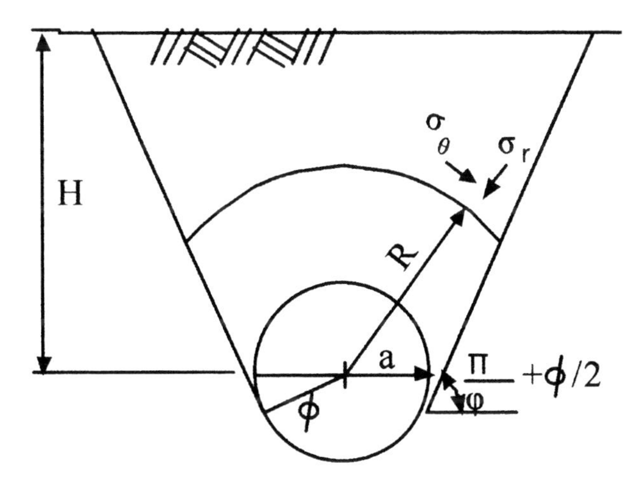

Figure 5.23 Roof barrier of a tunnel.

Then, substituting boundary conditions into the above simultaneous equations to obtain the plastic region leads to the following equation:

$$\ln \frac{R}{a} = \frac{\gamma_t}{2C}(H - R) \quad (5.6)$$

where R = plastic region, γ_t = average unit weight of the soil, H = depth to the centre of the tunnel, and a = radius of the tunnel.

5.4.4 Environmental applications

One of the more interesting uses of jet grouting is in the environmental field. There are many applications based on the ability of jet grouting to form bodies at considerable depth while only requiring small penetrating drill holes.

The main uses can be classified as follows:

1 *Encapsulation*: Achieving encapsulation of contaminants at depths where conventional excavation would be difficult as for the example shown in Figure 5.24. Additionally the grouted body is usually more impermeable than with conventionally grouted ground leading to more security in contaminant control.

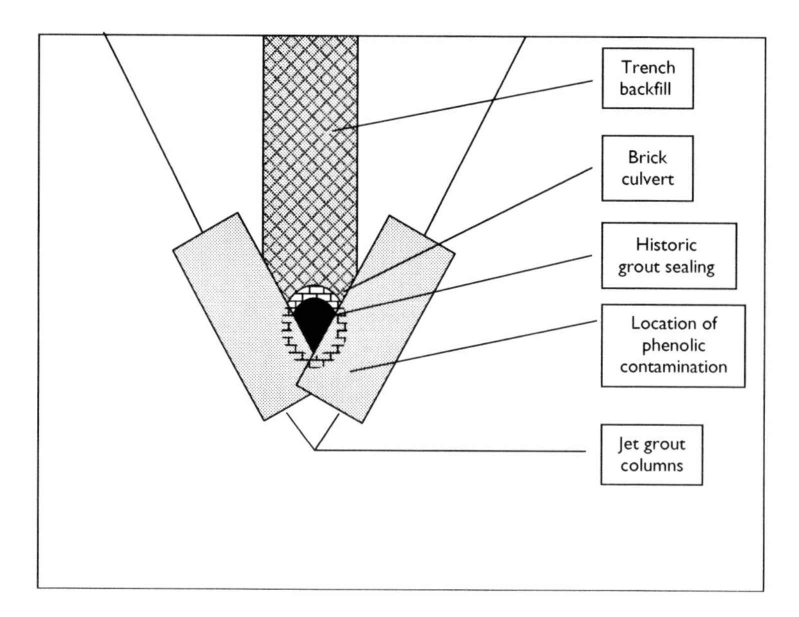

Figure 5.24 Encapsulation of contaminants at depth.

2 *Lateral barriers*: In difficult ground, jet grouting can provide an effective method of creating a barrier as was achieved in rock at Great Orton, Cumbria, England in 2001. Its main advantage is the ability to be selective in which zone is to be cut-off. This is especially advantageous for deep applications. The grout or water jet (depending on system utilized) scours weak and loose material, penetrating into fissures and fractures replacing the permeable infill material with relatively impermeable grout. The effective distance penetrated will depend on the system and rock type but has been shown on a number of projects to be effective up to 1 m from the hole position. Lateral barriers are typically specified in terms of permeability and for rock it is usually possible to achieve 5–10 times lower permeability than using conventional rock grouting. For soils the reduction when compared to permeation grouting can be as high as 10–50 times. As with all jet grout projects consisting of barriers or bodies constructed from interlocking columns, care must be taken during construction to minimize deviation from design locations and this should always be taken into account when designing the scheme.

3 *Active barriers*: In recent years, jet grouting has been used in the construction of permeable reactive barriers (PRB). These barriers contain materials that react with specific contaminants such that they are rendered harmless or less dangerous. Typical materials are zero valent iron (ZVI), granulated active carbon (GAC) or biologically active (BA). Design of these barriers is beyond the scope of this chapter. A guide has been published by the United Kingdom Environmental Agency in 2002. To construct these barriers, the reactive material is introduced in place of the grout. Because of this delivery process, the system is restricted to small particles that can pass through the nozzle.

5.5 Case histories

5.5.1 Groundwater control: Waste water treatment plant, Caister, East Anglia, England, 1999

An existing waste water treatment plant required to be expanded to meet increasing treatment demands. The site is close to the sea and is underlain by an extensive deposit of Alluvium consisting of very soft silty clays and peat to a depth in excess of 22 m. It was known that during the original construction, significant problems occurred with excavations in the very soft clays and peat below 1–2 m. The proposed extension required a treatment bed upto 7 m below ground level.

The adopted solution combined the use of sheet piles for perimeter retention and a 2 m thick jet grout base slab installed immediately below the base of the excavation to prop the sheet piles and to resist base uplift by bonding to the permanent pre-cast driven piles required to support the structure in the long term. The Crossjet system of jet grouting was selected because of the variable nature of the ground and consisted of nominal 2 m diameter columns installed on a 1.5 m triangular grid. Additional columns were required due to shadowing by the pre-cast piles and totalled approximately an additional 5 per cent in number.

Trial columns were carried out in advance of the main works to check the quality of the product and demonstrated the minimum strengths of 1 MN/m^2 required by the design. Inclinometers were installed immediately outside the tank to monitor lateral movements at depth, and geodetic surveying was carried out to check line and level at the top of the sheet piling. Maximum lateral movement reached 70–80 mm which compared favourably with the values predicted (Figures 5.25 and 5.26).

Figure 5.25 Ground water control, Caister, UK.

Figure 5.26 Ground water control, Caister, UK.

5.5.2 Underpinning: Vienna

This project, completed in 2002 consisted of the construction of a new commercial development in Vienna, Austria (Figures 5.27–5.29).

A two-storey basement was required to be constructed within a site adjacent to buildings on two sides and roads on the other sides. Ground conditions consisted of sands and silts overlying a clay bedrock at depth. The development required the maximum use of the site, and the use of jet grout underpinning to the two buildings allowed the basement walls to be constructed right to the site boundary. Two rows of jet grout columns using the double system were constructed around the site, toeing into the clay bedrock to provide ground water cut-off. Three rows of ground anchors were used to support the sides of the excavation.

Figure 5.27 Jet grout underpinning, Vienna.

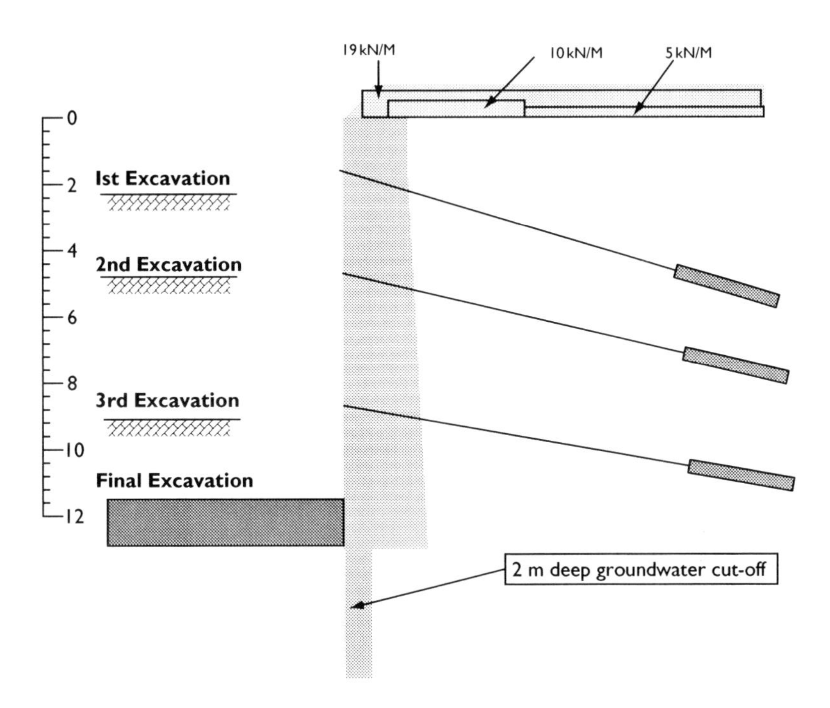

Figure 5.28 Jet grout cross-section, Vienna.

5.5.3 Environmental: Great Orton disposal site, Cumbria, England

Following a major outbreak of the Foot and Mouth disease in 2001, it was necessary to dispose of a large number of cattle carcasses. In some areas they were burnt but in Cumbria an abandoned airfield was used as a mass burial site. On completion of the burial operations there was a requirement imposed by the Environment Agency that a perimeter cut-off should be installed in association with ground water pumping to prevent leachate from the decomposing bodies causing off-site contamination. The jet grouting cut-off was to be constructed 5 m into the underlying mudstones and siltstones, and jet grouting at close centres was selected as providing the lowest permeability cut-off (Figure 5.30).

The cut-off was constructed by pre-drilling 300 mm diameter holes at 750 mm centres and then carrying out the jet grouting using the double system. The excess spoil generated by this process was then used during excavation of the cut off trench in the overlying boulder clay, thereby significantly reducing overall material usage and cost. The cut off was 2200 m in length and was completed in late 2001. Permeability of the cut-off achieved values in the range $1\text{–}5 \times 10^{-8}$ m/s (Figure 5.31).

Figure 5.29 Exposed jet grout underpinning, Vienna.

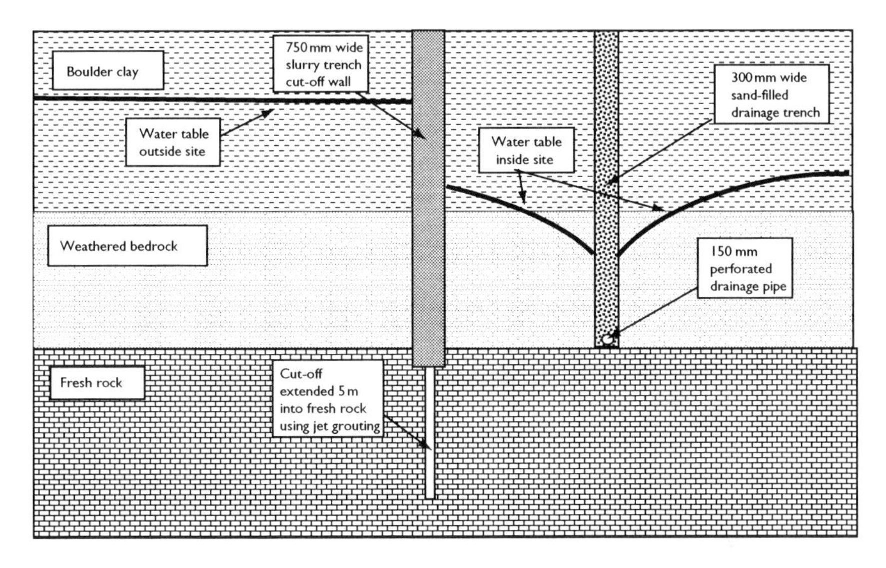

Figure 5.30 Proposed containment detail, Cumbria, UK.

Figure 5.31 Jet grouting construction, Cumbria, UK.

5.5.4 *Settlement control: Japan*

A tunnel was required to be constructed beneath a street under which were buried numerous services. In addition the adjacent buildings were sensitive to movement. The solution adopted was to construct a heading from spiles (horizontal piles) supported on jet grout columns toed into competent ground. In this way the jet grouting supported the tunnel drive and reduced settlements to acceptable levels. The small diameter holes required to install the columns were also of benefit in penetrating between the services. The crossjet system was chosen as the ground conditions were variable and with this system the column diameter could be guaranteed (Figure 5.32).

5.5.5 *Cofferdam sealing: Japan*

As part of the new construction of a station complex it was necessary to construct a 19.5 m deep excavation beneath an existing road. There were a number of large services that could not be diverted. The solution adopted was to construct the cofferdam around the services and then seal the gaps around and beneath the services using the superjet system. The superjet system was selected as a relatively large diameter was required to enable interlock of the columns beneath the 2.4 m diameter services. The columns

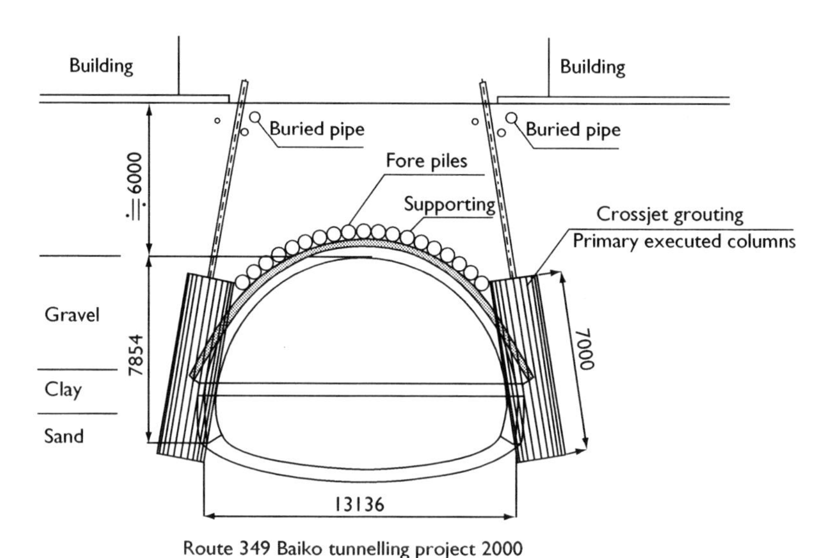

Figure 5.32 Settlement control for tunnelling, Japan.

were installed to a depth of 42 m to ensure that there would be no ground water flow into the excavation. Ground conditions consisted of medium dense sands toeing into hard clay at depth. Ground water was close to ground level. The use of the large-diameter superjet system minimized the risk of windows in the ground treatment at depth (Figures 5.33 and 5.34).

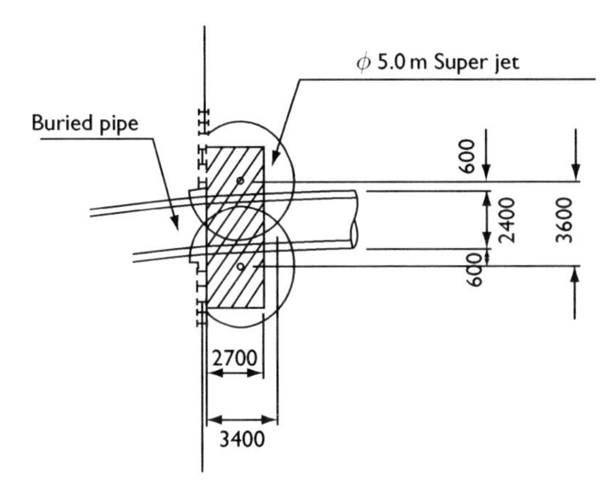

Figure 5.33 Cofferdam sealing, Japan.

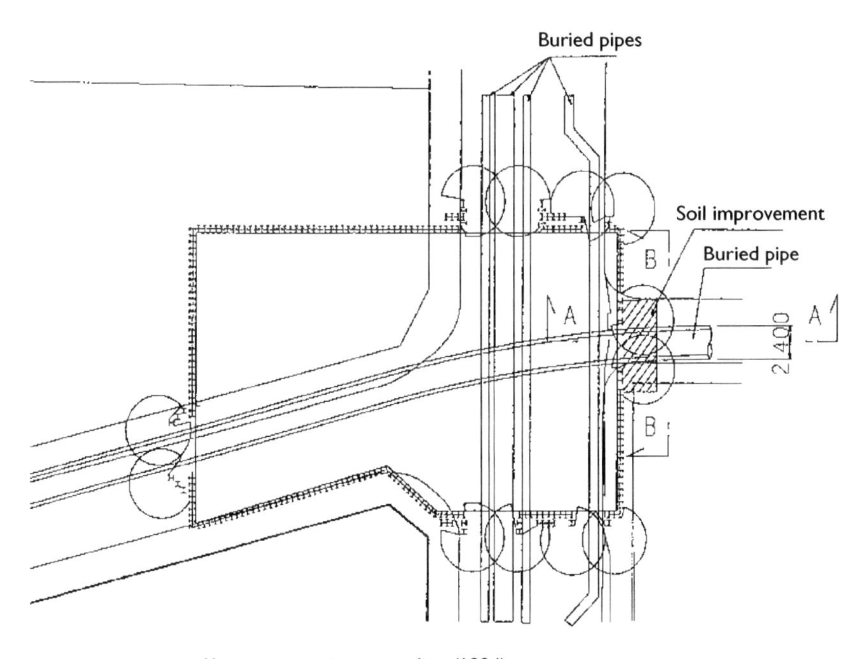

Figure 5.34 Cofferdam sealing, Japan.

Bibliography

Abramovich, G.N. (1963) *The Theory of Turbulent Jet*, MIT Press.

Environment Agency (2002) Guidance on the use of permeable reactive barriers for remediating contaminated groundwater, National Groundwater and Contaminated Land Centre Report NC/01/51, September.

Franz, N.C. (1972) Fluid additives for improving high velocity jet cutting, *Proceedings 1st Symposium on Jet Cutting Technology*, BHRA, A7-93.

Fritsch, M. and Kirsch, F. (2002) Deterministic and probalilistic analysis of the soil stability above jet grouting columns, *5th European Conference on Numerical Methods in Geotechnical Engineering*, NUMGE, Paris, Presses de LENPC.

Hermans, J.J. (1953) *Flow properties of disperse systems*, North Holland Publishing Company.

Kirsch, F. and Sondermann, W. (2002) Zur Gewölbestabilität über Soilcrete-Körpern. 9. Darmstädter Geotechnik Kolloquium, TUD.

Kirsch, K. (1997) Contractor's view on the risks involved with deep excavations in water bearing soils. XIV ICSMFE, Hamburg.

Lichtarowicz, A. (1995) Future of water jet technology basic research, *Proceedings, 4th Pacific Rim International Conference on Water Jet Technology*, pp. 13–26.

Noda, H. *et al.* (1996) Case of jet grouting for 10.8 m diameter shield. In: Yonekura *et al.* (eds), *Grouting and Deep Mixing*, Balkema, Rotterdam.

Pollath, K. (2000) Baugrube Schleuse Uelzen II. Deutsche Gesellschaft für Geotechnik. Baugrundtagung, Hannover.

Reichert, D. *et al.* (2002) Baugrube Domquarree. Deutsche Gesellschaft für Geotechnik. Baugrundtagung, Mainz.

Shavlovsky, D.S. (1972) *Proceedings of 1st International Symposium on Jet Cutting Technology*, BHRA.

Shibazaki, M. (1996) State of the art grouting in Japan, *Proceedings, 2nd International Conference on Ground Improvement Geosystems*, pp. 857–862.

Shibazaki, M. and Ohta, S. (1982) A unique underpinning of soil solidification utilizing super high pressure liquid jet, *Proceedings of the Conference on Grouting in Geotechnical Engineering*, ASCE, February, pp. 685–689.

Shibazaki, M., Ohta, S. and Kubo, H. (1983) *Jet Grouting*, Kajima Publishing Company, pp. 15–20.

Wachholz, T. *et al.* (2000) Planung, Konzeption der Schleuse Uelzen II und Berechnung der Baugrube. Deutsche Gesellschaft für Geotechnik. Baugrundtagung, Hannover.

Wichter, L. and Kugler, M. (2002) Fehlstellen in Dusenstrahlkörpern durch Inhomogenitäten im Baugrund. Deutsche Gesellschaft für Geotechnik. Baugrindtagung, Mainz.

Wittke, W. *et al.* (2000) Neues Konzept für die Schildeinfahrt in eine Baugrube im Grundwasser. Tunnelbau, Verlag Glückauf.

Wolbrink, R. *et al.* (2000) Neue Erkenntnisse bei der Herstellung von HDI-Dichtblocken am Beispiel U5 Berlin. Deutsche Gesellschaft für Geotechnik. Baugrundtagung, Hannover.

Yahiro, T. (1996) *Water Jet Technology*, Kajima Publishing Company, pp. 15–29.

Yahiro, T., Yoshida, H. and Nishi, K. (1983) *Soil Modification Utilizing Water Jet*, Kajima Publishing Company, pp. 7–20.

Yanaida, K. and Ohashi, A. (1980) *5th International Symposium on Jet Cutting Technology*, BHRA, A3-33.

Chapter 6

Compaction grouting

R. Rubright and S. Bandimere

6.1 Introduction

In the early 1950s grouting contractors in California began experimenting with the use of low-slump mortar-type grout. They discovered that they could inject the material under high pressure to densify the loose soil formations beneath distressed structures. The term they used to describe this unique grouting process which compacts soil was 'compaction grouting'. In July of 1980 the grouting committee of the American Society of Civil Engineers gave compaction grouting a more formal definition in the *Journal of the Geotechnical Engineering Division*:

> Compaction grout – grout injection with less than one inch (25 mm) slump. Normally a soil-cement with sufficient silt sizes to provide plasticity together with sufficient sand sizes to develop internal friction. The grout generally does not enter soil pores but remains in a homogenous mass that gives controlled displacement to compact loose soils, gives controlled displacement for lifting of structures, or both.

The compaction grout mass is generally spherical, but its shape is ultimately governed by numerous other factors including grout mix design, injection rate, the strength and texture of soil zones, the overburden or applied structural loads, etc. Compaction grout bulbs 3 ft (0.9 m) in diameter and larger are not uncommon in loose soil conditions (Figure 6.1). As the compaction grout bulb grows during injection, the soil nearest to the bulb undergoes severe deformation and stressing resulting in some local zones of distress at the interface of the soil and grout mass. Areas further away from the interface are more uniformly compacted since the stresses are more uniform and the strains are elastic rather than plastic.

At the time of its inception the idea behind compaction grouting was a radical departure from the permeation and fracture slurry grouting techniques which were in wide use. The engineering community was familiar with the idea of injecting chemicals or cementitious-type materials into rock

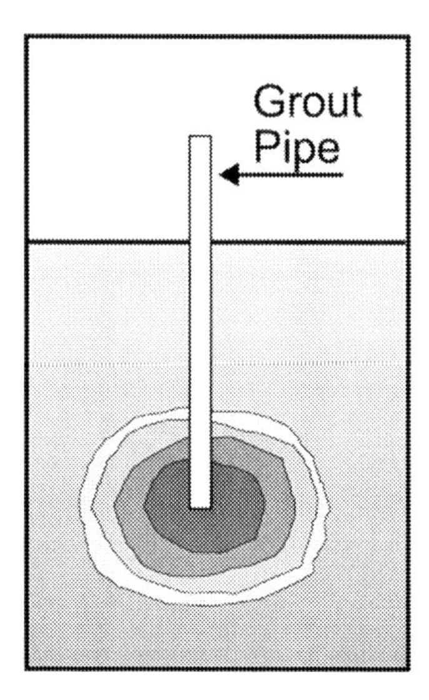

Figure 6.1 Compaction grout bulb formation.

or soil to fill the joints or interstitial voids for various purposes. The idea of injecting a thick grout into the ground for the purpose of compacting these alluvial type soils was a new concept, but it gradually proved itself useful and versatile. Despite the fact that new replacement techniques such as jet grouting and soilfrac grouting have been developed since compaction grouting, there has been no doubt that the technique provides a valuable service in the grouting industry. Compaction grouting has gained wide acceptance in the United States since its creation, and is now being used in many other parts of the world. In the United States compaction grouting has clearly replaced slurry injection or 'pressure grouting' as the preferred method of densification grouting.

6.2 History and development

Compaction grouting – the first 30 years written by James Warner in 1982, indicates that the development of the compaction grouting technique occurred in the early 1950s on the West Coast of the United States. For the first 30 years this system was used exclusively as a remedial technique. Graf (1969), Mitchell (1970), and Brown and Warner (1973), reported to the industry on how the compaction grouting technique was replacing conventional slurry grouting methods. Early applications of compaction grouting to correct structural settlement of buildings and floor slabs are documented by Warner (1978, 1982).

In the late 1970s the compaction grouting technique was gaining acceptance in other parts of the United States. In 1977 a new use for compaction grouting was found in the area of urban soft ground tunnelling. Baker *et al.* (1983) described the historical significance of settlement control for ground loss and loosening related to urban tunnelling (Figure 6.2). Compaction grouting has become a standard for subway construction in the United States where it has been used in Washington, Baltimore, Boston, Seattle and Los Angeles.

In the early 1980s compaction grouting was used for the first time as a site improvement technique for new construction. It was used in conjunction with dynamic deep compaction (DDC) to densify the soil beneath two 600 MW coal-fired electrical generating units in Florida. Shortly afterwards, it was used again in conjunction with dynamic deep compaction and vibro-compaction systems for a large nuclear submarine servicing facility in Georgia. In both cases it was used to densify potentially liquefiable loose granular soil zones at depths which were uneconomical if treated by other means.

Since that time its use as a site improvement tool has spread further throughout the United States. In the mid-1980s compaction grouting began to be used extensively for sinkhole repairs in Florida and other sinkhole-prone areas in the Northeast United States (Figures 6.6 and 6.7). Henry (1986, 1987) and Welsh (1988) describe this work in detail.

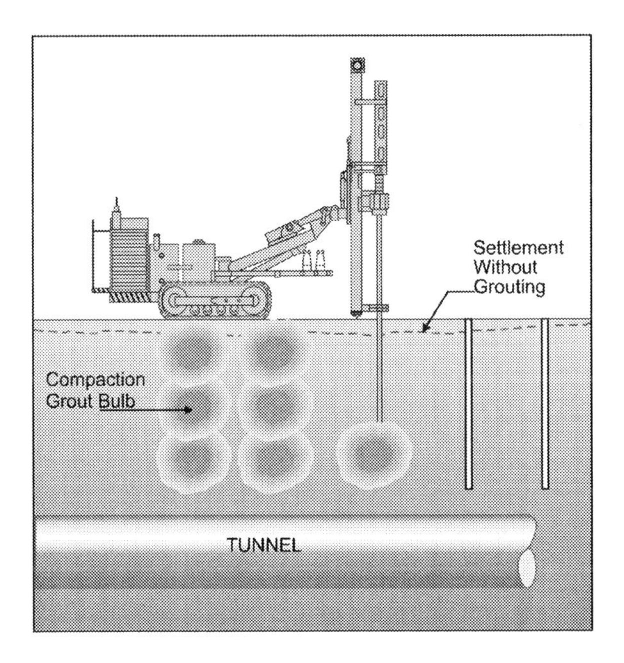

Figure 6.2 Use of compaction grouting above advancing tunnel in urban area.

In 1990 the American technology of compaction grouting was exported to Japan. The technology was met with some scepticism until the 1995 earthquake in Kobe, Japan. Following the quake, compaction grouting was used extensively to correct numerous structures that experienced settlement and tilting due to seismically induced settlements. Since that time, the American compaction technology has been successfully transferred to other countries and is finding acceptance on a world-wide level.

The most recent development of the 'low-mobility grouting' application involves a case near Denver, Co., where an apartment complex was built on post-tensioned concrete slabs over highly expansive clays. Water infiltration around the perimeters of several structures caused up to 8.25 in. (21 cm) of differential movement from heaving in less than 5 years from construction. A low-mobility (compaction) grout was used to lift the low portions of the structures to meet the heaved elevations. The long-term effect of this particular application remains to be seen, but it demonstrates the versatility of the low-mobility compaction grouting system.

6.3 Applications and limitations

In general compaction grouting is used to repair natural or man-made compaction deficiencies in various types of soil formations. Compaction grouting is most frequently used as a remedial measure beneath or adjacent to an existing structure. These situations usually arise from the need to non-destructively increase the weight/volume relationship of an *in situ* soil condition, while providing additional *in situ* structural elements. This can work as an overall composite structure for support purposes. The need for compaction grouting in these situations usually arises from the following conditions:

- Loose or deteriorating (i.e. organic degradation, etc.) natural soil conditions.
- Loose or voided fills either improperly placed at the time of construction or placed in an uncontrolled manner before construction was anticipated.
- Loose or voided soils caused by adjacent excavation activity, sinkhole activity, improper dewatering, broken utility lines or the like.
- Change of moisture content in a collapsible soil, i.e. loess.
- Need to increase a bearing capacity of a soil due to load changes (modification) of an existing structure.

A secondary use of compaction grouting is to re-level settled structures if desired. This is usually done in conjunction with a densification programme as discussed above.

Compaction grouting (Figures 6.3 6.6) can also be performed as pre-treatment or site improvement before a structure is built. Extraordinary

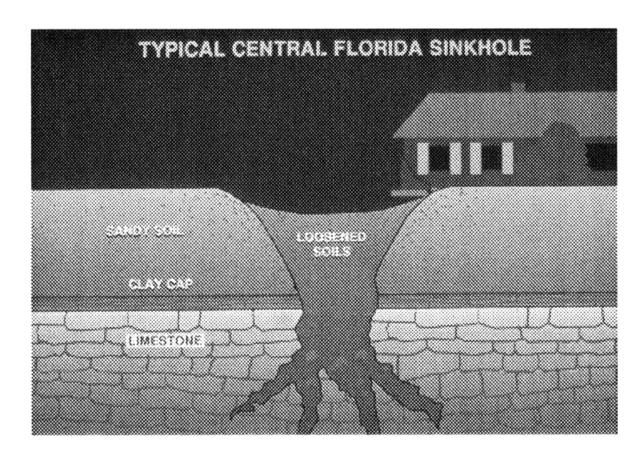

Figure 6.3 Schematic of Central Florida sinkhole.

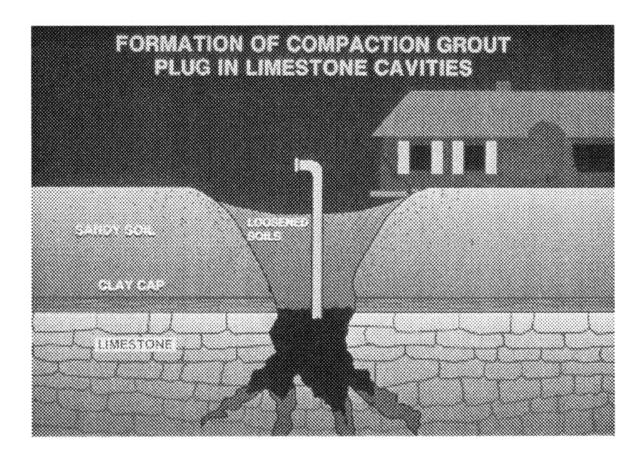

Figure 6.4 Formation of sinkhole plug with compaction grout.

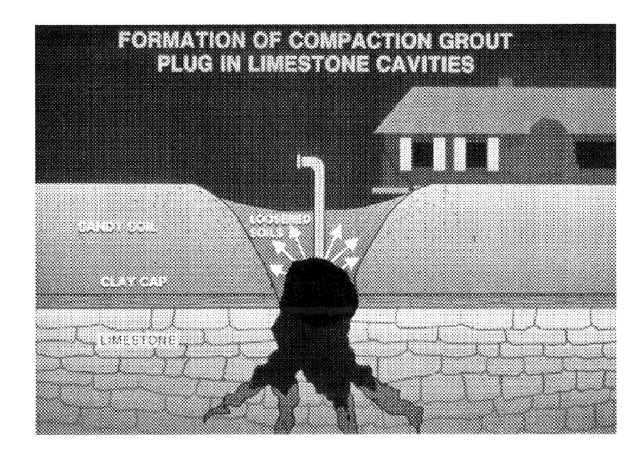

Figure 6.5 Compaction grout densifying loose overburden over sinkhole after plugging.

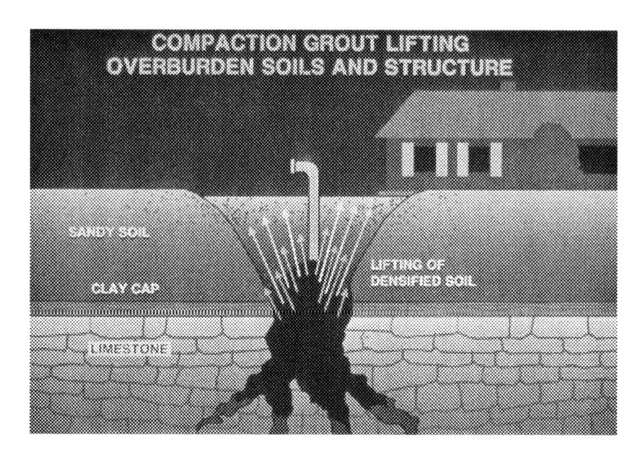

Figure 6.6 Compaction grout used to lift structure after plugging sinkhole and compaction of loose overburden.

circumstances need to exist for the process to be cost-effective as a pre-treatment technique. Some circumstances which have led to the use of compaction grouting for site improvement projects include:

- The presence of a thin but deep 'problem zone' which will cause unacceptable settlements due to new loads or future seismic events.
- The presence of loosened overburden soils in highly variable and pinnacled limestone areas where deep foundations cannot be economically founded in sound rock.

6.3.1 How compaction grouting works

The primary objective of compaction grouting (the densification of improvable soil masses) can be accomplished through a variety of geometrical configurations. The most common involves the placement of grout in a quasi-continuous cylinder along the axis of the injection pipe as it is withdrawn or inserted. All orientations of pipes have been used successfully with some suggestion that pre-mature surface heave is more likely to occur as the pipe departs from the vertical. Single or staggered injection bulbs can be successful when access or treatment zone geometry precludes other configurations.

The concept of a compaction grout pedestal or pile is another means of applying the technology. In thin layers of unimprovable soils a stout (length over diameter ratio less than 2) compaction grout pedestal as a structural element can be used. More commonly, however, the design concept of a structure element composed of compaction grout (sometimes with unconfined strengths as high as 5000 psi (345 bar ±) and a steel rebar in the centre for strength and continuity) working in conjunction with the densified surrounding soil is used to provide support for the applied load. It is difficult to analyse such a composite effect but experience has proved it to be successful on numerous occasions. One major advantage of this technique over mini-piling or underpinning is that a rigid connection to a foundation is generally not necessary due to the area effect of the grout. Compaction grouting is also frequently used to re-level structures which have settled. Surface heave is a common effect of the grouting operation and can be used under the right circumstances and with appropriate controls to correct problems. Shallow injection causes concentrated or localized heave while deeper injection has a broader effect. The existence of underground utilities in the area is one major determining factor for using deep injections for lift. To minimize the risk of damage to underground utilities, they should be lifted from a minimum depth of 10 ft (3 m) below their location. Simultaneous injection of multiple points has even been performed to raise larger portions of structures. Relevelling can be tricky, therefore expectations should be low until experience of the technique is obtained.

6.3.2 Applicable soil conditions

Soil conditions which lend themselves to effective compaction grouting fall into five general categories:

1 *Loose, granular soils above or below the groundwater table*: This category accounts for the greatest number of compaction grouting projects. The soil is mostly sands and/or gravel but can contain substantial amounts of silt and some clays, provided that the soil still drains and behaves mechanically like a granular soil. Typical soil SPT *N*-values range from 0 to 15. *N*-values can usually be improved by 10 or more points depending on spacing of pipes and other factors discussed later in this chapter. The formation can be a candidate for compaction grouting regardless of its thickness or the presence or absence of any surcharge provided that proper planning and construction techniques are used.

2 *Loose, non-saturated fine-grain soils*: Soils composed primarily of silt and/or clays can usually be improved by compaction grout if they are not saturated. Poorly placed cohesive fills are the most common occurrence under this category. Typical SPT *N*-values for candidate soils would be between 0 and 10. Treatment of thick saturated silt or clay strata should not be performed. On several such projects the compaction grouting has actually aggravated the problem by accelerating settlement of structure. It is believed that the compaction grouting squeezes the saturated soils causing a dramatic increase in pore pressures which cannot dissipate. This heightened pore pressure leads to a direct and sudden loss of shear strength. Initial heave during the pumping operation is followed within hours by net settlement of the affected structure.

3 *Collapsible soils*: Thallus formations created by a wind deposit mechanism in arid regions have been successfully treated with compaction grouting. Typically, these deposits create problems when poor drainage or broken plumbing leads to saturation of the soils, causing the soil microstructure to collapse. It has been learned that these soils can be densified or artificially collapsed by the compaction grouting method, thus avoiding the risk of water-related problems.

On a site where varying moisture contents might impact the overall effectiveness of the compaction grouting programme, it might be necessary to utilize the drilling operations to even out the collapse potential of the *in-situ* soils (i.e. vary the drilling method from dry/wet methods and means on a given site). Also, under certain conditions, soaking the holes for a period of time will weaken inter-granular bonds and ease the effects of compaction grouting.

4 *Soil voids*: Voids in soil or rock formations at depth are frequently filled with compaction grout rather than fluid grouts. Compaction grout is more controllable under these circumstances since it will not travel far beyond the intended treatment area and because the thixotropic nature of the material can be used to minimize the application of hydrostatic pressure. These are important considerations in applications such as sinkhole remediation and wash-out repair where the size and direction of the void are unknown (Bandimere, 1997).

5 *Thin, unimprovable soil strata contained by adequate surcharge*: This category appears to break all the rules established above, but experience has proved that it is a valuable exception worth noting. The soil can be dry or saturated silt, clay or organic (such as peat or wood waste) provided that it is not more than 6 ft (1.8 m) in thickness and is positioned beneath at least 6 ft (1.8 m) of competent or improvable soils or comparable structural load. The compaction grout does not improve this soil but rather bridges through it by creating a pedestal on which to support some of the overburden load. It is believed that the effect described in '*Loose, non-saturated fine-grain soils*' takes place in saturated situations but is quickly bypassed by the pedestal effect.

6.4 Design and construction

6.4.1 Design methods and parameters

The first step in designing a compaction grout programme is to have a clear understanding of the underground problem. A thorough soils investigation is essential for the success and cost-effectiveness of the compaction grouting programme. A thorough understanding of physical site constraints is also necessary. With this information at hand, objectives can be developed. 'We need to improve this loose silty sand strata between elevation −10 and −18 to an average *N*-value of >15 within these plan view limits,' might be the stated objective of a small remedial project. With clear objectives, it is then a matter of deciding how to achieve them.

Pipe spacing: Determination of pipe spacing is critical to the technical success and cost-effectiveness of the project. Pipe spacings ranging from 3 to 15 ft (0.9–4.6 m) have been used in appropriate circumstances. Tight spacings are used when trying to achieve a compactive effort near the surface with little overburden or structural load. Good compaction results near the surface have been documented in silty sands with grout pipes on 3 ft (0.9 m) centres with nothing more than a 6 in. (15 cm) concrete slab for confinement. Larger spacings such as 15 ft (4.6 m) are used when there is sufficient overburden pressure of 30 ft (9.2 m) or more, soil conditions are good, and improvement need not be very great (i.e. for liquefaction improvement). These are extreme cases. The majority of compaction grouting is done with pipes at spacings between 5 ft (1.5 m) and 7 ft (2.1 m). Remedial work with limited confinement (i.e. loose fills directly under a spread footing 3–4 ft (0.9–1.2 m) below grade) is usually spaced between 5 ft (1.5 m) and 7 ft (2.1 m) on centres. Area treatments with greater overburden pressure (i.e. densification of loose soils created by soft ground tunnelling operation) are usually between 8 ft (2.4 m) and 10 ft (3 m) on centre.

Injection sequence: The order in which pipes are injected is important for best technical results. A primary and secondary sequencing of pipes will allow for initial densification followed by secondary densification which is confined by the previous work. This is desirable because it gives the secondary grouting something to compress against as it does its job. Tertiary grouting is generally not seen as necessary provided that the secondary spacing is adequately tight for the circumstances. Compaction grouting can also be sequenced upwards or downwards within a single grout pipe. Upstage grouting or injection from the bottom up as the pipe is withdrawn is the most common system. It is less expensive and usually as effective as its counterpart, downstage grouting. Downstage grouting injects a compaction grout bulb at the shallowest injection point first. After the initial set, a grout pipe or grout hole is advanced through to the next lower stage and the process is repeated. The most useful place for the downstage grouting process is where minimal surface confinement exists, or where incremental lifting of a structure is required. The top bulb and resulting compaction appears to provide better confinement for subsequent stages. A variation of the two systems utilizes 'down-stage grouting' at the top of the treatment zone followed by 'up-stage grouting' for all other work.

Injection pressures: Surprisingly, the grouting pressure is usually a dependent variable rather than an independent variable in the design equation.

Compaction grouting equipment should generally be capable of 1200 psi (83 bar ±) at the injection point. Line losses are significant with low-slump grout and must be considered when sizing the pump. Injection pressures, measured at the point of injection, usually range between 100 psi (6.89 bar ±) and 400 psi (28 bar ±). It is important to understand that the stiff grout consistency causes rapid pressure dissipation in the ground. However, caution must be exercised when large volumes of material are placed at single locations in extremely loose or voided areas. The hydraulic pressure of large masses of unset grout, although less than a low-viscosity fluid, is still a force worthy of respect. Broken lagging boards, crushed concrete tunnel liners and unanticipated real estate purchases are the legacy of overzealous pump operators.

In routine compaction grouting, the ground will refuse to take grout at a specific pressure. Use of significant additional pressure will squeeze water out of the grout wherever a perfect seal does not exist. This will cause blockages in the delivery system. The usual refusal pressure on most granular soil projects is 400–600 psi (28–41 bar ±). Frequently the refusal pressure is not obtained before surface heave is observed.

Injection rate: The rate at which compaction grout is injected can be a more critical parameter than injection pressure although the two are related. Slow pumping rates of the order of 0.5–1.0 ft^3 (0.01–0.02 m^3) per minute are used in poorly draining soils and close to the surface. Medium rates of the order of 1.0–4.0 ft^3 (0.02–0.11 m^3) per minute are used in free draining or dry soils with reasonable cover and fast pumping rates of 4–12 ft^3 (0.11–0.34 m^3) or more are used in 'safe' voided or loose situations involving significant cover.

Injection volume: In conventional densification applications the objective should be to distribute the compaction grout evenly through the targeted strata without causing pre-mature surface heave. It is best to select a standard 'target volume' for each injection stage. This target volume should be adjusted as the programme progresses. An increase in volume is warranted if heave is not observed until the pipe is near the surface. A decrease in volume is appropriate if heave is observed at deep stages. Target volumes per individual stages should be estimated during the design process. They should be a fraction of the affected area. For example, if the compaction grout pipe grid is on 7 ft (2.1 m) centres, and stages are set at 2.5 ft (0.8 m), then the volume might be:

$$7\,\text{ft} \times 7\,\text{ft} \times 2.5\,\text{ft} \times 10\% = 12.25\,\text{ft}^3 \text{ or } 4.9\,\text{ft}^3/\text{linear ft}$$

$$2.1\,\text{m} \times 2.1\,\text{m} \times 0.8\,\text{m} \times 10\% = 0.35\,\text{m}^3 \text{ or } 0.35\,\text{m}^3/\text{linear m}$$

The percentage used in this equation represents a 'compaction factor' or 'replacement factor', and usually ranges from a minimum of 3 per cent to a maximum of 12 per cent for normal non-voided natural soil deposits. Replacement percentages as high as 20 per cent have been documented in extremely poor (and presumably voided) fills consisting of cobbles and boulders or waste debris. Target volumes should not be used as a criteria on projects where voiding is suspected. Information regarding replacement values is of interest not only in calculating target volumes but also in calculating project costs and schedules.

Heave: Surface and/or structural heave is the most common limiting factor in compaction grouting. It is a phenomenon which can be observed on most compaction grouting projects. Heave indicates that compaction stresses have exceeded the confining stresses and that the soil mass is fracturing rather than compacting. Any further injection, at least at the injection stage, is virtually useless. For this reason, the injection design should call for movement to the next stage once heave is observed. Injection at higher stages is worth continuing since lateral stress can still provide additional compaction, but generally additional heave will be noted quicker than before.

Before pumping begins it is important to establish acceptable limits of surface or structural heave. If possible the cumulative heave should be set high enough to observe the smallest possible heave on multiple injection stages. One half inch (0.5 in.; 2 cm) of total heave with a tenth of an inch (0.1 in.; 0.2 cm.) as refusal criteria at each stage is a good starting point. Remember that structures have frequently suffered settlements of this much or more resulting in the need for compaction grouting and therefore, some heave is frequently beneficial to the site or structure.

Material: The design of the compaction grout material is critical to success and is frequently misunderstood by engineers or owners not familiar with the process. The objective is to design a mix which has enough granular components to create internal friction in the mix. This encourages the grout to stay in a quasispherical, homogenous mass at the point of injection. If the mix lacks internal friction it behaves like a low slump 'grease' which is still free to lens and travel uncontrollably in the ground. On the other hand, if the fines content is too low and internal friction too high, the grout requires too much pressure to pump which usually results in water bleed and dry packing or sand blocking in the pipeline. This phenomenon can also be observed just inside the tip of the injection pipe.

The most common misconception about compaction grout material is that unlike most building materials, strength is not an important issue. Remember that in most compaction grouting applications the object is to densify the surrounding soils rather than to create a structural element in the ground. Often cement is not even used in the grout mix. Moist, properly graded silty sand without cement is frequently adequate to accomplish the densification objective. Furthermore, grouting without cement makes clean-up and long pumping shifts easier and less expensive. The compaction grout slump is usually set at 1–3 in. (2.5–7.5 cm). Higher slumps are sometimes used especially in suspected void situations.

Sand/soil: The major component of the compaction grout mix is sand or a sandy soil. In the early days of the process, almost all grout was site batched using a sandy top soil material or a silty sand with 15–30 per cent passing the No. 200 sieve. This material can be used alone or with cement as an additive. Since it is often difficult to find such a naturally graded material, it is now frequently manufactured using a combination of more readily available components. Clean masonry sand can be supplemented with cement and fly ash to add fines. Typically three parts of sand are used to one part cement and/or fly ash. Bentonite is usually added in small quantities to increase pumpability.

Cement: Type I or II Portland cement is usually used in the mix. It can vary from as little as 0 to as much as 900 pounds (408 kg) per cubic yard depending on the need for fines, availability of fly ash, the need for strength and the anticipated pot life required for the operation. Use of set retarder is common.

Fly ash: In most cases fly ash is a desirable substitute for cement in an effort to get an appropriate fines content in the mix. Fly ash allows much more working time before set and costs less than cement. When it is available, it is frequently used to replace 50 per cent or more of the cement content.

Bentonite: Use of bentonite should be kept to an absolute minimum in the mix. Overuse results in a greasy consistency which is poor for the reasons explained above. Unfortunately all the problems associated with pumping of compaction grout go away with excess bentonite use. One seasoned superintendent once said 'If a crew is smiling, they're probably using too much bentonite'. Typically bentonite composes less than 3 per cent of the combined weight of cement and fly ash and except in extremely unusual circumstances it should not exceed 5 per cent.

Coarse aggregate: Pea gravel and other rounded, small aggregate (under 0.5 in. or 1.3 cm) have been used successfully in applications with large pumps and grout lines of 2 in. (5 cm) in diameter or greater. Angular rock materials (i.e. crushed limestone) work very well, but typically require a slight variation to fine content in the mixes relationship, and will usually increase the wear on hoses, pump cylinders, etc. Excellent technical results can be achieved as long as the same basic objectives are kept in sight.

6.4.2 Construction procedure

Regardless of how thorough the planning has been, there will always be a need for procedural adjustments in the field. An experienced superintendent or engineer should be positioned at the injection point so he can observe the pressure readings, withdrawal rates, and heave data. He should also be provided with a remote control for the pump, which has minimum control parameters of 'forward pumping, off, and reverse pumping'. The supervisor or technician should have a written outline of the compaction grout work plan. It should list grouting sequences, refusal criteria such as pressure, heave and/or volume, as well as mix design and pumping rate. A written log should record the hole number, date and time of each injection stage, pressure, volume and heave. The heave should be monitored by an appropriate survey technique. Laser level monitoring is particularly effective because it leaves the site clutter-free and provides continuous monitoring with automatic sensors, thus decreasing the risk of unobserved movement. Laser monitoring usually needs to be supplemented with occasional optical level readings to pick up accumulated elevation change over several days. Another flexible device for monitoring levels is a manometer, or water level.

Grout pipe installation: Compaction grout pipes should have a minimum inside diameter of 2 in. (5 cm), but should not be larger than 4 in. (10 cm). The pipe should be flush jointed or of single-piece steel construction and must be placed snugly in the ground to avoid leakage or unintended push-out caused by the grout pressure. These effects are particularly troublesome near the surface. The grout hole is then advanced downward through subsequent grout stages without a steel pipe. In downstage grouting it is recommended that the grout casing be pulled and then redrilled to the next stage depth.

Usually pumping sequence dictates that a time frame of 8–10 h is allowed to lapse before redrilling occurs. This also gives the grout stage a chance to gain an initial set so subsequent drilling operations do minimal or no damage to that particular grout stage. Drilling operations for either upstage or downstage grouting are typically handled separately from the pumping operations. While any grouting project requires careful coordination between the drilling and pumping operations, it is particularly crucial that the two operations work very closely with each other when doing downstage grouting. Otherwise it is almost certain that some stages will either be repeated or skipped. A precise logging system will optimize the grouting operation.

Pipe extraction is best handled with a specialized casing extraction (jacking) system operated and handled by the pumping crew (see Section 6.5.4). In some cases, drill rigs or cranes are used but the advantages and disadvantages of their use must be understood. Drill rigs and cranes require large access and overhead areas; therefore when the treatment soil is permeable, grout injection rates can be increased, making the drill or crane-assisted system potentially more productive and cost effective.

Grout injection: Before beginning injection, the grout lines should be pre-wetted to avoid excessive moisture loss from the initial grout front. Grout presence and quality should be confirmed at the injection point before connecting the hose to the grout pipe. Soils found to be dry of optimum moisture content can be pre-wetted to assist grout take by reducing shear stresses in the ground, water loss from the grout as it initiates its compactive effort at the tip of the grout casing.

Monitoring involves the recording of all pertinent information required to analyse volume and flow-rate, as well as the pressure behaviour of any grout interval or stage. The pressure behaviour is the variation of pressure vs time of any grout stage. This should include system accuracy checks by performing volume output tests with varying back pressures on the pump. Field calibrations should be performed anytime the accuracy of the gauge is in question. Gauges should be mounted on the injection header as near as possible to the point of hook-up to the grout pipe. Gauges should have at least a 4 in. (10 cm) face diameter and they should be fitted to a proper gauges saver. Grout volume calibrations should be performed by using large enough containers so variations can be accurately determined. Typically, a known volume form works very well for measurements by volume or weight.

6.4.3 Quality assurance testing

When properly performed, compaction grouting is a self-proofing process. The compactive effort is applied at the point of greatest need since that is also the point of least resistance. Before overburden stresses are overcome, the grout take will be greatest in the loosest soil zones, assuming all cut-off criteria are applied evenly over the entire site. Quality assurance testing can be performed on the finished work if desired, although this is uncommon in remedial work because of difficult access for test rigs and highly variable soil conditions. Site improvement work is more likely to undergo *in-situ* quality assurance testing. This testing usually takes the form of standard penetration test (SPT) or cone penetration test (CPT). In-place density tests can also be run if access to the densified strata is available, but this is rare.

Test cylinders of the compaction grout can and should be taken if strength is an important factor in the project. Be careful not to over-employ this testing method if it is not important to the project.

Standard slump tests should be taken routinely at the pump and at the end of the discharge line. It is important to note that the slump of the material can change by 2 or 3 in. (51 or 76 mm) between the pump and the discharge end of the hose. This phenomenon occurs as water is forced into the soil particles under pressure, thus reducing the amount of free water available for interstitial lubrication.

6.5 Plant

6.5.1 Compaction grout pumps

In the early days of compaction grouting, the equipment was purpose built and only capable of small discharge volumes. Valving was unsophisticated and the equipment frequently malfunctioned. Concrete pumping equipment has improved. Today, most compaction grouting is performed with modified, state-of-the-art concrete pumps. These concrete pumps are generally capable of pumping material at the required high pressures, but not at the low flow rates which are sometimes necessary for compaction grouting. For this reason most concrete pumps must be modified to pump compaction grout at lower flow rates. The most critical aspect of compaction grout pumping (and of concrete pumping) is to avoid mechanisms which allow the pressurized grout material to bleed-off water. The most likely place for this problem to occur is at the pump intake chamber. Most new generation concrete pumping equipment has solved this problem with sophisticated valving.

One crucial aspect of concrete pump modifications is the option to lower its output capability by either sleeving or replacing the material piston cylinders. Most concrete pumps optimize their output by using 6 in. (15 cm) or larger material piston cylinders. Compaction grouting on the other hand, does not require high volume capability. Instead, it requires high sustained pressure control with variable output. Concrete pumps with 80–90 h.p. engines and 4 in. (10 cm) material cylinders work very well, but few pumps are configured this way. Therefore, to do the compaction grouting with the control and pressure capabilities this process requires, it becomes necessary to modify the pump.

6.5.2 Mixing

Delivery and mixing of the compaction grout has developed into three distinct methods. Each method has characteristics which make it more or less desirable for a given project.

The first mixing method is the batch-style pugmill mixer. The compaction grout materials are usually added by hand, but sometimes by machine, to a conventional pugmill-type mixer ranging in size from 5 ft^3 (0.1 m^3) to 1 yd^3 (0.8 m^3). The mixer must be adequately powered to handle the stiff consistency of the compaction grout. Batch mixing is labour intensive at the job site but is well suited for low-volume situations and projects where the grout intake is highly unpredictable, but in the low to medium-volume range of 1–5 yd^3 (0.8–3.8 m^3) per shift.

The second mixing method is the continuous mix batch plant. These units utilize belt-driven feed systems for the aggregate and cement and a continuous screw auger to blend the water with the solid components. These units can be built to accommodate output ranges from 0 to 50 yd^3 (38.2 m^3) per hour or more. The materials are usually loaded mechanically into the batch plant. These units are equipment intensive and work best when the material demands of the project are in the medium to high range of 10–200 yd^3 (7.6–153 m^3) per shift.

The third mixing method is commercial ready-mix concrete trucks. A majority of compaction grout projects are now supplied in this manner since it requires no on-site labour or equipment. Proper mix design and use of retarders can usually allow 'truck time' of 3 h or more for the compaction grout material. Batch plant proximity to the job site, fleet size and stand-by time charges must be considered when ready-mix is an option.

6.5.3 *Drilling*

Much compaction grouting work is done in confined spaces and therefore requiring small equipment. A variety of such equipment has been developed over the years and is best known to contractors involved in the business. For hole depths of 25 ft (7.6 m) or less in relatively soft soils, hand-held rotary percussion equipment can be utilized. Soft to medium soils are good candidates for hydraulic or pneumatic driving equipment. In more difficult conditions, small pneumatic or hydraulic track drills capable of fitting through standard doorways are available. All of these systems utilize driving and/or simultaneous drilling of the casing as the hole is advanced so as to give the desired tight fit between the grout pipe and soil. Pre-augering and later insertion of the grout pipe is undesirable in this class of work since compaction close to the surface is frequently a prime objective.

Medium-sized track and wheeled rigs capable of duplex drilling are well suited for remediation work with reasonable access. The same conditions discussed above should apply for these tools.

With the increasing popularity of compaction grouting as a site improvement technique, larger equipment such as cranes with pile hammers and large bore, high derrick drills are becoming more prevalent. Some soils allow conventional driving of a heavy grout pipe for this application. In non-cohesive soils a water jetting system is sometimes utilized with vibratory or impact hammers to advance the grout pipe. Pre-augering of an undersized heavy casing has been successfully used. This latter method may not be appropriate if compaction at near surface depths of 10 ft (3 m) or less is desirable.

6.5.4 *Extraction equipment*

A range of purpose-built extraction equipment has been developed in the industry for withdrawing the tight-fitting compaction grout casing. Hydraulic cylinders configured to pull the casing from the ground are the most popular extraction systems for remedial work. Drill rigs can be used but this frequently delays progress and slows the overall completion of the project. On site improvement work, a crane with a vibratory hammer or impact hammer is sometimes used in a single-stroke application to both insert and withdraw the grout pipe. Regardless of which extraction technique is used, it is important that it be sized properly and that the injection procedure is conducted properly so as to avoid lock-in of the grout pipe.

6.5.5 *Miscellaneous equipment*

Compaction grout hoses, delivery lines and grout pipes should have at least a 2 in. (5 cm) inside diameter. Grout pipes/casings with larger inside diameters require heavy equipment to handle them during operations. Steel line should be used where possible to minimize friction loss. There is no practical limit to pumping distance other than that dictated by the friction loss generated in the transmission pipe. Any bends in the pipeline should be sweeps of not less than 1 ft (0.3 m) in radius. Pressure gauges should be mounted at the pump and at the injection pipe and should be fitted with appropriate gauge savers. Metering of material can be done by measurement at the source such as ready-mix tickets or dry material tickets, stroke counters on the pump or with magnetic field flowmetres. The latter has not proved to be very reliable due to the nature of the material and the pulsing of the pump cycle.

6.6 Case histories

As discussed in Section 6.3 the applications of compaction grouting are numerous and varied. The case histories selected for this chapter are typical of some of these applications but they also leave many applications untouched.

6.6.1 *One Woodway Plaza, Houston, Texas: 1997*

Over several years, a four-storey office building had undergone major settlement of all of its foundation elements. Built in the early 1970s on spread footings, deep grade beams and drilled caissons, the structure sits on 30 ft (9.1 m) of construction rubble fill, beneath which are competent soils. A significant part of the rehabilitation programme involved compaction grouting 33 ft (10 m) deep to stabilize the fill and lift spread footings back to their original elevation. Work was accomplished at night, with more than 3600 yd^3 (2752 m^3) of grout pumped through 467 low-headroom, interior locations. The settled footings were raised up to 8 in. (20.3 cm). In addition, drilled caissons were underpinned with micropiles, and an anchored retaining wall was constructed to stabilize a failed MSE (mechanically stabilized earth) wall (Figure 6.7).

Figure 6.7 Compaction grouting stabilized the rubble fill underlying a four-storey building and successfully restored the building to near original elevation.

6.6.2 La Reina Building, Hollywood, California: 1995

The La Reina Building is a six-storey glass and steel office complex founded on large spread footings. The building sits some 80–90 ft (24–27 m) directly above the alignment of a new, twin-tube subway tunnel. Compaction grouting was used to protect the building against settlement resulting from foundation soils being loosened during tunnelling. As the tunnelling machine passed beneath the building, grouting was initiated just following the advancement of the tunnel shield and expansion of the tunnelling pre-cast segments. The complex array of 150 precisely angled compaction grout pipes were positioned within 5 ft (1.5 m) of the tunnel crown. Gyroscopic survey of installed pipe tip locations and as-built CAD drawings, aided the critical sequencing of tunnelling and grouting (Figure 6.8).

6.6.3 LRT Extension, Morena Segment, San Diego, California: 1996

In the Mission Valley area of San Diego, three light rail transit bridges are supported by individual piers. The piers bear on 9 ft (2.7 m) diameter caissons up to 130 ft (40 m) deep. These caissons are founded in dense sands and gravels underlying potentially liquefiable soils. Although the caissons are founded below the zone of liquefaction, they rely on support from the surrounding soils for lateral stability. Prior to bridge construction, compaction grouting was performed to depths of 45 and 115 ft (13.7 and 35 m) around six abutments and 68 caissons to densify and reinforce the soils, mitigating their liquefaction potential and thereby ensuring the long-term

Figure 6.8 Compaction grouting through pre-placed pipes prevented tunnelling induced settlements during subway construction beneath this building on Hollywood Boulevard.

protection needed for the caissons and the bridge superstructure in the event of an earthquake (Figure 6.9).

6.6.4 Hampton Inn, Albuquerque, New Mexico: 1995

In the 5 years since construction, a Hampton Inn had settled almost 2.5 in. (6 cm). Data indicated that the moisture contents of the upper 20 ft (6 m) of soils were now higher than at the time of construction. Modelling tests performed indicated a soil collapse potential of 5–7 in. (13–18 cm). Compaction grouting was performed to varying depths at 150 locations to

Figure 6.9 Pre-treatment with compaction grouting for long-term liquefaction mitigation protected the integrity of caissons supporting light rail piers.

target the higher moisture content soils, with the majority of the work done within the occupied building in limited headroom. Total grout take for the project represented 14.5 per cent of soil volume for the treated zone. Post-grouting survey results indicate that movement of the structure has slowed to a rate of less than 0.0625 in. (1.6 mm) per year (Figure 6.10).

Figure 6.10 Remedial compaction grouting successfully treated collapsible soils beneath this hotel, reducing future settlement potential to an acceptable level.

6.6.5 Paiton Private Power Complex, Paiton, Indonesia: 1995

Before completion of blocks 7 and 8 of the power complex, extensive soil improvement was needed. Compaction grouting offered an exceptional outcome in relation to the design criteria of sufficient bearing capacity able to withstand an earthquake. To prevent an operation stoppage due to damages on buildings and pipelines, certain settlement should not be exceeded during operation of the plant (Figure 6.11a).

When examining the degree of soil improvement, SPT-values of $N > 15$ were required in sand after treatment, in order to reduce the existing liquefaction potential in the soil and to ensure an allowable bearing pressure of 140 kPa. SPT-values of $N > 25$ were required for the boiler platform of block 8 with a foundation pressure of 280 kPa. Due to the very heterogeneous ground, and in order to control the soil improvement for the very silty soils, an adjustment factor was necessary. For each SPT-test that did not reach the values of $N > 15$ or 25, the content of fines was established, and the SPT-values were classified (Figure 6.11b). The subsoil and soils with a relatively high content of fines (Figure 6.11c) showed considerable improvement. No extraordinary heave was detected.

Figure 6.11 (a) Excavation trench at power complex site. Left: Non-compacted soil collapses to form a natural slope of roughly 45°. Right: Compacted soil dense enough to stand nearly vertical (Wegner, 1997; courtesy of Technische Universität Darmstadt).

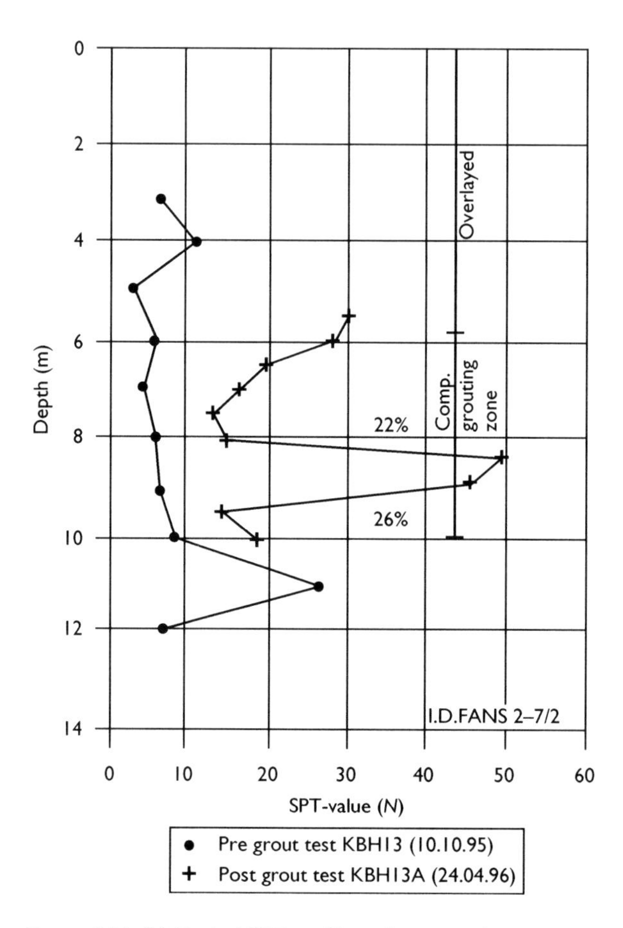

Figure 6.11 (b) Typical SPT-profile with pre- and post-test and the contents of fines for SPT $N < 15$ (Wegner, 1997; courtesy of Technische Universität Darmstadt).

	Content of fines – passing sieve # 200 mm							
	<15%		15–35%		35–50%		>50%	
	Before	*After*	*Before*	*After*	*Before*	*After*	*Before*	*After*
Minimum	2.0	15.0	2.0	5.0	3.0	5.0	2.0	4.0
Average	6.5	20.1	4.7	11.2	5.5	10.6	3.9	7.5
Maximum	15.0	40.0	13.0	18.0	10.0	20.0	12.0	24.0

Figure 6.11 (c) Average of achieved SPT-values with increasing fines content of the soil.

Bibliography

Baker, W.H., MacPherson, H.H. and Cording, E.J. (1980) Compaction grouting to limit ground movements: instrumented case history evaluation of the Bolton Hill subway tunnels, Baltimore, MD *Report No. UMTA – MD-06-0036-81-T*, US Department of Transportation, Washington, DC.

Baker, W.H., Cording, E.J. and MacPherson, H.H. (1983) Compaction grouting to control ground movements during tunneling, *Underground Space*, Vol. 7, Permagon Press, pp. 205–212.

Bandimere, S. (1997) Compaction Grouting, State of the Practice, 1997, *1997 GeoLogan GeoInstitues Convention*, Utah State University, Logan, Utah.

Brown, D.R. and Warner, J. (1973) Compaction grouting, *Journal of the Soil Mechanics and Foundations Division, ASCE*, Vol. 99, No. SM8, Proceedings Paper 9908, pp. 589–601.

Graf, E.D. (1969) Compaction grouting technique, *Journal of the Soil Mechanics and Foundations Division, ASCE*, Vol. 95, No. SM5, Proceedings Paper 6766, pp. 1151–1158.

Henry, J.F. (1986) Low slump compaction grouting for correction of Central Florida sinkholes, *Proceedings, National Water Well Association Conference*, Bowling Green, KY.

Henry, J.F. (1987) The application of compaction grouting to karstic foundation problems, *Proceedings, 2nd Multidisciplinary Conference on Sinkholes*, Orlando, Florida.

Mitchell, J.K. (1970) In-place treatment of foundation soils, *Journal of the Soil Mechanics and Foundations Division, ASCE*, Vol. 96, No. SM1, Proceedings Paper 7035, pp. 73–110.

The Committee on Grouting of the Geotechnical Engineering Division. ASCE (1977) Slabjacking-State-of-the-Art, *Journal of the Geotechnical Engineering Division, ASCE*, Vol. 103, No. GT9, Proceedings Paper 13239, pp. 987–1005.

Warner, J. (1978) Compaction grouting – a significant case history, *Journal of the Geotechnical Engineering Division, ASCE*, Vol. 104, No. GT7, Proceedings Paper 13897, pp. 837–847.

Warner, J. (1982) Compaction grouting – the first thirty years. *ASCE Specialty Conference, Grouting in Geotechnical Engineering*, New Orleans, Lonisiana. pp. 694–707.

Warner, J. and Brown, D.R. (1974) Planning and performing compaction grouting, *Journal of the Geotechnical Engineering Division, ASCE*, Vol. 100, No. GT6, Proceedings Paper 10606, pp. 653–666.

Wegner, R. (1997) Compaction grouting-soil improvement in complicated soil conditions for the foundations of a large thermal power complex in Indonesia, *4th Darmstadt Geotechnical Conference*, 13 March.

Welsh, J.P. (1988) Sinkhole rectification by compaction grouting, *Proceedings of Geotechnical Aspects of Karst Terrains*, Geotechnical Special Publication No. 14, ASCE. New York, NY, pp. 115–132.

Chapter 7

Soil fracturing

E. Falk

7.1 Introduction

The phenomenon of hydraulically fracturing soils was initially observed as an undesirable by-product of traditional grouting measures. The uncontrolled propagation of soil cracks which were rapidly filled with suspension did not achieve the objective of homogeneously filling cavities in unfixed soils. In the past, high-suspension pumping rates which no longer allowed the suspension to penetrate the pore system of the soil continuously were regarded as a grouting defect. It was crude oil technology using hydraulic soil fracturing for increasing the permeability and thus the yield of oil fields that provided the impetus for systematically applying geotechnical methods for utilising deliberately produced cavities in the soil. In the meantime, fracture grouting has been used for systematically improving soil properties. The load-bearing capacity and permeability of both granular and cohesive soils can be modified by incorporating a cement or solid matter skeleton. The repeated application of this method also permits the controlled raising of buildings with very different support systems. The most spectacular use of the method is found in connection with the complex tasks of compensating for settlements which, as a result of tunnel driving, endanger the continued existence of buildings above the tunnel.

7.1.1 Characteristics of fracture grouting

High-viscosity grouting agents are introduced through valves installed in the ground in such a way that the sum of the reachable cavities in the surrounding soil accommodates only a small percentage of the amount of liquid introduced. As far as equipment is concerned, this method requires mixers for producing suspensions rich in solid matter content and pumps which achieve a sufficiently high-pressure build-up inside the suspension which is accumulating in the soil. After the fracturing pressure in the soil has been exceeded, cracks open up in the soil which are widened immediately by the subsequent grout. By injecting small amounts of solid substance per

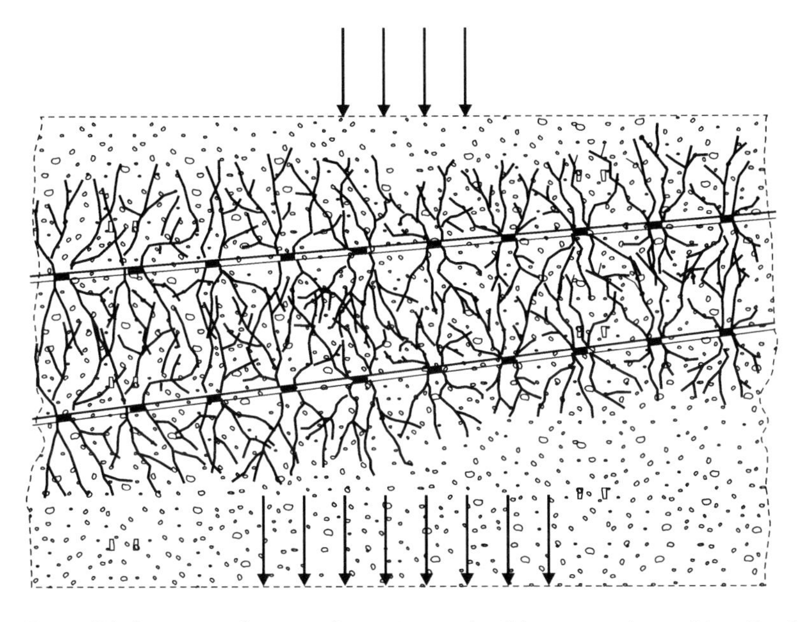

Figure 7.1 Supporting framework consisting of solid matter veins and lamellae for transferring concentrated stresses.

grouting operation and by repeatedly pressurising individual grouting valves, it is possible to achieve a grout framework of hardened solid matter veins and lamellae (Figure 7.1).

The membranes as described have an irregular shape and a medium thickness which can range from just a few millimetres to several centimetres. Soils with large cavities may need to be pre-treated in order to achieve a pressure build-up and, subsequently, to be able to carry out the process of hydraulic fracturing. In order to control the development of 'Fracs' in the ground regarding their length, grout volumes are strictly limited and the flow characteristics of the grout are controlled by the use of accelerators.

7.1.2 Construction and technical aspects

The use of hydraulic fracturing in construction technology has to meet some important pre-conditions:

1 Basic project assumptions: The achievable modifications of the soil parameters have to be assessed just as realistically as the geometric relations of foundations subjected to loads, which permit systematic raising of a structure. Two to five-fold increases in stiffness can be achieved.

2 Performance description: The performance description on which the works contract is based has to take into account the actually expected works progress, with observation times and work interruptions playing a part for organisational reasons.

3 Time factor involved in the course of the construction work: A permanent soil improvement by hydraulic fracturing is achieved when the number of grouting phases is large. In consequence, the desired effect only occurs after the passage of a considerable amount of working time. For compensation grouting it is particularly important to include in the time schedule a suitable period for installing the grouting system and for pre-treating the soil up to the point in time when it is ready to be lifted.

4 Measuring technology: The method is controlled by means of a highly developed measuring technology which makes it possible to observe both surface movements as well as any deformation in the subsoil. The reliability of the measured values and their early evaluation are essential pre-conditions for a successful application of the method.

5 Whereas, with hydraulic fracturing, the soil is improved in small stages and with the objective of achieving a permanent increase in the lateral soil resistance, only those tests can provide conclusive evidence which, both in terms of time and space, reflect the geometry of the actual subsequent application.

6 Application limits: After very promising results have been achieved in connection with raising buildings and compensating for settlements during the last two decades, there exists a tendency to exceed previously known application limits. However, an essential element of a successful application consists in observing maximum injection rates per soil unit and working day. As far as economics are concerned, it has to be taken into account that a sensible decision on the usefulness of the process application can be taken only if the prevented damage and its frequently complex consequential effects are realistically assessed.

The purpose of the following chapter is to make available relevant principles on which decisions on the subjects mentioned can be based and which can be adapted to the respective specific project conditions. The following examples should be regarded as an overview of those areas of application where hydraulic fracturing has been used in recent years.

7.1.2.1 *Case studies*

7.1.2.1.1 MODIFICATION OF SOIL STIFFNESS

(a) Homogenisation of foundation soil underneath a machine foundation (Figure 7.2).

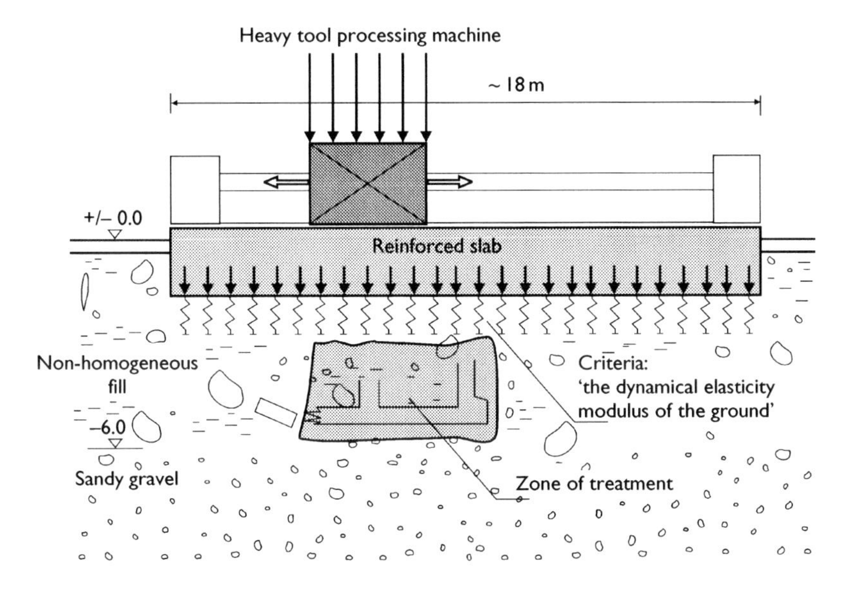

Figure 7.2 Consolidation of a demolished, inadequately refilled basement underneath the foundation of a precision machine, with the objective of homogenising the dynamic deformation modulus of the soil.

7.1.2.1.2 STABILISATION OF LONG-LASTING OR CURRENT DIFFERENTIAL SETTLEMENTS

(a) Different settlement rates of a church which had continued over centuries and which, eventually, constituted a safety problem (Figures 7.3a,b).

7.1.2.1.3 COMPENSATING FOR DIFFERENTIAL SETTLEMENTS WHICH HAD ALREADY TAKEN PLACE

(a) Straightening of structures which were affected by settlements in the vicinity of deep excavation pits or by tunnel driving (Figure 7.4).

7.1.2.1.4 ACTIVE COMPENSATION OF DIFFERENTIAL SETTLEMENTS AS THEY OCCUR

(a) Compensating for settlements above cavity structures which, in traditional excavation methods, are often built in several phases (Figures 7.5a,b,c).

(b) Conditioning and compensation for settlements which occurred as a result of driving with tunnel-boring machines (TBM).

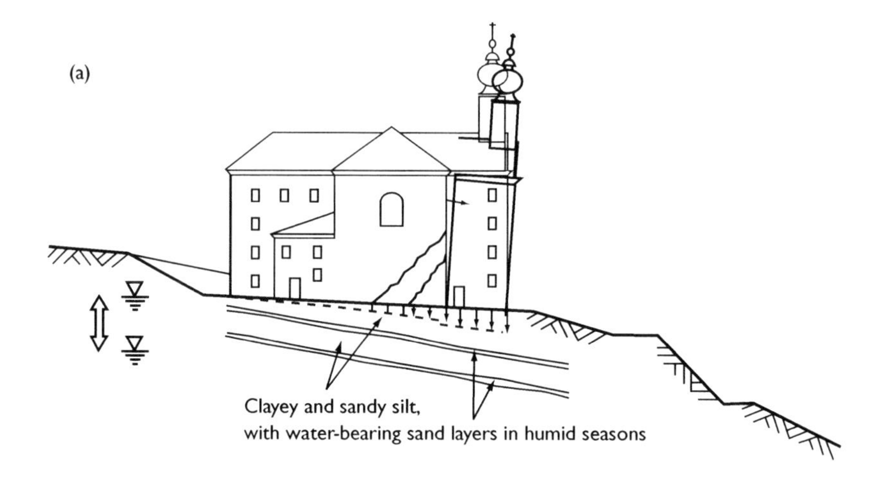

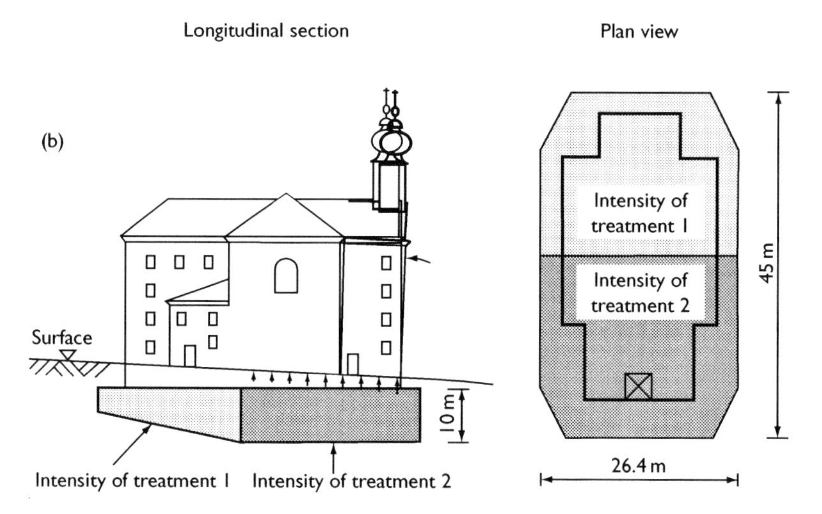

Figure 7.3 (a) Shrinkage processes caused by cyclic drying of the soil lead to differential settlement occurring in stages; (b) The intensity of the stabilisation process by hydraulic fracturing is adapted to the existing loads.

7.1.3 Existing references

As far as available documentation is concerned, it should be noted that, because of the spectacular results, a large part of the spectacular cases of the application of compensation grouting have found access to international technical conferences, but the basic project assumptions still form the subject of discussions by experts. This is the reason why the attached references

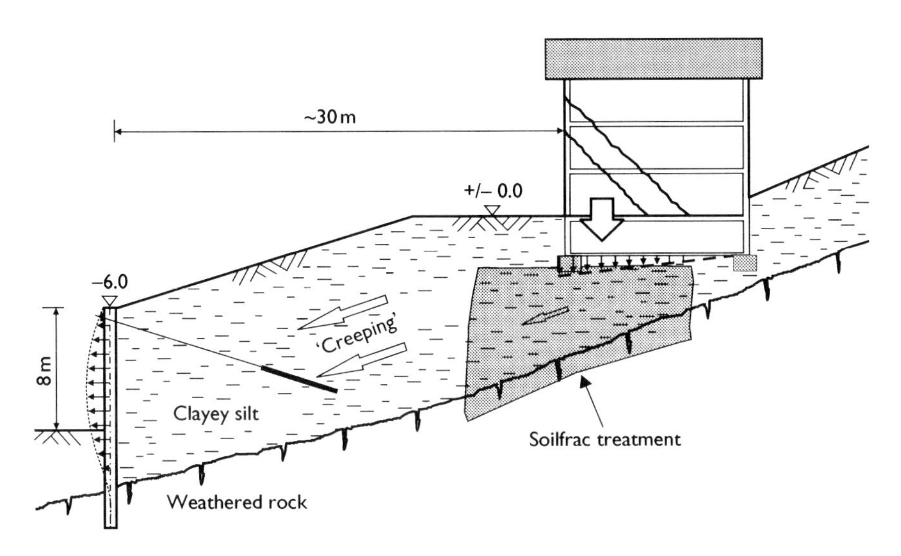

Figure 7.4 Stopping creep movements after deformation in the retaining structures had led to cracks in existing structures.

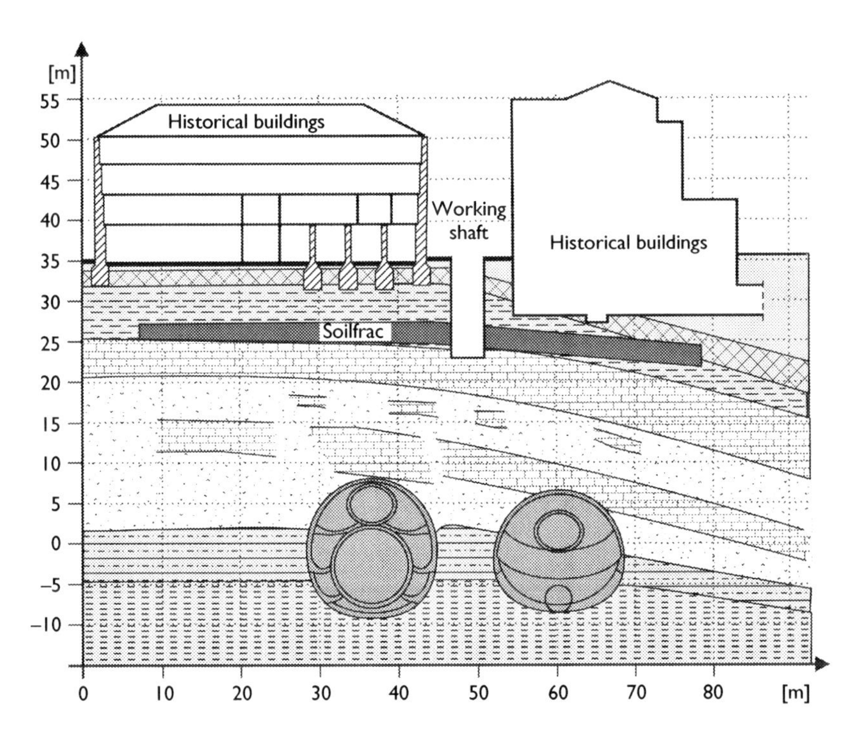

Figure 7.5 (a) Compensating for settlements above two station buildings for an underground railway which was built in stages over a period of approximately 2 years.

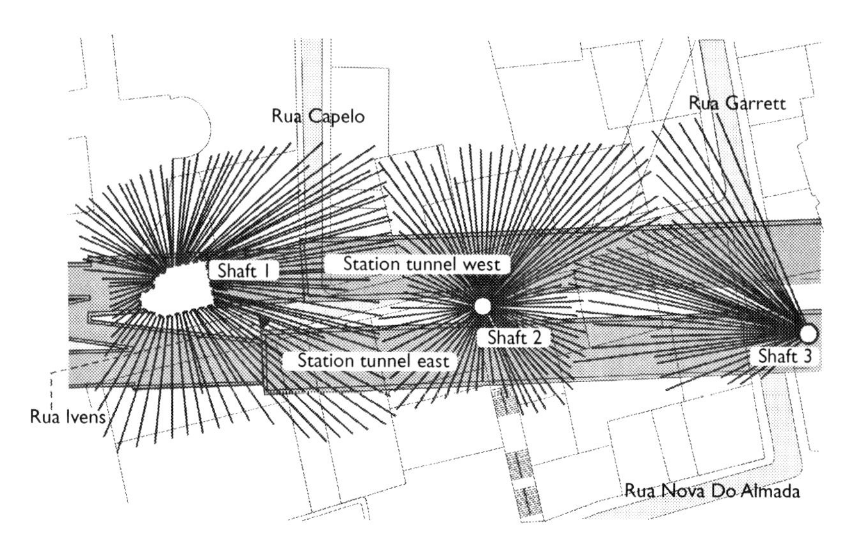

Figure 7.5 (b) Arrangement of shafts and drillings for the installation of tubes à manchettes for compensating for settlements in a densely built-up area of approx. 15 000 m^2.

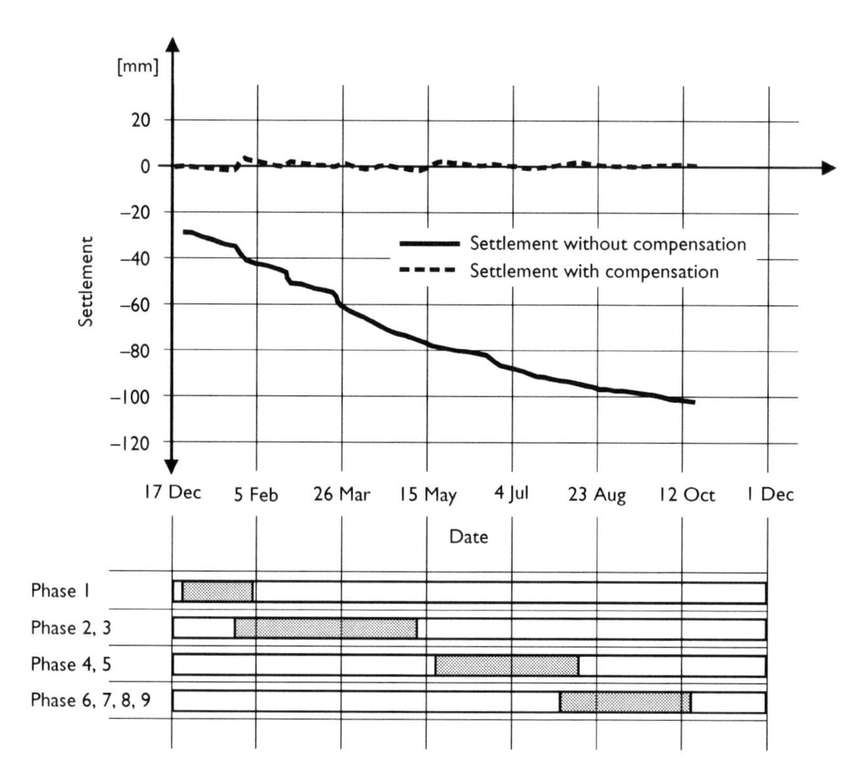

Figure 7.5 (c) Illustration of the average compensation success, indicating the compensation procedure in small stages during the individual excavation phases.

contain publications which deal with the further development of mathematical equations, in particular involving numerical methods, current improvements in measuring technology and well-founded views on the interaction resulting from forced deformation between the soil and the structure.

7.1.4 Basic project assumptions

In principle, it has to be stressed that hydraulic fracturing consists in imposing soil deformation by injecting grout rich in solid matter. The criterion for effectively applying the method consists in being able to control the movements in the soil as well as the interaction between such movements and the structure concerned! The more or less complex requirements demand a close inter-disciplinary cooperation between geotechnology, structural analysis, structural process technology and measuring technology. A pre-condition for effectively applying the method therefore consists in the data of different disciplines being compiled by a structural engineering department which is competent in these disciplines and which is capable, responsibly, to take decisions within the framework of the application of an observational method.

7.2 Historical development

After the suitability of the phenomenon known from the traditional grouting technology and of the equipment developed in oil fields to solve geotechnical problems had been recognised, the lifting method using cement suspension was first used in Essen for the purpose of raising a coke furnace, and described by Bernatzik (1951). Essential steps for improving the method and for widening its range of application consisted in adapting the valve pipe technology (approx. as from 1970) and in integrating electronic data processing which allowed the measured values to be shown in real time. As from 1986, it has been possible to compensate actively for settlements inside parallel settlement troughs.

In the meantime, the method has spread to many geographic locations. It is known to have been used in Germany, Italy, Austria, The Netherlands, Portugal, Spain, Belgium, Great Britain, USA, Canada and Puerto Rico.

7.2.1 Further technical developments

7.2.1.1 Settlement prognosis and risk analysis

Ever since experience with inner-city tunnel construction in recent decades highlighted the considerable extent to which environmental damage can influence the total costs of such measures, systematic settlement analysis has occupied an important place within overall project planning. This includes a detailed examination of the condition of existing structures in the area at risk and an evaluation of possible damage within the framework of a comprehensive risk analysis. It is on the basis of these data that the decision is taken as to the stages at which additional measures are applied economically to supplement settlement-reducing measures in tunnel driving itself.

7.2.1.2 *Sleeve pipe technology*

The improvement in the grouting system consisting of long-life valve pipes and double packers makes it possible to pressurise individual sleeves many times, to use long pipe lengths and, if necessary, to carry out controlled drilling operations. High-quality pump control devices do not only allow a large number of decisions to be made on parameters, but also allow these to be recorded and automatically presented. High-performance drilling methods and the use of flexible drilling assemblies allow the use of shafts and also the adaptation of existing working areas such as excavations or tunnels, which permits the method to be used in confined inner-city spaces as well (Figure 7.6).

7.2.1.3 *Soil fracturing combined with other methods*

To meet difficult project requirements and to find economical overall solutions, the method has already been developed and used in combination with

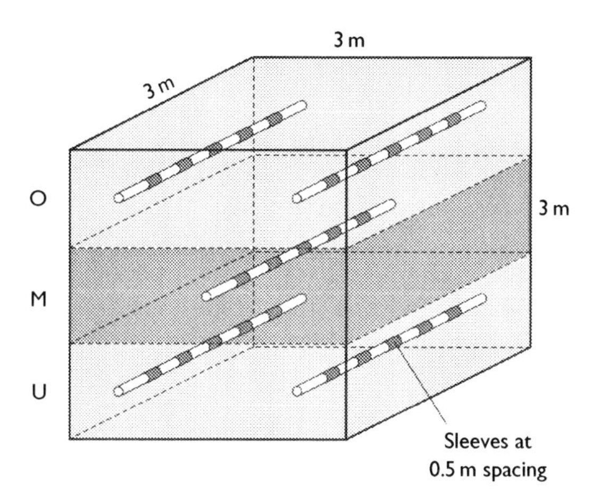

Figure 7.6 Sleeve pipes can be arranged on several levels and can serve different purposes. This figure shows a detail of a lifting mat used in compensation grouting with two horizons 'O' and 'U' for the purpose of distributing stress and a central lifting layer 'M'.

several geotechnical methods: compaction grouting, jet grouting, umbrellas, hydraulic lifting, floating pile foundations.

7.2.1.4 Data processing

Recording and storing of data obtained in grouting itself and from the monitoring-based observations are not sufficient for ensuring a professional application of the method. Only the use of professional visualisation programs and the combination of data by means of individual software modules make it possible for the site manager to take decisions, at the required speed, on grouting programmes which have to be modified continuously (Figure 7.7).

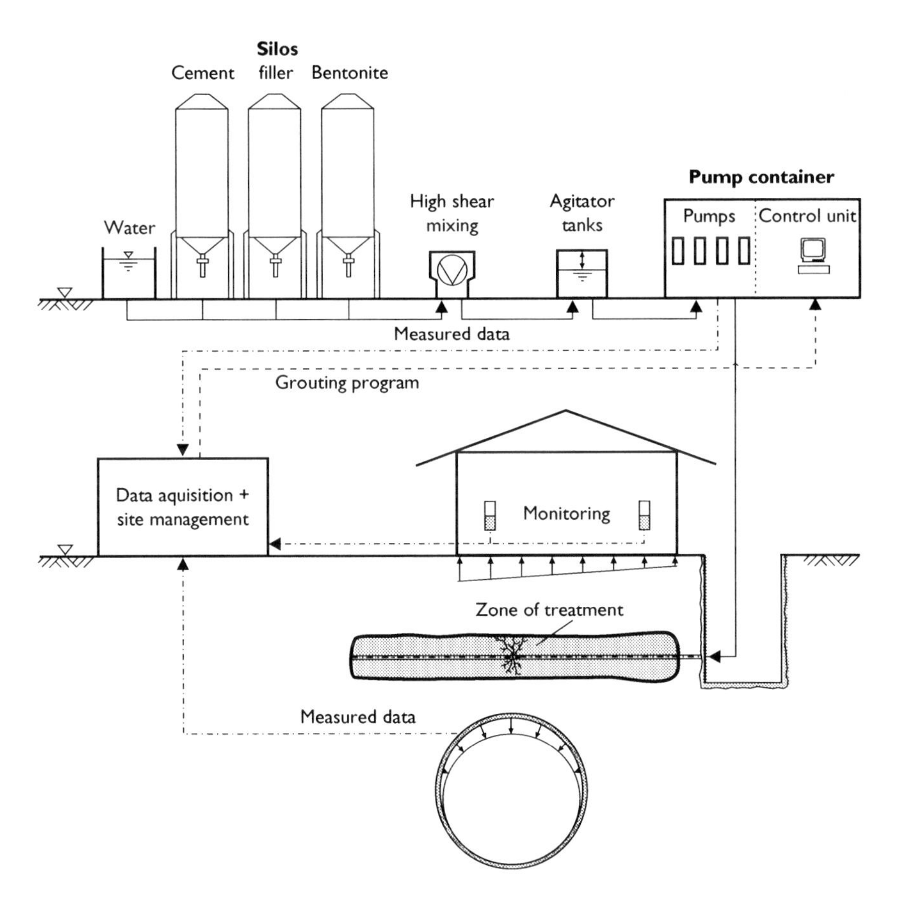

Figure 7.7 Data on deformation measurements and parameter recording are used by the site manager for determining actual grouting programmes.

7.3 Equipment

The following describes important parts of the equipment suitable for systematically applying the hydraulic fracturing method. The requirements are formulated exclusively in terms of characteristic functional requirements and do not deal with market-specific details.

7.3.1 Installation of grouting system

7.3.1.1 Drilling technology

The sleeve pipes can be installed under the protection of mud-flush drilling or fully cased boreholes. Both the rotary impact drilling method using a down-the-hole hammer and ramming methods with an external hammer are used. In soft soils it is also possible to apply worm drilling. The drill spoil material is conveyed by air pressure flushing or water flushing. In special cases it is possible to use directional drilling where long drilling lengths are required and in areas which are difficult to access. In the case of the double packer technology, it is advisable to limit the pipe lengths to approx. 50 m, although borehole lengths in excess of 70 m have also been successfully achieved. It should be noted that in the case of extended drilling lengths, the general application problems increase disproportionately. In the case of several controlled 100 m boreholes, the double-packer technology can be replaced by single-valve technology (Figure 7.8).

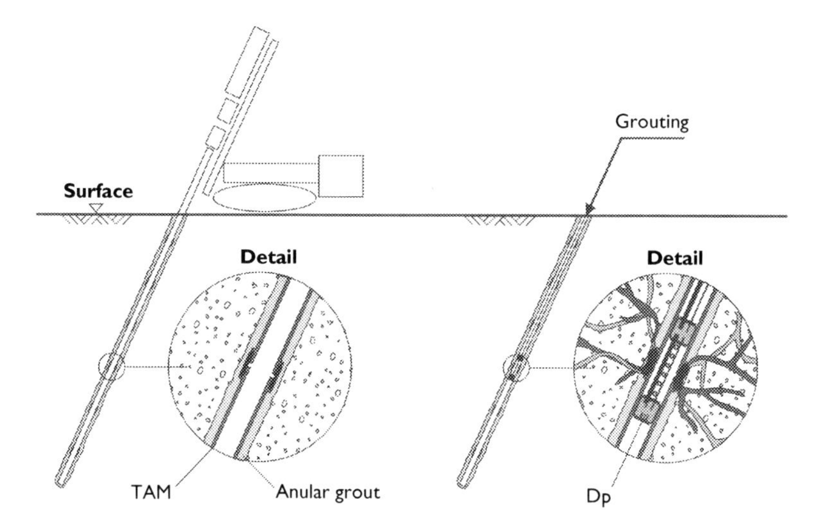

Figure 7.8 Sleeve pipes TAM are installed under the protection of a stable and cased borehole and sealed by an annular grout. Individual valves can be pressurised by double packers (Dp).

7.3.1.2 *Borehole set-ups*

Drilling masts for economically producing boreholes in excess of 20 m comprise lengths of at least 4 m, but preferably approx. 5 m. Any drilling shafts, ditches and working tunnels should be designed to be at least 1 m wider than the length of the drilling masts. An orderly discharge of cement-containing flushing water has to be included in the works schedule, just like the safe introduction and lifting of the drilling and grouting equipment. All safety technical requirements have to be observed during the different operating phases and are normally combined in a safety schedule which is pointed out to those participating in the project prior to the commencement of work. If shafts have to be located in the direct vicinity of the area where the raising operations take place, it may be necessary to provide a dilatation joint, if necessary even underneath the water table (Figure 7.9).

7.3.2 *Preparation of grout suspensions*

7.3.2.1 *Storage system*

The storage system must allow separate storage of the major components required for mixing suspensions. It is necessary to provide steel or plastic tanks for storing water in drinking-water quality, silos with a capacity of at least 20 tonnes for bonding agents (cement, lime, fly ash) and filler material (limestone, slag, bentonite) as well as small containers for additives.

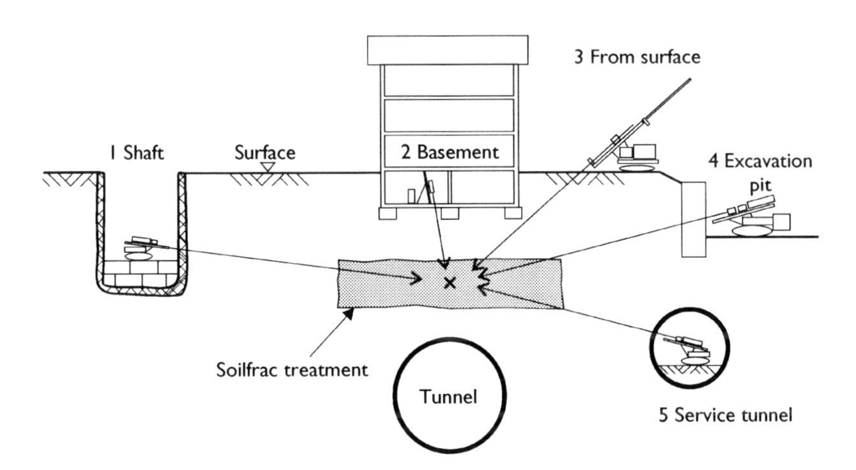

Figure 7.9 As far as the drilling geometry is concerned, a large number of options is available for reaching certain regions in the soil to be grouted.

7.3.2.2 Mixing technology

It involves the use of so-called colloidal mixers or high-frequency mixers which still allow the homogeneous mixing of suspensions with a water/solids ratio of 0.45. Even with grouts with high solid content and bentonite added, the mixing capacity has to ensure an adequate supply of material to the proposed number of pumps (Figure 7.10).

7.3.2.3 Stored quantity of suspension

Not too large a quantity of suspension, say 500–1000 l, is stirred in agitator tanks with electronic filling level control means (multirangers) to ensure that, on the one hand, grouting units are supplied continuously and that, on the other hand, larger suspension quantities are prevented from being 'stirred dead', should there be any interruptions in the grouting operation.

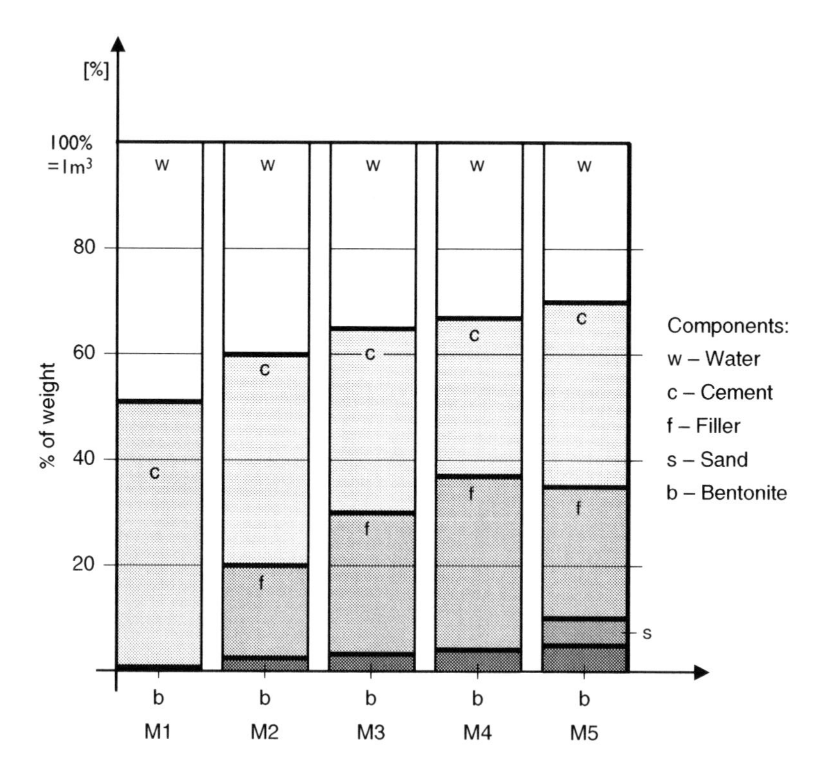

Figure 7.10 Examples of suspension compositions which can be used in different soils and operating phases.

7.3.3 Grouting technology

7.3.3.1 Grouting pumps

Modern grouting pumps are provided with different control options, thus permitting an automatic reaction to the development of pressure at a constant or variable pump rate. The pumps have to be suitable for a pressure range of 0–100 bar and a pump rate of 1–20 l per minute; any parameters set within these ranges are expected to be kept constant even for suspensions with a high solid content. In modern grout modules it is a common practice to combine 2–8 pumps of similar design. The essential pump parameters are either graphically recorded immediately or stored electronically and, via software programs, processed in databanks and printed out. In each and every case, the type of data saved has to ensure that, in each individual grouting operation, the pressure and quantity ratios must be clearly associated with the respective location of the grouting operation. This is the reason why modern grouting data recording programs and visualisation programs are coupled; they permit an early interpretation of data (Figures 7.11a,b).

7.3.3.2 Sleeve pipes

Pipes available on the market consist of PVC or steel; their diameters range from 1 to 4 in. and the distance between valves amounts to 0.33–1 m. In special cases, the steel pipes are reinforced and the rubber valves are protected by steel rings in such a way that either the pipes can be rammed in directly or that the function of a reinforcement of the soil, for example the shape of a pipe umbrella, is supported. In principle, by surrounding the grouting pipes with a so-called skin-forming mixture (stable, low-strength mixture, but clearly stiffened) a direct connection between the individual grouting valves is prevented.

7.3.3.3 Packer system

Double packers are used to close the grouting pipe before and after the aimed grouting aperture (Figure 7.8).

The packer elements consist of wire mesh and have a rubber surface, and are inflated either with compressed air or water. The pressure applied has to be clearly higher than the maximum grouting pressure expected. After completion of the grouting operation, the grouting pipes are cleaned by high-pressure hoses integrated into the packer system or by separate cleaning systems. The quality of the individual components and the care taken in using these are the key to an efficient process application and to being able to use the system over longer periods of time.

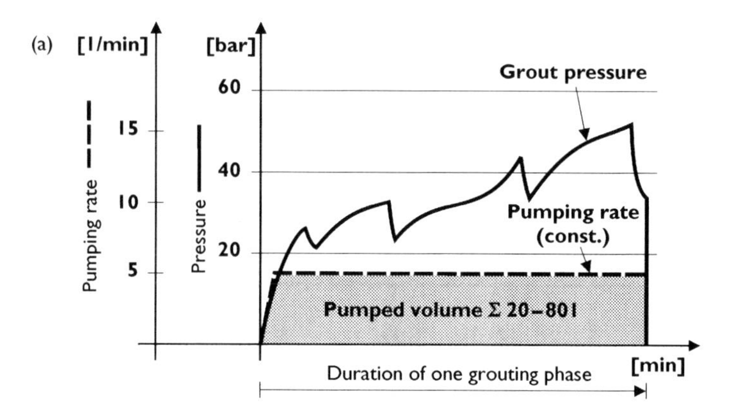

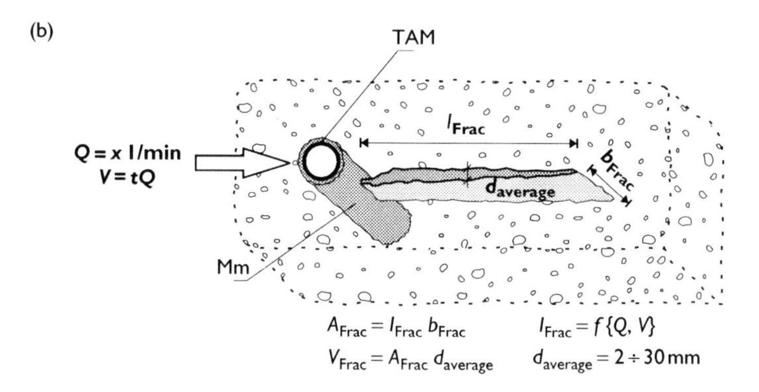

Figure 7.11 (a) Typical injection graph showing a drop in pressure after the occurrence of 'fracs'; (b) A frac usually has a flat, oblong shape and starts from an initial crack in the skin-forming mixture. However, frequently there occur secondary fracs whose geometric description can be given in a statistical form only.

7.3.3.4 Data recording

A compensation grouting operation generates a huge amount of data which have to be suitably administered and made available to the site management which has to come up with a suitable decision. For instance, there have been applications in the past in the course of which more than 100 000 individual grouting operations and the continuously measured values of more than 100 measuring elements had to be recorded over a period of more than 1 year. Associating the data in terms of time and place was just as important as combining the effects which start from the construction of cavities and from the grouting measures and influence individual regions of the structures positioned above. The object is therefore to aim at a simplified presentation of the structures and the soil in a form which shows the interaction between cavity, grouting operation and structures. When designing the measuring system, care has to be taken to ensure that the actual measuring accuracy and the frequency of data recording indicate movement tendencies. Important external influences such as temperature fluctuations have to be filtered out. Valuable tools in the application of compensation grouting have been found to be visualisation systems which, for example, at any time, show the development of pressure ratios during grouting or the distribution of quantities inside physically limited units.

7.4 Planning and theory

Injecting solid matter into the soil leads to deformation on all sides. The directions in which individual injections spread largely depend on the homogeneity conditions in the soil. From the point of view of statistics, however, it has to be assumed that the greatest part of the volume introduced into the soil leads to deformations whose amounts are distributed proportionally relative to the respective stress conditions. Only a small part of the movement rates is caused by the compression of the existing soil, as the highest effective injection pressure cannot greatly exceed the magnitude of the highest existing standard stress. However, locally and over a short period of time, higher forces can become effective if tensile forces within the grout skeleton, which have already been set, and cohesion forces in the soil are activated (Figure 7.12).

Experience has shown that if a limit speed in respect of the injected quantity per unit time within a limited soil volume is exceeded, said adhesion forces are overcome and, in consequence, the time-dependent deformation resistance is clearly reduced. This conclusion is based on the

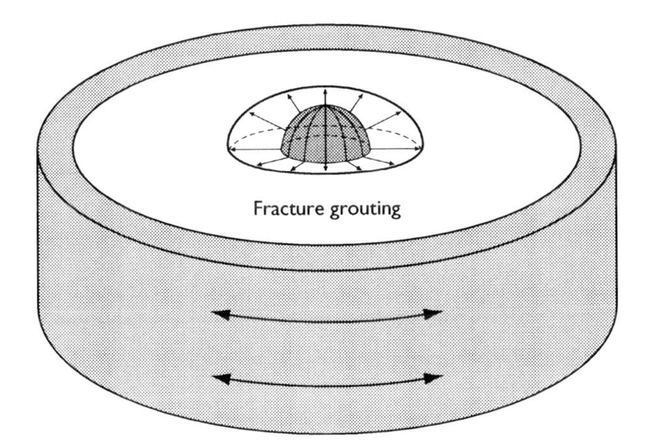

Figure 7.12 The model image of a 'tension ring' comprises the sum of all forces which permit a central lifting injection.

observation that the efficiency of injected quantities is reduced in the course of a lifting operation, if, in order to achieve a greater lifting speed, the quantity injected per working day is increased excessively. The above-mentioned limit balance between effective injection pressures applied (which is not the injection pressure measured at the pump) and the annularly acting pressure forces in the surrounding soil depends on the respective soil characteristics to such an extent that global recommended values for reliable injection quantities cannot reasonably be given. Instead, it is advisable to monitor limited injection areas by making use of the available measuring technology, and to obtain clarity on the achievable efficiency and lifting speed by varying the injection parameters (Figure 7.13).

Clear information on the movement rates and their directions as a result of the injection of solids cannot only be obtained by observing surfaces. Information from the soil is essential, with extensometers and inclinometers being able to provide useful services. For the purpose of checking the accurate arrangement of geotechnical measuring instruments it is safe to observe the rule that, in principle, all quantities of solid material introduced into the soil have to be identifiable as movement rates in the soil.

7.4.1 Soil improvement by hydraulic fracturing

A permanent improvement in cohesive soils is achieved by producing a continuous supporting framework consisting of a hardened solid matter skeleton. Since it is only the homogenisation of stress conditions in the soil and the closure of the solid matter skeleton which permit a supporting effect which is independent of the original soil, the improvement curve of a multi-phase injection application is not to be regarded as being linear by any means. Any lifting which does not occur immediately after the end of the injection operation is a clear indicator for achieving the desired improvement. The slab stresses of up to approx. 1000 kN/m^2 usually occurring in structures, if at all, can only be transferred into the ground by soils improved by soil fracturing. If the values are higher, it is possible to transfer stresses in the range of 2.0–3.0 MN/m^2 with negligible deformation rates. Borderline cases are soft soils. Whereas effective consolidation has already been achieved in peat layers, there are no examples yet for very soft and structureless types of soil, although in such cases, too, the selection of a suitable injection agent promises success.

Soil improvements by hydraulic fracturing are always carried out with the intention of generating a controlled stress flow in the ground. Because of the low material-strength values of the soil and the intention to include said soil in a supporting system, it is frequently necessary to increase the existing soil stiffness by 2–5 times the existing value.

Direction of Deformation

Phase	Horizontal (H)	Vertical (V)
Pre-treatment	100%	0%
Multi-stage injection before actual heaving	95 ÷ 100%	0 ÷ 5%
Heaving phase	75 ÷ 95%	5 ÷ 25%

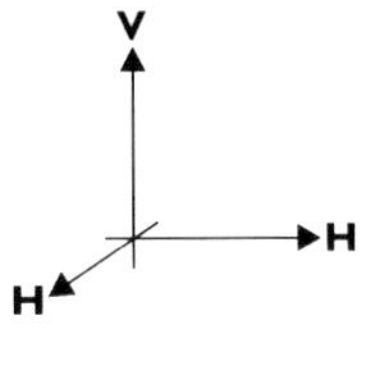

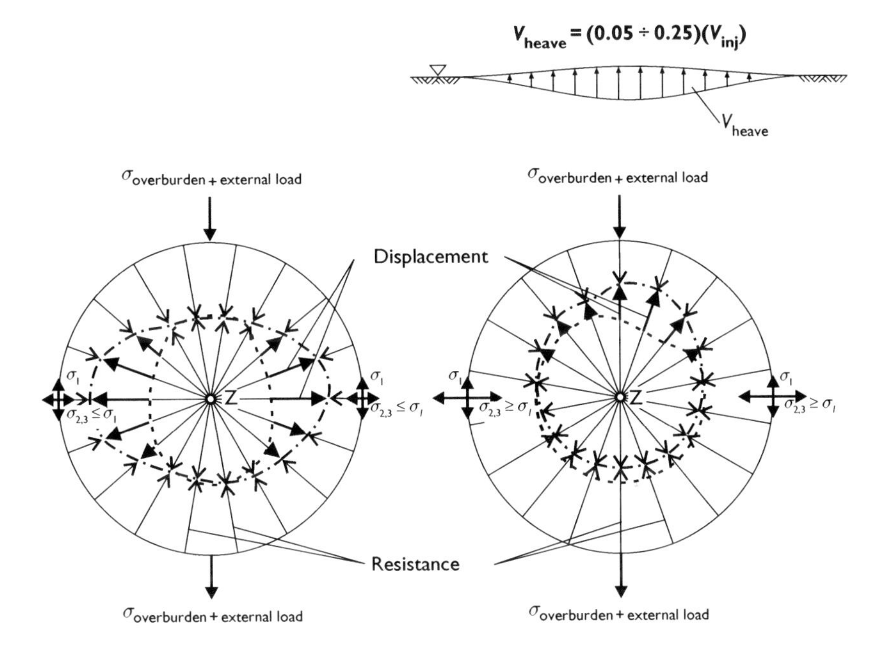

Figure 7.13 Inhomogeneities in the soil and the stress distribution influence the deformation direction which is the result of an imposed addition of material and all-round resistance.

As the deformation method described involves the displacement of large soil masses by small individual amounts, the respective plans must take account of the principle that slim deformation elements should not be considered in planning such methods. Figure 7.14 shows basic geometric relations which result from experience with different lifting operations. Needless to say, in such considerations the different load intensities of adjoining foundation faces and the different depths at which they are located also play an important part.

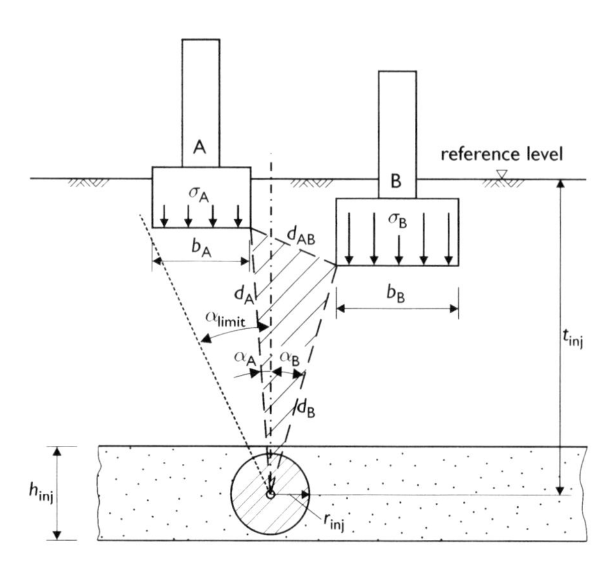

Figure 7.14 Recommendations for geometrical situations and load intensities to permit controlled leveling: $d_{limit} = 10–25°$; $d_A/d_B \leqslant 1.5$ with $\alpha_A + \alpha_B \leqslant 20°$; $b_A/d_A \geqslant 1$ resp. $b_B/d_B \leqslant 2$; $\sigma_A/\sigma_B \leqslant 3$ with $\alpha_A + \alpha_B \geqslant 20°$. Situations exceeding these values need additional verification or practical trial.

Table 7.1 List of parameters with significance for project considerations (1 – absolutely necessary; 2 – important for calculations; 3 – significant for numerical models)

Designation	*Short symbol*	*Unit*	*Importance*
Grain-size distribution	–	mm	1
Coefficient of non-uniformity	C_u	–	1
Moisture content	w	–	1
Porosity	n	–	1
Void ratio	e	–	1
Relative density	D	–	1
Consistency index	I_c	–	1
Liquidity index	I_L	–	1
Coefficient of permeability	k	m/s	2
Young's modulus	E	kN/m^2	2
Shear strength parameters	φ', c', φ_r	°, kN/m^2, °	2
Undrained shear strength	c_u	kN/m^2	3
Model of rheology			3

7.4.2 *Soil description*

In principle, hydraulic fracturing is suitable for improving all types of soil with an adequate consistency. However, to be able to quantify the requirements which have to be met, it is necessary to have specific information on the initial soil properties. Table 7.1 contains a selection of the necessary parameters and an evaluation of the effect on the quality of the prediction. In principle, it has to be said that such a method is used almost exclusively in difficult non-homogeneous and anisotropic soil conditions. Even if it would be possible, accurately, to describe individual soil layers, it would be almost impossible to describe the problems generated by the layering effect and by even greater irregularities such as karst fillings. However, it is precisely these irregularities which are usually the actual causes for the occurrence of considerable differential settlements. Therefore, the most important objective of the project preparation phase consists in recognising the nature of the causes of differential settlements and in formulating a concept which makes the provision of further information part of executing the actual work.

7.4.3 *Mathematical models*

Simple evaluations of the geometric relationships occurring in the course of lifting injections can be made by using a 'block model' in which the zones of different treatment intensities are modelled in the form of an idealised tensile stress ring, and the lifted zone in the form of a centrally positioned lifted piston.

The presently available numerical calculation methods allow the modelling of many movement phenomena in the soil and their interaction with structures positioned above. For the time being, a considerable impediment as regards simulating lifting injections has been found to be the continuous change in the input parameters, which results from the treatment of the soil. The respective publications describe the present-day status of calculation possibilities. Two-dimensional models are more economical to produce and are more reliable than three-dimensional simulations of complex situations whose results are sometimes also influenced by special characteristics of calculation programs.

7.4.4 *Monitoring*

The concept for comprehensive deformation measurements always includes the two completely different sets of problems of the structure on the one hand and of the soil underneath on the other hand. As far as structures are concerned, monitoring the height of supporting components is of primary importance. In addition, floors and existing cracks can be provided with

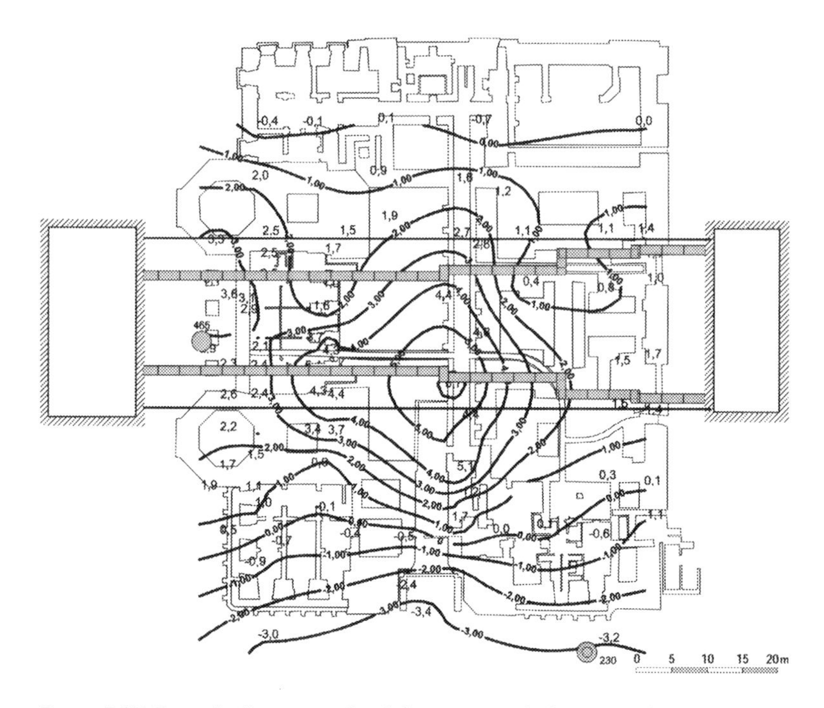

Figure 7.15 Example illustrating the deformations of a historic railway station, which were continuously monitored by means of a complex measuring system.

instruments whose values are read automatically or visually. Measuring instruments such as inclinometers, extensometers, incremental extensometers, settlement piezometers and similar devices can be arranged in the ground with the objective of determining spatial deformation rates and their directions in the direct vicinity of the area treated (Figure 7.15).

7.4.5 Basic information required for planning purposes

To be able to assess the expenditure involved in applying the method of hydraulic fracturing, it is necessary to make assumptions regarding the required percentage of solid matter of the treated soil volume and the type of suspension to be used.

7.4.6 Performance elements

The plan for the application of hydraulic fracturing is normally followed by a performance description which is used as a basis for the works contract to be concluded with the contractor. Table 7.2 contains a list of normally

Table 7.2 Performance elements of compensation injection

Performance element	*Important specifications*
Licences for drilling areas and construction site equipment, access to structures for exploration purpose	Duration of use/number of measuring points to be provided
Exploring installation situation	Type of exploration Safety measures
Recording details of structures	Type of structure
Preparing and maintaining drilling surfaces	Shaft depth and diameter Types of ditches/expenditure for adapting existing spaces Access situation
Setting up drilling equipment	Mean and maximum drilling length Drilling method, borehole dia. Diameter of valve pipe to be built in, spaces available
Execution of drilling operations installation of sleeve pipes	Borehole diameter, diameter and wall thickness of pipes, material properties, distance between valves, valve characteristics
Filling the annular space	Suspension mixture data
First injection	Number and sequence of individual injections, quantity injected per injection operation
Mobilisation of grout station	Number of injection units to be operated separately
Multiple injection; pre-consolidation and 'conditioning'	Number of injection operations and quantity to be introduced per injection operation, minimum and maximum pump rates
Pre-lifting and injection up to point when the structure is ready to lift the structure	Number of injection operations and quantity to be introduced per injection operation, minimum and maximum pump rates
Lifting 'compensating injection'	Listing lifting stages and lowest and highest settlement rates to be compensated for per shift
Installing an automatic measuring system	Type and number of measuring elements to be installed
Providing and maintaining the measuring system for the duration of the project and/or for the period of continued observation	Assessing the required functional periods, required accuracies, measuring frequencies

Table 7.2 (Continued)

Performance element	Important specifications
Data recording and visualisation	Providing, in a standard form, the injection parameters and data from monitoring, frequency and type of data sets to be handed over
Technical site management	Interpreting the measured data jointly with planner, setting up individual injection programmes/evidence of experience with similar construction situations

occurring working phases and a selection of necessary specifications. In addition, there are proposed units according to which the very different services have to be evaluated, depending on the type of project.

Note: When applying soil fracturing in connection with very complex infrastructure measures, it may be advisable to include, in the performance description, regularly occurring interruptions in the construction work; in addition, a distinction has to be made between phases when the equipment is ready for injection and phases of pure measuring technical observation on the one hand and downtimes on the other hand.

7.5 Case histories and limits of application

Hydraulic fracturing can be used for different objectives and types of settlement reducing operations. The spectrum ranges from pure prevention via repairs after settlement damage has already taken place to the simultaneous compensation for extensive settlement troughs in inner-city areas. A number of examples illustrates the many different opportunities for this method and should provide creative ideas for the planning engineer.

7.5.1 Prevention

Soil improvement by hydraulic fracturing takes place to homogenise the stress conditions in large soil regions or to increase the load-bearing capacity of soils to enable them to receive additional loads. It is important to point out the economical nature of this method as compared to deep foundation systems because the readiness of the soil/structure to be lifted under existing loads precisely indicates the point in time at which a sufficient degree of soil improvement has been reached for the load actually occurring in the slab joint.

7.5.1.1 Increasing the load-bearing capacity of foundation soils

In the case of the already overloaded foundation of a three-storey dwelling on strip foundations, it is planned to improve the soil down to a depth of 8 m to enable the building to accommodate, free of settlements, additional loads resulting from the construction of additional storeys (Figure 7.16).

7.5.2 Passive settlement reduction

If differential deformation rates are expected, a pre-treatment with the objective of improving the soil can considerably reduce the absolute extent of such deformation and, if necessary, allows an active correction of actually occurring movements. Normally, the costs of such a measure are considerably lower if the installation of the sleeve pipes is already included in the construction schedule.

7.5.2.1 Limiting deformation when securing a construction pit

The most economical variant of securing a construction pit by means of an anchored diaphragm wall was likely to result in an extent of horizontal

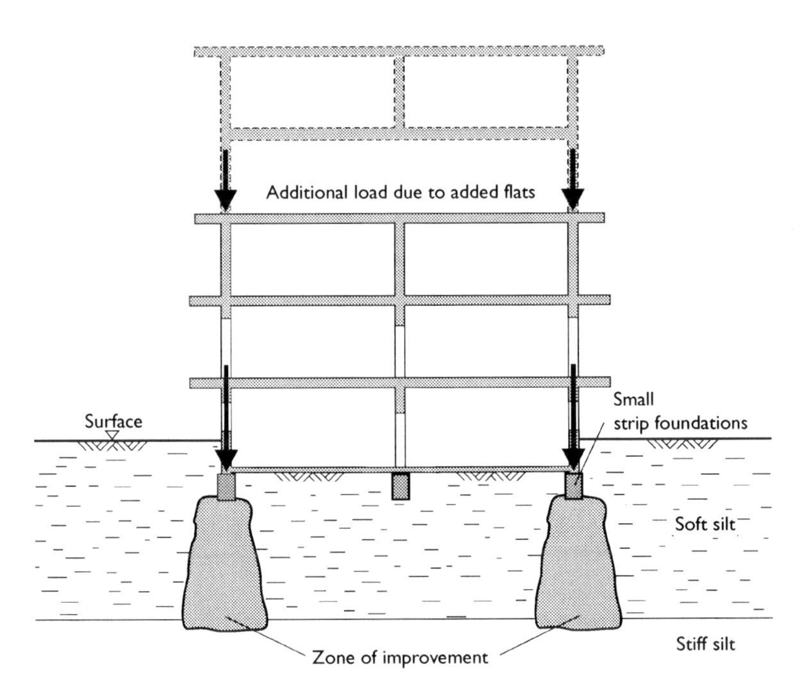

Figure 7.16 Increasing the load-bearing capacity in the area where additional loads have to be carried due to the addition of two additional storeys.

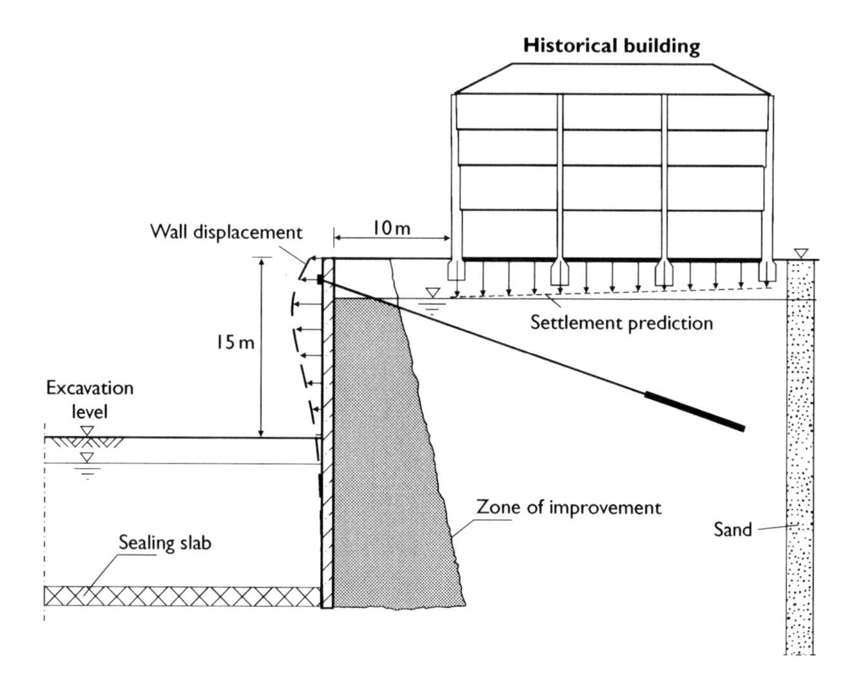

Figure 7.17 Passive settlement reduction by carrying out specific soil improvements in the area subject to deformation.

deformation which was unacceptable for the adjoining buildings. As an alternative to secondary underpinning of foundations by jet grouting, it was considered to improve the block of soil behind the diaphragm wall by means of soilfrac, which, if needed, would also allow volume losses to be compensated for during the excavation phase (Figure 7.17).

7.5.3 *Back levelling*

There are several traditional methods for performing repairs to settlement-damaged buildings, including the use of piling systems, jet grouting and hand digging.

Most methods have complications which are connected with the use of working surfaces inside the building. In addition, lifting a building by hydraulic presses is a very complicated operation which, if applied accurately, requires a very sophisticated control technology and a sufficiently intact and stable structure. By adapting the soil fracturing technology to small-scale applications, it is possible to obtain a technically and economically interesting alternative even in those cases where, normally, access to the building interior is necessary for observation purposes only.

7.5.3.1 *Stabilising a layer of peat*

A wedge-shaped layer of peat existing underneath the foundation slab had not been properly identified during the construction of a multistorey dwelling, which meant that a flat foundation was produced on a concrete slab. The resulting differential settlements led to maximum inclinations of 1:85, as a result of which the usability of the building was put in question. On this occasion it was successfully shown for the first time that organic soils, too, can be improved by hydraulic fracturing, even at higher costs (Figure 7.18) than volume stable soils.

7.5.4 Compensation injections

Ever since electronic measuring systems have allowed a real-time presentation of deformation measurements, it is possible, by means of soilfrac, to actively compensate for simultaneously developing settlement troughs. In this context, it is essential to asses realistically the total settlement to be expected as well as the highest settlement rates occurring during a limited period of time. Needless to say, minimising settlements by driving a suitable tunnel still constitutes the essential element of an environmentally acceptable construction method in inner-city areas.

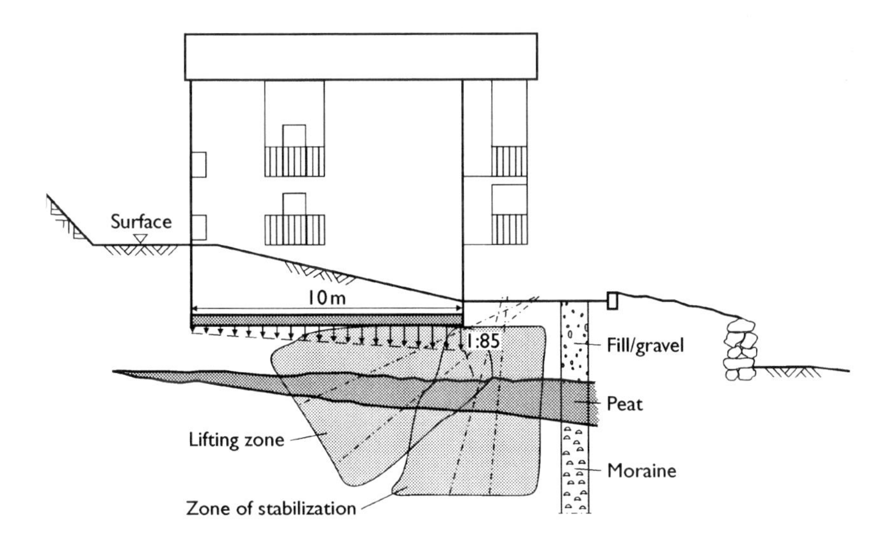

Figure 7.18 Lifting an apartment building.

7.5.4.1 *Compensating for settlements above convential underground excavations*

7.5.4.1.1 COMPENSATION DURING SEQUENTIAL EXCAVATIONS

Over a period of 18 months, historic structures had to be protected from differential settlements caused by station tunnels with internal diameters of up to 20 m. Total settlements in excess of 100 mm were compensated for in lifting stages of 2 mm per lifting operation, so that the position of the building foundations prior to the tunnel driving operation remained almost unchanged (Figures 7.5a,b,c).

7.5.4.2 *Compensating for settlements above tunnels drilled by tunnel-boring machines*

Producing tunnels by tunnel-drilling machines differs from using mining methods above all in respect of the time characteristics. Because of the higher driving speed and occasional abrupt interference in the region of earth pressure support, greater attention has to be paid to the passive effect of a compensation injection. As a result of the number and type of sleeve pipes, it is possible, additionally, to use their reinforcing effect on the soil. Furthermore, an extensive measuring system (widely distributed measuring elements) contributes considerably towards controlling shield machines when installing a suitable communication system.

7.5.5 *Application limits*

The large number of practical applications shows that soil fracturing can be applied under very different conditions with technical and economic advantages. As far as improving soft soils are concerned, the limit is reached in very soft and organic soils. If the method is to be applied in refuse tips, it is necessary to carry out a detailed preliminary chemical investigation. As far as treatment geometry is concerned, it should be pointed out once again that the treatment in question mainly modifies the stress flow in the soil. The concept of the works schedule should therefore be formulated in the sense of an improved and controlled stress transmission and should distance itself from the idea of structures in the soil with defined dimensions. The assumption that the dimension of the treated soil member must permit the deflection of the stresses occurring can be the simple guiding principle.

In cases of doubt, large-scale tests have always been found to be useful which, in most cases, can be integrated into subsequent main measures and thus do not represent substantial additional costs.

7.6 Test methods and monitoring methods

The objectives of tests in connection with hydraulic fracturing can consist in obtaining information on the injection technology itself, in evaluating the quality of soil improvement or in assessing the effect on the structures concerned. Usually, the intention is to draw conclusions on the achievable lifting rates, the associated time requirements and the material quantities involved.

7.6.1 Field tests

Experience has shown that field tests are of value only if their dimensions correspond in every way to those of the subsequent application. However, the execution of such tests should be conceived as part of the concrete application rather than as a kind of isolated basic test. In such cases, there is no need to assess whether the test area selected comprises representative soil and load conditions. What is important is to install a set of geotechnical instruments providing reliable information and to record all relevant parameters. The documentation of such a test has to permit the evaluation of all parameters in terms of their time sequences. As far as geotechnical instruments are concerned, apart from inclinometers and extensometers, all systems are suitable which are able to record deformation and changes in stress in the soil, the requirement being that the deformability of the measuring elements themselves and that of the annular grout should be similar to that of the surrounding soil; in the case of pressure gauges, this is not always the case with every product (Figures 7.19a,b).

7.6.2 Laboratory tests

The suspensions provided as a skin-forming mixture and for the injection itself can vary considerably in respect of the number and type of components used and their composition. In view of the different reactions of similar bonding agents with different production origins, it is absolutely essential to carry out basic tests at the start of each project. Especially the effect of additives can be greatly influenced by the local properties of cement and filler materials. The tests are carried out with the objective of determining the flow properties, the bleeding of water and the setting behaviour as well as the stability of suspensions. Although it is important to determine special technical characteristics, the most important premise remains the processibility of the suspensions, using the equipment available on site. While it is a common practice to vary the suspension composition in the course of the phasewise execution of the soil fracturing work, it is advisable to set up a site laboratory. The objectives of suitability tests can be the production of 'soft mixtures' to achieve low-end strength values of granular soils as well as the production of 'harder mixtures' with the release of small amounts of water for the purpose of improving soft soils.

7.6.3 *Measuring technology*

The measuring technology used must allow the measures applied in the soil and their effects on the soil to be very clearly associated with the structures concerned. It is important to ensure that any blurred measured values do not feign any movement tendencies which could lead to pre-mature changes in the works programme. Therefore, in cases of doubt, reference arrangements have to be used under controlled conditions. Such calibrations under

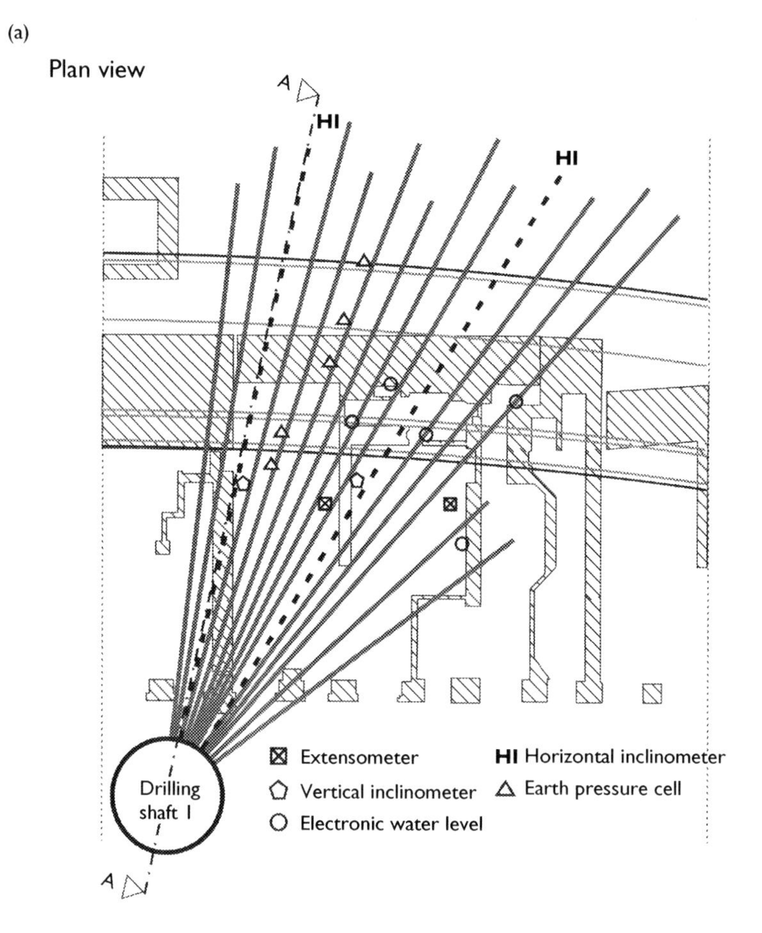

Figure 7.19 (a) Layout of a large-scale test which was planned as part of the subsequent settlement compensation measure.

(b)

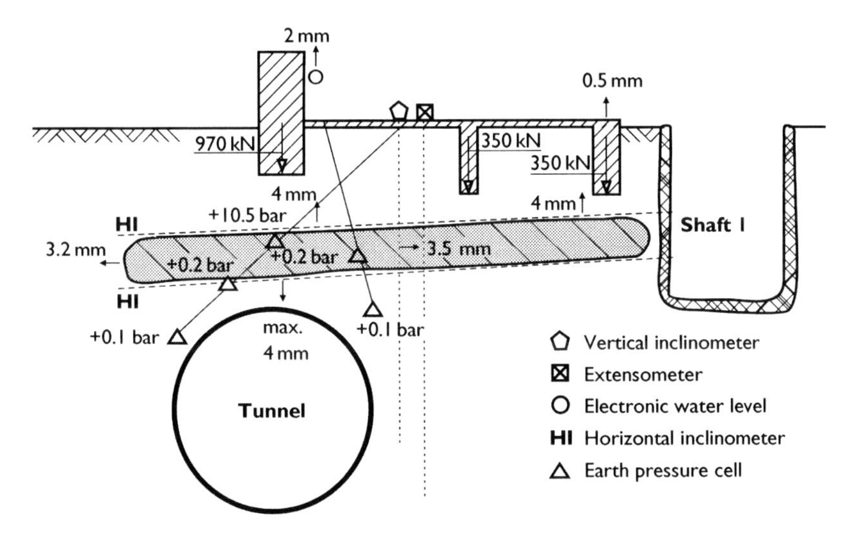

Figure 7.19 (b) Test results which, on the one hand, show the suitability of the measure for specifically lifting individual foundations and, on the other hand, the negligible influence on the tunnel shell.

realistic conditions can concern all types of electronic and visual-settlement-measuring systems such as hose scales based on the filling level principle, pressurised water level systems, automatic levelling instruments and theodolites, rotary laser, precision levelling instruments and special types such as floor-level-measuring instruments, inclination instruments and fissuremeter devices.

7.7 Environmental aspects

Although hydraulic fracturing mainly manifests itself in the form of movements in the soil and on structures, it is its ability of being able to control the modification of the soil and of the respective condition of the structure on which attention is focused. When agreements are drafted in connection with complex structural methods, it is important, at an early stage, to clarify ownership conditions and the interests of those who may be affected. The chemical environmental compatibility of the individual injection components has to be proved. In cases of doubt, additional tests of the actually used combination of materials and products have to be carried out.

7.7.1 Contractual situation

Agreements regarding the realisation of projects involving hydraulic fracturing are concluded on the basis of mutually agreed projects. As some of the effects of the measures applied can often not be specified at that particular point in time, the agreement must permit the consistent use of the observation method. The type of reaction to possible scenarios in the individual construction phases has to be planned and contractually permitted, of course with the intention of safeguarding the rights of third persons and the economic execution of the project in the interest of solving an existing problem. While taking into account the measures associated with strategies against undesirable effects, it should be noted that a performance schedule can list additional measures for limiting or warding-off injection effects. Equally, the acceptance of crack formations and arching taking place in the foot path region while a building is being lifted can be included in the working agreement as part of the overriding project objective.

Bibliography

Bernatzik, W. (1951) *Anheben des Kraftwerkes Hessigheim am Neckar mit Hilfe von Zementunterpressungen*, Der Bauingenieur, Heft 4.

Boeck, Th. and Scheller, P. (2000) *4. Röhre Elbtunnel – Sicherung der Bebauung am Nordhang der Elbe*, Hannover: Deutsche Baugrundtagung.

Brandl, H. (1981) Stabilization of excessively settling bridge piers, Florence: *Proceedings of the Xth ICSMFE, S.* pp. 329–336.

Chambosse, G. and Otterbein, R. (2001) Central Station Antwerp Compensation grouting under high loaded foundations, Istanbul: *Proceedings of the fifteenth ICSMFE.*

Droof, E.R., Tavares, P.D. and Forbes, J. (1995) *Soil fracture grouting to remediate settlement due to soft ground Tunnelling*, San Francisco: Rapid Excavation and Tunnelling Conference.

Falk, E. (1997) Underground works in urban environment, Hamburg: *Proceedings of the XIVth ICSMFE.*

Falk, E. and Schweiger, H.F. (1998) *Shallow Tunneling in Urban Environment – Different Ways of Controlling Settlements*, Felsbau 16, No. 4.

Gittoes, G., Sir William and Simic, D. (1995) *Design of Large Diameter Shield Driven Tunnels for the Southern Extension of the Lisbon Metro*, Forschung und Praxis.

Grabener, H.G., Raabe, E.W. and Wilms, J. (1989) *Einsatz von Soilfracturing zur Setzungsminderung beim Tunnelvortrieb*, Essen: Taschenbuch für Tunnelbau, Deutsche Gesellschaft für Erd- und Grundbau Verlag Glückauf GmbH.

Harris, D.I. (2001) *Protective measures*, London: Response of buildings to excavation – induced ground movements, Ciria Conference.

Mair, R.J., Harris, D.I., Love, J.P., Blakey, D. and Kettle, C. (1994) *Compensation grouting to limit settlements during tunnelling at Waterloo Station*, London: Tunnelling '94, IMM, pp. 279–300.
Schweiger, H.F. and Falk, E. (1998) Reduction of settlements by compensation grouting – Numerical studies and experience from Lisbon underground, Brazil: *Proceedings of the world tunnel congress '98 on tunnels and metropolises*, Sáo Paulo.

Chapter 8

Lime and lime/cement columns

B.B. Broms

8.1 Introduction

8.1.1 The lime column method

Lime and lime/cement columns, where quicklime and/or dry cement are mixed *in situ* with soft soil as shown in Figure 8.1, are common in Sweden and Finland, to stabilize soft clay and silt as well as organic soils. The method has gradually been improved and new applications have been found. Lime and lime/cement columns have mainly been used to increase the stability and to reduce the settlements of road and railroad embankments and to increase the stability of trenches for sewer lines, water mains and heating ducts. New efficient machines have been developed for the installation of the columns. The diameter and the length have gradually increased and the time required for the installation of the columns has been reduced significantly as well as the costs. New methods have been introduced to check *in situ* the shear strength, the bearing capacity and the stiffness of the columns.

Lime/cement columns have also been used to stabilize organic soils, where unslaked lime alone has not been effective. Lime columns have the advantage that the permeability and the ductility are normally high. In addition the ground temperature is increased by the heat generated during the slaking. The increase of the shear strength, caused by the reduction of the water content, is usually significant.

Since a large number of factors affect the behaviour of lime and lime/cement columns, it is necessary to determine for each site the effect of different stabilizers (e.g. lime, cement, gypsum, industrial waste and different ashes) on compressibility, shear strength and permeability of the stabilized soil. Extensive field and laboratory tests are usually required.

The stabilized soil has a high angle of internal friction even at undrained conditions since the stabilized soil is only partially saturated. The bearing capacity and the shear resistance of the columns are to a large extent governed by the axial load and by the confining pressure.

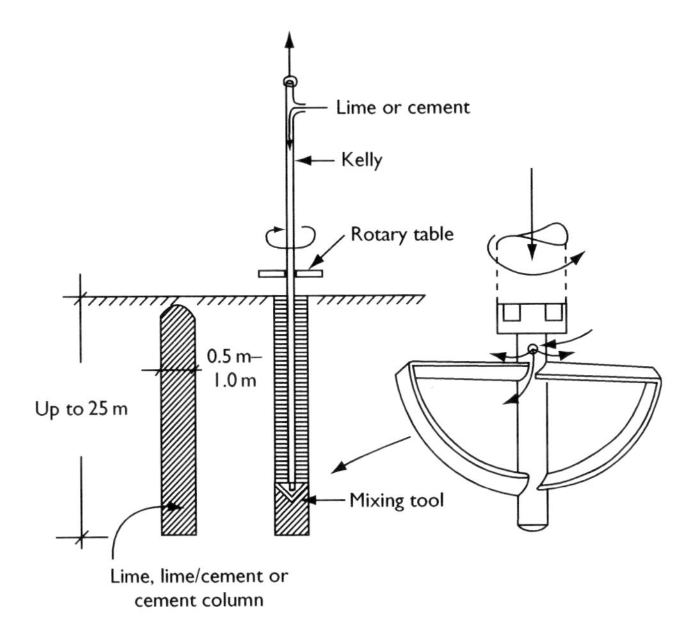

Figure 8.1 Installation of lime and lime/cement columns.

8.2 Shear and tensile strength

8.2.1 Shear strength at dry and at wet soil mixing

A higher undrained shear strength can generally be obtained by dry than by wet method at the same cement content (Chida, 1981) due to the reduction of water content and water/cement ratio (w/c-ratio). The w/c-ratio of the slurry at the wet method is usually 0.6 to 1.3 (Okumura, 1996) which will reduce the undrained shear strength. Sandros and Holm (1996) have reported that the undrained compressive strength of a silty clay was 1.7 to 3.2 times higher by dry than by wet method of soil mixing.

Due to the difficulty to mix dry cement with the soil in dry method especially when the cement content is high, the variation of shear strength and stiffness across and along the columns is usually less by wet method than by dry method. It is easier to mix cement slurry with soil compared with dry method especially when the water content of the soil is low and the initial shear strength of soil is high. The wet method is common in Japan while in Sweden and Finland only the dry method is used.

8.2.2 *Undrained shear strength with lime*

The short-term increase of the shear strength with unslaked lime (CaO) depends mainly on the reduction of the water content during the slaking, on the increase of the plastic limit and on the reduction of the plasticity index of the soil by cation exchange. The long-term increase is mainly governed by the pozzolanic reactions of the lime and the cement with the clay fraction in the soil.

The strength and deformation properties of the stabilized soil are similar to those of a partially saturated overconsolidated clay. The undrained shear strength is usually determined by unconfined compression tests, by undrained triaxial or by direct shear tests (UU-tests). Sometimes consolidated undrained triaxial tests (CU-tests) are used. The undrained shear strength, as determined by direct shear tests, is often lower than the shear strength determined by triaxial tests when the shear strength is high due to the high angle of internal friction $\phi_{u,col}$. Yoshida (1996) has reported that the shear strength as determined by triaxial tests with cement was 1.57 times higher than the shear strength determined by direct shear tests.

The undrained shear strength, $\tau_{fu,col}$ of the soil stabilized by lime/cement or cement, which governs the stability of embankments, slopes, trenches and excavations, increases with increasing normal pressure.

$$\tau_{fu,col} = c_{u,col} + \sigma_f \tan \phi_{u,col} \tag{8.1}$$

where $\phi_{u,col}$ is the undrained angle of internal friction of the stabilized soil, $c_{u,col}$ is the undrained cohesion and σ_f is the normal total pressure acting on the failure plane through the columns. A value $\phi_{u,col} = 30°$ is typical for lime and lime/cement columns, when the stabilized soil is not fully saturated.

An undrained shear strength $\tau_{fu,col}$ equal to $0.5q_{u,col}$ is usually assumed in the design of the columns, where $q_{u,col}$ is the unconfined compressive strength. However, this assumed shear resistance could be too high when the soil, which is stabilized by lime, lime/cement and cement is not fully saturated. The ratio $2\tau_{fu,col}/q_{u,col}$ increases from 1.34 at $\phi_{u,col} = 25°$ to 1.91 at $\phi_{u,col} = 45°$. The assumed undrained shear strength $\tau_{fu,col}$ at $\phi_{u,col} = 0°$ can thus be 34 to 91 per cent higher than the undrained shear strength $c_{u,col}$ as determined by unconfined compression tests at $\phi_{u,col} = 25$ and $45°$, respectively.

The design of lime, lime/cement and cement columns is usually based on the undrained shear strength determined 28 days after the installation of the columns. This shear strength is too low for lime columns since a large part of the long-term increase occurs later than 28 days after the mixing. It is therefore proposed that the design strength of lime columns should be based on the measured or the estimated undrained shear strength after 90 days.

A high friction angle $\phi_{u,col}$ has been reported for soils stabilized by lime and lime/cement due to dilatation, when the confining pressure is low. A negative pore water pressure is expected when the soil is fully saturated. The increase of the pore water pressure when the soil is loaded is expected to be small, when the soil is only partially saturated and that $\phi_{u,col} = \phi'_{col}$.

The shear strength of laboratory samples is often higher than the *in situ* shear strength when the undrained shear strength is high. In a few cases the *in situ* shear strength has been higher than the shear strength of laboratory samples, probably due to the confining pressure *in situ* and the high ground temperature.

Unslaked lime affects mainly the clay fraction due to the reduction of the water content and the increase of the plastic limit by ion exchange. The increase of the shear strength with 5 to 10 per cent unslaked lime is typically 10 to 20 times for normally consolidated inorganic clays. The maximum undrained shear strength is usually 200 to 300 kPa. For quick clays the relative increase can be 10 to 50 times. There is usually no further increase of the shear strength when the lime content exceeds 10 to 15 per cent.

The long-term increase of the undrained shear strength of the stabilized soil depends mainly on the puzzolanic reactions in the soil and on consolidation as illustrated in Figure 8.2.

The highest undrained shear strength is usually obtained with unslaked lime when the clay content is high. The increase of the shear strength is relatively slow for quick clays and for soft marine clay with a high salt content. The long-term shear strength is often high probably due to the improved mixing when the shear strength of the remolded soil is low.

Consolidation also increases the shear strength. The increase of the volume caused by the slaking when water is drawn from the soft soil around the columns or from pervious silt and sand layers, affects the lateral pressure around the columns and the shear strength.

8.2.3 Undrained shear strength with lime/cement

Often a higher undrained shear strength can be obtained with lime/cement than with lime alone especially for organic soils. The shear strength increases in general with increasing lime/cement content and with increasing content of fines. The maximum undrained shear strength with lime/cement is about 500 kPa with 10 to 20 per cent lime/cement (50/50). The increase of the shear strength is often small for organic soils when the water content exceeds 200 per cent (Babasaki *et al.*, 1996). Also the pH-value affects the shear strength.

Green and Smigan (1995) have reported values between 30 and 40° on $\phi_{u,col}$ by direct shear tests with lime/cement (50/50 and 80/20). A friction angle of 41 and 33° was obtained by triaxial tests. Kivelö (1996) determined a value of 45° by triaxial and by direct shear tests (UU-tests) as shown in

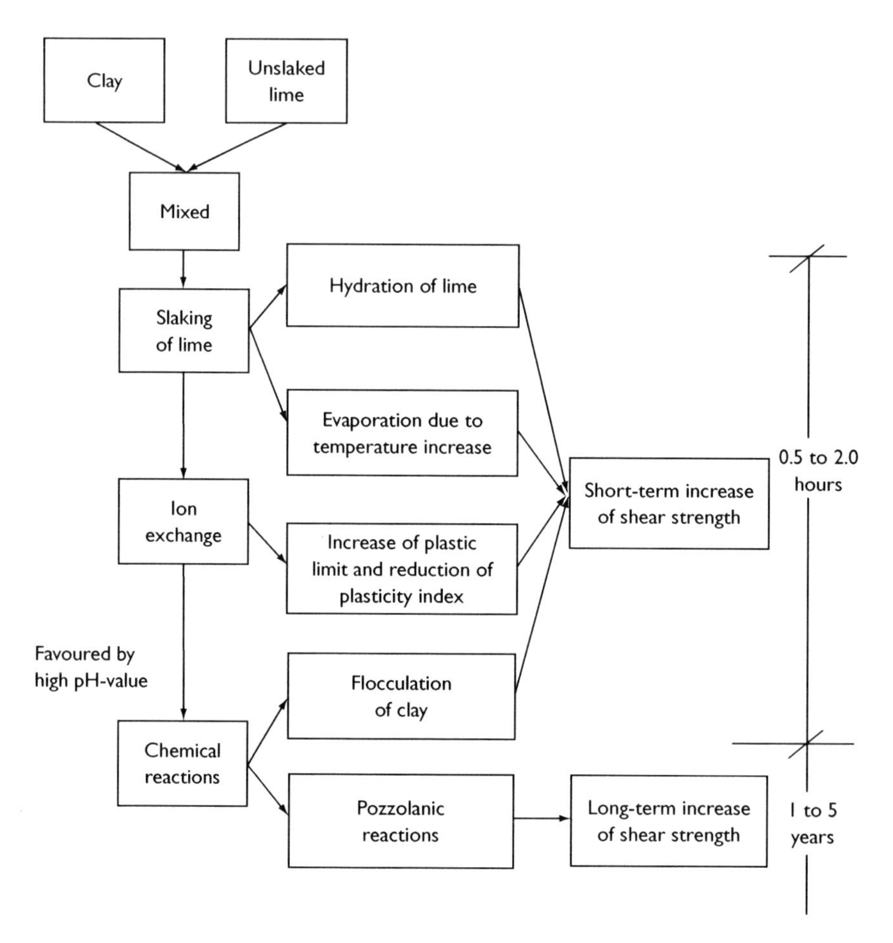

Figure 8.2 Stabilization of soils with unslaked lime.

Figure 8.3. The cohesion was reduced to zero when the deformations were large. The friction angle was not affected much.

Axelsson and Larsson (1994) obtained an average angle of internal friction ($\phi_{u,col}$) of 42° for cores from lime/cement columns. A friction angle ($\phi_{u,col}$) of 43 to 45° was determined by Björkman and Ryding (1996) by direct shear tests. The friction angles as determined by undrained and drained triaxial tests (UU and CD-tests) were 41 and 40°, respectively.

8.2.4 Undrained shear strength with cement

The undrained shear strength increases in general with increasing cement content and with decreasing liquid limit. The increase of the shear strength

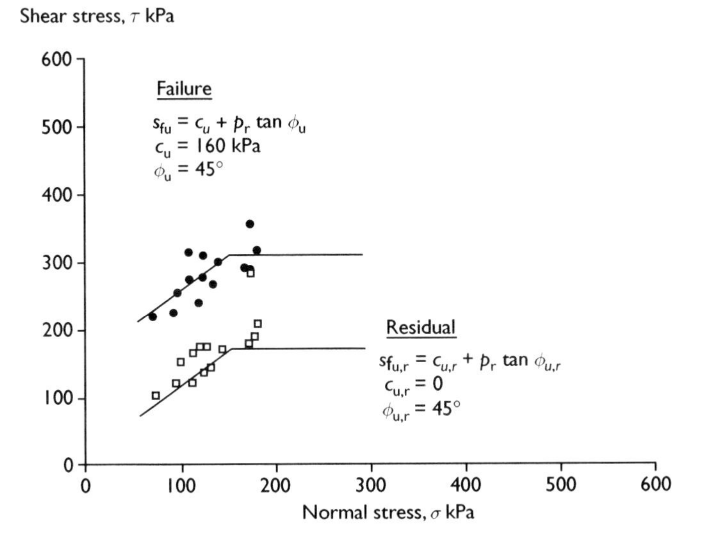

Figure 8.3 Undrained direct shear tests (after Kivelö, 1996).

decreases with increasing plasticity index and with increasing clay content (Woo and Moh, 1972). Åhnberg *et al.* (1994) have reported that the shear strength increases with increasing cement content between 6 and 16 per cent. The increase of the shear strength is also affected by the soil type, the initial water content and by the water/cement ratio. The increase of the shear strength with cement is often small when the water content of the soil exceeds 200 per cent (Babasaki *et al.*, 1996). The increase has been small for organic soils when the ignition loss exceeded 15 per cent even when the cement content exceeded 20 per cent. The shear strength has also a tendency to decrease with decreasing pH-value. Below 5 the increase is usually small.

Also the cement type affects the shear strength as reported by e.g. Bergado *et al.* (1996). Åhnberg *et al.* (1995) found that the shear strength with fast setting cement was higher than the shear strength with Standard Portland cement.

Cement had to be mixed thoroughly with the soil compared with lime to obtain a high shear strength and a high uniformity. The increase of the shear strength with cement depends mainly on the pozzolanic reactions in the soil while for lime the increase depends to a large extent on the reduction of the water content, on flocculation and on the increase of the plastic limit by cation exchange.

The pozzolanic reactions are initially faster with cement than with lime especially with finely ground cement. The pozzolanic reactions with lime continue for many months. The increase of the ground temperature is less with cement than with lime, which slows down the pozzolanic reactions and the increase of the shear strength with time.

The shear strength of samples mixed in the laboratory with cement has been up to 2 to 5 times higher than the shear strength of samples obtained from columns (Ansano *et al.*, 1996) due to the difference in mixing in the laboratory and in the field, and the difference in ground temperature.

8.2.5 Undrained shear strength with gypsum, fly ash and blast furnace slag

Kujala (1983a) has reported a friction angle $\phi_{u,col}$ of 23° determined by triaxial tests and 40° by direct shear tests with gypsum and lime. Kujala and Nieminen (1983) found that the friction angle was about 10° higher with lime/gypsum than with lime.

Also fly ash has been used for soil stabilization (Wu *et al.*, 1993). An increase of 40 times has been observed after 28 days with 30 per cent fly ash mainly due to the reduction of the water content. Also the increase of the plastic limit and the reduction of the plasticity index have been contributing factors. Fly ash usually decreases the permeability. Often a high shear strength can be obtained with cement and blast furnace slag also for soils with a high initial water content.

8.2.6 Drained shear strength

The drained shear strength τ_{fd} of the stabilized soil, which governs the long-term stability, is estimated by the following equation:

$$\tau_{fd} = c'_{col} + [\sigma_f - u_{col}] \tan \phi'_{col} \tag{8.2}$$

where σ_f' $[\sigma_f - u_{col}]$ is the normal effective pressure on the failure plane, c'_{col} is the effective cohesion and ϕ'_{col} is the effective angle of internal friction.

The drained shear strength (ϕ'_{col} and c'_{col}) can be determined by drained triaxial and by direct shear tests (CD-tests) or by consolidated undrained triaxial tests (CU-tests) with pore pressure measurements. The shear strength as determined by drained triaxial tests, is often lower than the shear strength determined by undrained direct shear tests when the normal pressure is low. The effective friction angle ϕ'_{col} can be assumed conservatively to 30° for lime columns, 35° for lime/cement columns and about 40° for cement columns.

8.2.7 Drained shear strength with lime

The effective friction angle ϕ'_{col} usually increases with increasing lime content and with time. Balasubramaniam and Buensuceso (1989) have reported values of 38 and 35.8° with 5 to 10 per cent lime. With 15 per cent lime the friction angle was 40.1°. Rogers and Lee (1994) found that ϕ'_{col} increased a few degrees with increasing lime content. Göransson and Larsson (1994) determined values of 31 to 36° by triaxial tests (CD-tests). Brookes *et al.* (1997) reported values of 30 to 42° on ϕ'_{col} by triaxial tests (CD-tests) with Gault Clay from the UK at a lime content of 5 to 15 per cent. The highest friction angle was observed with 5 per cent lime. The friction angle ϕ'_{col} was 23 to 25° for the unstabilized Gault Clay. For the London Clay, ϕ'_{col} was 31 to 41° with 5 to 15 per cent lime. The friction angle for the unstabilized weathered clay was 17 to 23°.

8.2.8 Drained shear strength with lime/cement and cement

The friction angle ϕ'_{col}, which has a tendency to increase with time (Åhnberg *et al.*, 1995), is often lower for lime columns than for cement and lime/cement columns. The highest values have been observed for clayey silt and the lowest values for organic soils. A friction angle of 34 to 36° has been determined by drained triaxial tests for organic clay, 34 to 39° for clay and 40 to 44° for clayey silt, which have been stabilized by lime/cement. Åhnberg (1996) found from drained triaxial tests that ϕ'_{col} was 34 to 44°. The higher angles are for clayey silt while the lower are for clayey 'gyttja'. The high friction angles observed at drained tests could at least partly be caused by dilatancy when the stabilized soil is loaded. The behaviour of both lime/cement and cement columns has been similar to that for an overconsolidated clay.

Björkman and Ryding (1996) found for samples from lime/cement columns that ϕ'_{col} was 40° by triaxial tests. Triaxial tests with 0.5 m diameter column segments by Steensen-Bach *et al.* (1996) indicated values on ϕ'_{col} of 37.5 to 39.3°. Ekström (1992) has reported a value of 40° for ϕ'_{col} for silty clay and 25 to 29° for clay, which has been stabilized by lime/cement or cement. Values of 40 to 45°, have been reported for cement columns. Ekström (1994b) observed at triaxial tests (CU-tests) a friction angle ϕ'_{col} of 38 to 40° with lime/cement.

Huttunen *et al.* (1996) found for peat that ϕ'_{col} was 36.8 to 41.3° with cement and granulated blast furnace slag. Huttunen and Kujala (1996) have reported that ϕ'_{col} was 37.1 to 60.6° for peat with cement and gypsum (Finnstabi) at a stabilizer content of 250 kg/m^3. The water content was 174 to 198 per cent. The unconfined compressive strength was 130 to 234 kPa.

8.2.9 Drained shear strength with gypsum

Kujala (1983b) has reported, that ϕ'_{col} was 40° at triaxial tests with gypsum and lime and $\phi'_{col} = 23°$ with only lime. Brandl (1981) obtained values of

35 to 40° by direct shear tests with clays stabilized by lime and by lime and gypsum. The normal pressure was low.

8.2.10 *Effective cohesion c'_{col}*

Åhnberg (1996) found that the cohesion c'_{col} varied between 50 kPa for clays stabilized with lime to 1600 kPa for clayey silt stabilized with cement. The effective cohesion c'_{col} was 46.2 to 83.6 kPa for cement and gypsum (250 kg/m^3) and 37.4 to 48.5 kPa for cement and granulated blast furnace slag. The cohesion was higher for clayey 'gyttja' than for clayey silt and clay stabilized by lime/cement.

8.2.11 *Water/cement ratio*

The shear strength of organic soils stabilized with cement is to a large extent governed by the water/cement ratio (w/c-ratio) as discussed by Åhnberg *et al.* (1995), Kukko and Ruohomäki (1985), Rathmayer (1997) and by Nagaraj *et al.* (1996). Holm (1994) has reported that the shear strength increases approximately linearly with decreasing cement content. Ansano *et al.* (1996) found that the increase of the shear strength depended mainly on the *w/c*-ratio

$$\tau_{fu} = \frac{\tau_{fou}}{(w/c)} \tag{8.3}$$

where τ_{fou} is a reference undrained shear strength after 28 days at a *w/c*-ratio of 1.0 including the initial water content of the soil. The reference undrained shear strength varies with soil type, organic content and with the pH-value of the pore water. It is expected from equation (8.3) that the shear strength of organic soils will increase by 100 per cent when the water/cement ratio is reduced 50 per cent i.e. from 8 to 4. When the water/cement ratio is 1.2 to 1.5 the expected unconfined compressive strength is about 1 MPa. An unconfined compressive strength of 2 to 4 MPa is often obtained at a *w/c* ratio of 1.0 (Ansano *et al.*, 1996; Matsuo *et al.*, 1996).

8.2.12 *Residual shear strength*

The undrained and the drained shear resistance are reduced when the peak strength is exceeded. This reduction had to be considered, when the peak strength of the stabilized and of the unstabilized soil is utilized in design. The reduction increases in general with increasing peak shear strength and with decreasing confining pressure. It is proposed that the effective cohesion should be neglected when the residual shear strength is used in design

($c'_{col,res} = 0$). The friction angle ϕ'_{col} is not affected much as discussed by Steensen-Bach *et al.* (1996).

8.2.13 Failure strain and ductility

The strain ε_f at the peak shear strength of the stabilized soil could be as high as 10 per cent or higher for lime columns when the unconfined compressive strength of the stabilized soil is low, less than about 200 kPa. The failure strain for lime/cement and cement columns could be as low as 0.5 to 2 per cent when the unconfined compressive strength exceeds about 200 to 300 kPa as shown in Figure 8.4 (Åhnberg *et al.*, 1995). Unami and Shima (1996) have reported a failure strain of 1 per cent for cement at an unconfined compressive strength of 1 MPa. A failure strain of 0.77 per cent has, for example, been reported by Holm (1994) at a peak shear strength of 5.0 MPa.

The failure strain was 0.7 to 0.8 per cent for lime/cement columns with an unconfined compressive strength of 520 to 760 kPa (Kivelö, 1996). Ekström (1994b) found that the failure strain at unconfined compression tests with lime/cement columns was 1.8 to 2.2 per cent at a shear strength of 130 kPa. A strain of only 0.1 to 1 per cent has been reported by Tatsuoka *et al.* (1996), by Tatsuoka and Kobayashi (1983) and by Terashi and Tanaka (1981) for soils stabilized by cement.

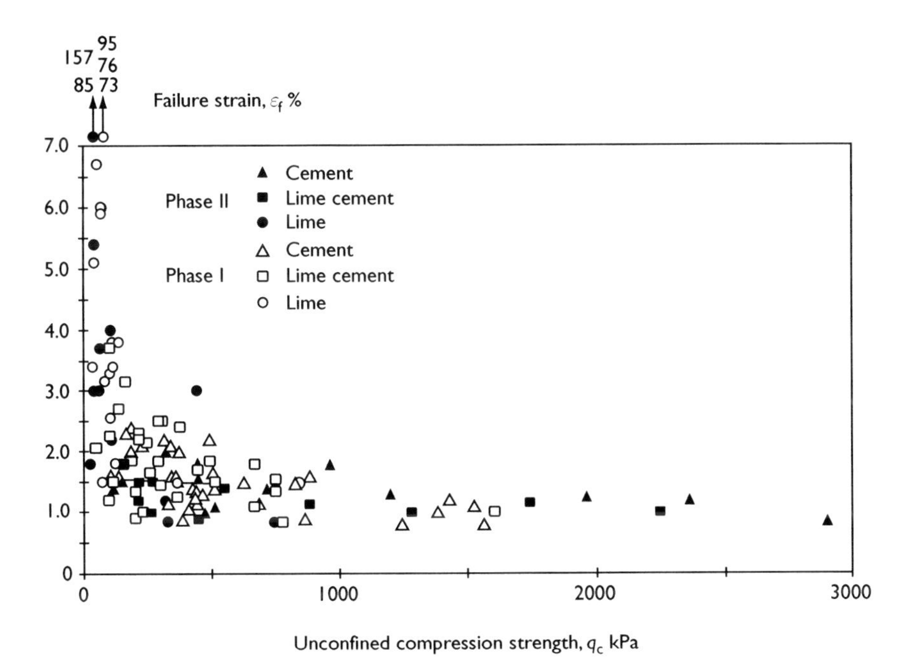

Figure 8.4 Failure strain at unconfined compression tests.

It should be noted that the strain gradient affects the failure strain, which for eccentrically loaded columns is probably 2 to 3 times higher than the failure strain at unconfined compression tests. The failure strain has also a tendency to decrease with time as the shear strength of the columns increases (i.e. Brandl, 1995; Sandros and Holm, 1996).

The axial strain at the peak shear strength increases in general with increasing confining pressure. Tatsuoka and Kobayashi (1983) have reported that the failure strain increased from less than 1.1 per cent to more than 15 per cent when the confining pressure was increased from 20 to 686 kPa. The failure strain increased from 1.5 per cent at a confining pressure of 50 kPa to about 8 per cent at a confining pressure of 400 kPa as reported by Terashi and Tanaka (1981).

Balasubramaniam and Buensuceso (1989) found for soils stabilized with lime, that the failure strain increased from 2.5 per cent at a confining pressure of 50 kPa as determined by triaxial tests to about 10 per cent at a confining pressure of 400 kPa. Rogers and Lee (1994) found that the failure strain increased from about 2 per cent to about 13 per cent with 3 per cent lime, when the confining pressure was increased from 0 to 600 kPa. Ekström (1994b) has reported for lime/cement columns, that the failure strain increased from 7.5 per cent when the confining pressure was 60 kPa to more than 11 per cent when the confining pressure was 200 kPa.

The failure strain is often higher at the wet than at the dry method at the same cement content since a higher shear strength is obtained by the dry than by the wet method. The failure strain also increases with increasing water content since the shear strength of the stabilized soil is reduced. The failure strain is usually higher for organic soils than for inorganic soils.

The failure strain is often higher for lime columns than for lime/cement and cement columns because of the often slow increase of the shear strength with lime. The difference of the failure strain between lime and cement columns is usually smaller at drained than at undrained tests. The failure strain is about the same at the same shear strength (Ekström, 1994b). The failure strain is also low for soils stabilized by fly ash (Brandl, 1995) even when the unconfined compressive strength of the stabilized soil is low, less than about 100 kPa. Kujala and Nieminen (1983) have reported that the strain at the peak shear strength was reduced from 4.1 per cent with lime to 1.4 per cent with lime and gypsum.

The failure strain is usually sufficient when the undrained shear strength is less than 100 to 150 kPa so that the stability of embankments and slopes can be evaluated from the peak shear strength of the stabilized soil. However, this failure strain could be too low with gypsum and fly ash especially when the shear strength of the stabilized soil is high and the confining pressure is low. Progressive failure could also affect the bearing capacity of lime/cement and cement columns and of columns stabilized by gypsum and fly ash when the shear strength is high.

8.2.14 Increase of shear strength with time

The shear strength and the bearing capacity of lime, lime/cement and cement columns increase with time as well as the modulus of elasticity and the compression modulus. The increase of the shear strength is initially faster with cement than with lime. The additional increase of the shear strength after 30 days is often small for cement while for lime the increase continues for several years. Brandl (1981, 1995) and Nagaraj *et al.* (1996) report that the shear strength increases with $\log t$. Also Sherwood (1993) indicates that the shear strength increases with $\log t$ with cement when the clay content is low. Okumura and Terashi (1975), Bredenberg (1979), Åhnberg and Holm (1986) and Kujala *et al.* (1993) have reported, however, that the undrained shear strength increases with $\sqrt{t}$.

8.2.15 Increase of the shear strength with lime with time

A 10 to 20-fold increase of the shear strength can usually be expected for inorganic clay with quick lime. Typically 50 per cent of the final shear strength is obtained after 1 month and 75 per cent after 3 months. About 90 per cent of the long-term shear strength is expected after 1 year. The increase of the undrained shear strength is initially relatively slow with lime compared with lime/cement and cement. Some increase of the shear strength can occur even after 200 days with lime (Eriksson and Carlsten, 1995). The increase of the shear strength with lime is usually faster in the field than in the laboratory due to the high ground temperature *in situ* and the high confining pressure. The ground temperature affects mainly the pozzolanic reactions in the soil.

The long-term increase of the shear strength depends partly on the type of clay mineral present in the clay fraction. The increase is often small with lime when the clay content is low, and the soil consists mainly of silt. The pozzolanic reactions, which are similar to those caused by hydration of cement, are slow. The pozzolanic reactions increase with increasing ground temperature and with increasing pH-value ($pH > 12$) since the solubility of silicate and aluminate increases with increasing pH-value. The rate is reduced when the water content is high.

8.2.16 Increase of the shear strength with lime/cement and cement with time

The shear strength increases initially faster with lime/cement and cement than with lime. The increase is usually very rapid the first month after the installation of the columns especially when finely ground cement is used as binder. The increase of the shear strength is usually small with lime/cement and cement after 2 to 3 months. After 3 months the shear strength could be lower for many inorganic soils with lime/cement and cement compared with lime even when the shear strength with lime is low 1 month after the mixing.

The increase of the shear strength with lime/cement is initially slow for clays when the sulphide content is larger than 1 to 3 per cent. The shear strength increases gradually with time and could eventually be about the same as for clays with a low sulphide content. Often the increase is slower in the laboratory than in the field.

The shear strength next to the columns increases with time due to consolidation and diffusion. The diffusion rate for the Ca-ions is very low, about 10 mm/year (Holeyman *et al.*, 1983). Åhnberg *et al.* (1995) found that the concentration of Ca-, K- and SO_4-ions had increased up to a few centimetres around the columns, 3 to 6 months after the installation.

8.2.17 Increase of the shear strength with gypsum, fly ash and other additives

The shear strength increases very fast with gypsum and lime the first 2 to 3 months after the mixing (Nieminen, 1978). The long-term undrained shear strength is often higher than the shear strength with only lime. Kujala and Nieminen (1983) and Kujala (1983b) report that the shear strength with lime and gypsum has been 2 to 4 times the shear strength with lime only, 1 month after the mixing. After 1 year the shear strength with lime and gypsum was about twice the shear strength with only lime. The maximum shear strength was 200 kPa.

Soils with an initial water content of up to 140 per cent have been stabilized successfully with lime and gypsum (Holm *et al.*, 1983a; Holm and Åhnberg, 1987). The needle-shaped ettringite particles, which are formed when the SO_4-ions in the gypsum react with the clay fraction, contribute also to the shear strength (Nieminen, 1979; Holm *et al.*, 1983a; Kujala, 1983b). The volume of the stabilized soil is increased and the water content is reduced, since the amount of water bound by the ettringite is very large. The increase of the *in-situ* shear strength with lime and gypsum can at least partly be attributed to the volume increase and the resulting increase of the lateral pressure around the columns by the ettringite.

There is a risk that the long-term shear strength and the bearing capacity could be reduced when the pH-value is reduced since ettringite is stable only when $pH > 10$. It is therefore recommended that gypsum should not be used to increase the long-term stability of embankments, slopes and excavations.

Cement and gypsum as well as fly ash and cement have been used in Finland to stabilize peat (Ravaska and Kujala, 1996). Kuno *et al.* (1989) report that slaked lime and gypsum could increase the shear strength of soils with an initial water content as high as 300 to 400 per cent. Nieminen (1979) found that the cement content could be reduced with fly ash, slag and gypsum.

Peat has been stabilized in Japan and India by fly ash and gypsum (Ansano *et al.*, 1996; Mishra and Srivastava, 1996). Fly ash has been effective in many silty and clayey soils with a low clay content (Nieminen, 1979; Holm and Åhnberg, 1987). However, the short-term strength could be reduced with fly ash while the long-term shear strength could be higher than the shear strength with only cement. Blast furnace slag and fly ash have been found to contribute to the pozzolanic reactions.

8.2.18 Deterioration of lime, lime/cement, cement columns and of columns stabilised by other additives

Information about the long-term performance of lime, lime/cement and cement columns is very limited. The shear strength is affected by, for example, the pH-value of the ground water. The shear strength and the bearing capacity could be reduced when pH < 4.5 to 5.0 and the concentration of Ca-ions in the flowing ground water is low.

Volume changes caused by the ettringite can also be detrimental. Mitchell (1986) has reported that the bearing capacity of pavements constructed on expansive soil, which have been stabilized by unslaked lime, has been reduced. The pavement behaved initially satisfactory but started to deteriorate after about 2 years. The reduction of the shear strength was attributed to the gradual increase of the volume by the ettringite, caused by sodium sulphate (Na_2SO_4) and gypsum ($CaSO_4$) present in the soil. The expansion reduced the bonding and thus the shear strength.

A volume increase can also occur with time, of lime and lime/cement columns when the flowing ground water contains, for example, sodium sulphate. The bonds between the individual soil particles in the soil are broken by the large volume increase by the ettringite. The long-term bearing capacity is reduced if the pH-value of the stabilized soil is reduced (Kujala, 1983b). The long needle-shaped ettringite particles could also increase the shear strength.

8.2.19 Tensile strength of lime, lime/cement and cement columns

The tensile strength of soils stabilized by lime, lime/cement and cement has been determined by the Brazilian tension test. The tensile strength has been low for lime/cement and cement columns, typically 10 to 20 per cent of the unconfined compressive strength (Terashi *et al.*, 1983b). Terashi *et al.* (1980) and Okabayashi and Kawamura (1991) report that the tensile strength has been about 15 per cent of the unconfined compressive strength. The tensile strength as evaluated by bending tests has been as high as 25 per cent of the unconfined compressive strength and thus higher than the tensile strength determined by direct tension tests. The tensile strength increases in general with increasing lime content and with increasing time after the mixing.

8.3 Compression modulus, modulus of elasticity and shear modulus

8.3.1 Compression modulus, M_{col}

The compression modulus as determined by consolidation tests is normally used to estimate the settlements. The initial compression modulus M_i is constant up to about the apparent consolidation pressure as shown in Figure 8.5. Thereafter the compression modulus is reduced to $M_{min.}$ The compression modulus increases approximately linear with increasing consolidation pressure when the load is increased further as illustrated in Figure 8.5.

The apparent pre-consolidation pressure is about 0.8 to 1.6 times the undrained shear strength according to Åhnberg *et al.* (1995a). Terashi and Tanaka (1983a), and Okumura and Terashi (1975) have reported that the compression modulus is about 1.3 times the unconfined compressive strength. Kohata *et al.* (1996) found for cement columns that the apparent pre-consolidation pressure was 1.2 to 1.7 times the unconfined compressive strength.

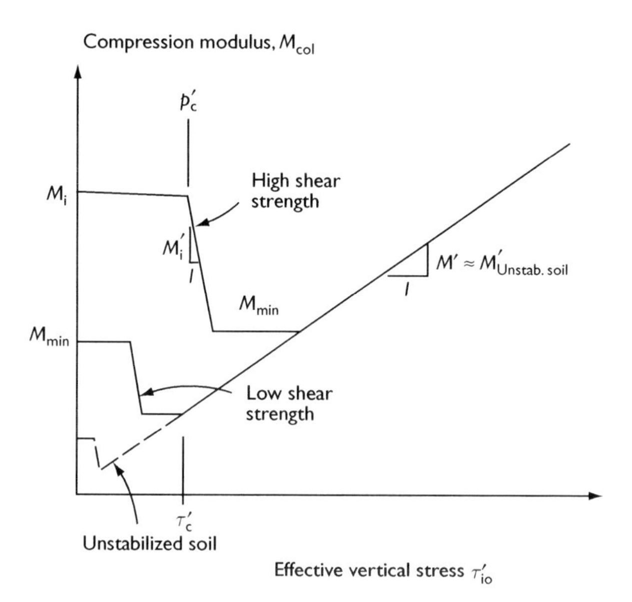

Figure 8.5 Compression modulus M_{col}.

The equivalent pre-consolidation pressure p'_c for lime/cement and cement has been 1.0 to 1.45 times the unconfined compressive strength, when p'_c is 200 kPa. At $q_{u,col}$ equal to 500 kPa the equivalent pre-consolidation pressure has been 0.15 to $0.3q_{u,col}$. The apparent pre-consolidation pressure is $1.7q_{u,col}$ at a c/p'-ratio of 0.30 and about $2.0q_{u,col}$ at a c/p'-ratio of 0.25. It is proposed that a value of $1.7q_{u,col}$ can be used to estimate the apparent pre-consolidation pressure p'_c.

The compression modulus $M_{o,lab}$ as determined by CRS-tests up to p'_c is typically 40 MPa for cement columns about 2 months after the installation and 10 to 30 MPa for lime/cement columns (Ekström, 1992). Åhnberg *et al.* (1995a) have reported that $M_{o,lab}$ as determined by oedometer tests (CRS-tests) is 60 to $300c_{u,col}$ up to p'_c. A somewhat lower ratio has been obtained for lime and gypsum. The compression modulus is often lower for lime columns than for lime/cement and cement columns at the same undrained shear strength. The *in-situ* compression modulus $M_{o,col}$ has been up to 5 times the oedometer value $M_{o,lab}$. The $M_{o,col}/c_{u,col}$-ratio increases in general with increasing shear strength.

Rogbeck and Tränk (1995) have recommended that a compression modulus of 50 to $150c_{u,col}$ can be used for lime/cement columns. The lower value, $50c_{u,col}$, is for organic soils and the higher value, $150c_{u,col}$, is for silty clay. Ekström (1992) has reported that $M_{o,col}$ is 50 to $150c_{u,col}$ for lime columns with an undrained shear strength of 100 to 400 kPa, 100 to $200c_{u,col}$ for lime/cement columns at $c_{u,col}$ between 200 and 400 kPa and 150 to $250c_{u,col}$ for cement columns. Hansbo (1994) has back-calculated an $M_{o,col}/c_{u,col}$-ratio of 100, 70 and 100 for lime/cement columns as determined by unconfined compression tests, column penetrometer tests and pressuremeter tests, respectively.

8.3.2 Modulus of elasticity, E_{50}

The $E_{50}/c_{u,col}$-ratio has been larger for cement than for lime and lime/cement columns. This ratio has also been larger for samples obtained from columns than for samples prepared in the laboratory. Lahtinen and Vepsäläinen (1983) have back-calculated an average modulus of elasticity of 15 to 25 MPa for lime columns. Similar results have been reported by Vepsäläinen and Arkima (1992) for cement columns.

Ekström (1994b) determined an $E_{50}/c_{u,col}$ ratio of 150 for samples from lime/cement columns and $130c_{u,col}$ for cement columns when the shear strength exceeded 50 kPa. For laboratory samples E_{50} was $75c_{u,col}$ with lime and $200c_{u,col}$ with lime/cement and cement. Baker *et al.* (1997) determined E_{50} to be 100 to 170 MPa for lime/cement columns, which corresponds to 250 to $350q_{u,col}$. Ansano *et al.* (1996) found that the $E_{50}/q_{u,col}$-ratio varied between 140 and 500 for cement columns and between 50 and 300 for soils stabilized by fly ash, gypsum and cement. Steensen-Bach *et al.* (1996) have reported values of 45 to 105 MPa for lime/cement columns as determined by undrained triaxial tests. The undrained shear strength was 127 to 225 kPa. The drained modulus was only 30 to 50 MPa.

The $E_{50}/c_{u,col}$-ratio can be assumed to be 200 for lime/cement columns and 250 to 300 for cement columns according to Carlsten and Ekström (1995, 1997) and SGF (2000). The modulus of elasticity is often estimated in Japan by standard penetration tests (SPT) to $0.7N$ in MPa, where N is the penetration resistance in blows/0.3 m (Babasaki and Suzuki, 1996). Ekström (1994b) found that the modulus as determined by unconfined compression tests with samples prepared in the laboratory, was always higher than the modulus for the columns. The difference can possibly be caused by cracks and fissures during the coring and the handling of the cores. Ekström (1994a,b) recommends that E_{col} should not be determined with laboratory samples.

Pre-loading is used to reduce the settlements and the variation of the modulus of elasticity. Carlsten and Ekström (1995, 1997) recommend, that the characteristic modulus of elasticity E_c can be assumed to $300c_{u,c}$ when the soil has been pre-loaded, where $c_{u,c} = 0.5q_{u,col}$ is the characteristic undrained shear strength of the stabilized soil.

8.4 Permeability

8.4.1 Permeability of the stabilized soil

Since the permeability of the laboratory samples often is different from the *in situ* permeability, where cracks and fissures in the columns effect the results, it is desirable to determine the permeability from the measured excess pore water pressures in the columns. The permeability can also be determined from the settlement rate of, for example, embankments and fills.

8.4.2 Permeability with lime

The permeability increases normally with lime. The permeability of samples prepared in the laboratory with lime has been low, 1.5 to 5 times the permeability of the unstabilized soil (Åhnberg *et al.*, 1994). Ekström (1992) reports that the *in situ* permeability has been 5 to 10 times higher than the permeability of laboratory samples. The difference between the permeability of the columns and the permeability of laboratory samples has been attributed to cracks and fissures in and around the columns caused by shrinkage.

Rogbeck and Tränk (1995) report that the permeability of lime columns could be 700 to 1000 times the apparent permeability of the unstabilized soil. Also Mitchell (1981) has reported that the permeability with unslaked lime could be increased up to 1000 times. Brandl (1981) indicates that an increase of the permeability by one or two orders of magnitude is possible. Pramborg and Albertsson (1992) have reported that the average permeability increased 225 times with lime. Bengtsson and Holm (1984) found with lime that k_{col} was about $100k_{soil}$, where k_{soil} is the permeability of the unstabilized soil. Åhnberg *et al.* (1995a) found that the permeability was higher for clayey silt than for clayey 'gyttja' and clay, and that the increase depended on the soil type. The permeability was higher for clayey silt than for clayey 'gyttja' and clay. Terashi and Tanaka (1983a) observed that the permeability for marine clay decreased with increasing lime content.

8.4.3 Permeability of lime/cement columns

The permeability of lime/cement columns has been determined in the laboratory by constant and by falling head permeability tests as well as by triaxial and oedometer tests with samples from the columns or with small laboratory samples. The permeability has been found to decrease with time.

Carlsten and Ekström (1995, 1997) recommend that a permeability ratio of 400 to 800 can be used in design of lime/cement columns. These recommended values appear to be too high as indicated by field tests. Lime/cement columns may not be as effective as drains due to the large hydraulic lag when the length of the columns is large.

Baker *et al.* (1997) report that the *in situ* permeability of lime/cement columns can be up to two orders of magnitude higher than the permeability of laboratory-mixed samples. The permeability as determined in the field was 51×10^{-9} to 350×10^{-9}m/sec for lime/cement columns. The permeability of the samples prepared in the laboratory was only 2×10^{-9} m/s. The permeability was observed to decrease with increasing curing time and with increasing confining pressure. The permeability has also been observed to decrease and with time even when the confining pressure is low.

Pramborg and Albertsson (1992) report that the *in situ* permeability of lime/cement columns has been 200 times the permeability of the unstabilized clay. Rogbeck and Tränk (1995) found for lime/cement columns that the permeability was about 750 to 1000 times the initial permeability. The variation of the permeability can thus be large.

Arnér *et al.* (1996) have reported, that the drainage and consolidation were small for lime/cement columns for an 8 m high test embankment at Norrala, Sweden and that the permeability of the columns was about the same as the permeability of the unstabilized soil. They found that the excess pore water pressures had dissipated fully after 5 months for the 5 to 6 m thick clay layer below the embankment in spite of the low permeability of the 6 to 8 m long columns. A permeability ratio $k_{col}/k_{v,soil}$ equal to 40 is recommended for lime/cement columns.

8.4.4 *Permeability of cement columns*

The permeability of cement columns is low. In some cases the permeability has even been lower than the permeability of the unstabilized soil (Okumura, 1996; Terashi and Tanaka, 1983a). Due to this reduction of the permeability, cement columns cannot be considered as drains. Mitchell (1981) has reported that columns stabilized with cement, fly ash or by a mixture of by-products are almost impervious. Tielaitos (1993, 1995) reports that the permeability of peat, which has been stabilized by cement decreased from about 10^{-5} m/sec to about 10^{-6} to 10^{-8} m/sec.

Suzuki (1982) has reported that the permeability decreased with increasing cement content between 15 and 20 per cent and that the permeability was reduced by two to three orders of magnitude. The permeability has also a tendency to decrease with time, as reported by, for example, Terashi and Tanaka (1981), and with decreasing water content. At a water content of 100 per cent, the permeability of a clay with an initial water content of 95 per cent and a liquid limit of 90 was reduced by two orders of magnitude, when the cement content was increased from 5 to 15 per cent.

8.4.5 *Permeability with gypsum*

Gypsum increases the permeability due to the volume increase caused by the formation of ettringite when gypsum reacts with the lime and the clay in the columns.

8.5 Choice of stabilization method

8.5.1 *General*

Both the undrained and the drained shear strengths are increased with lime, lime/cement, cement and by different waste products. Mainly inorganic soils with a water content less than 100 to 120 per cent can be stabilized with lime. Lime/cement or cement is required to stabilize organic soils. Cement in combination with granulated blast furnace slag has often been effective to stabilize peat, 'dy', 'gyttja' and several other organic soils. Also sand has been effective, when the water content of the soil is high.

8.5.2 *Stabilization of inorganic soils*

Unslaked lime (quicklime) is usually effective in normally consolidated or slightly overconsolidated inorganic clays with a medium to low plasticity index and a high sensitivity ratio. However, the plasticity index had to be larger than 10 per cent for lime to be effective. The increase of the shear strength is typically 10 to 20 times. The maximum shear strength is usually 200 to 300 kPa.

Lime affects mainly the clay fraction through ion exchange. Unslaked lime also reduces the water content. Cement affects mainly the silt and the sand fractions. It is thus expected that lime/cement and cement will be more effective than lime to stabilize silty or sandy soils as well as organic soils when the clay content is low. Lime together with industrial waste products has also been used successfully (Kamon and Nontananandh, 1991).

A high shear strength can usually be obtained with lime/cement and cement compared with lime. The increase of the shear strength with lime is often large for clays with a high salt content since the pozzolanic reactions are increased by the chloride ions (Cl-ions). The increase of the shear strength has, however, been small when the clay content is low (Carlsten and Ekström, 1995, 1997). A high ground temperature also contributes to a high shear strength.

Sodium and potassium sulphate have a tendency to reduce the shear strength due to the formation of ettringite. The initial increase is often slow with lime compared with lime/cement and cement. It is preferable to stabilize clay and silty clay with quicklime when the water content is less than about 90 to 100 per cent and the organic content is low.

The main advantage with lime is the high permeability of the stabilized soil and that lime columns function as drains. The ductility is high when the shear strength is low, less than 100 to 150 kPa. Lime is not very effective in organic soils or in silty soils especially when the salt and the clay contents are low.

8.5.3 *Stabilization of organic soils*

Lime/cement and cement are recommended for organic soils when the required shear strength cannot be obtained with unslaked lime. Lime/cement should be considered, when an undrained shear strength of at least 25 to 50 kPa cannot be obtained with lime alone. The largest increase of the shear strength with lime/cement is expected for clays with a low sensitivity ratio and a water content between 40 and 80 per cent. The undrained shear strength before stabilization should be less than 10 to 20 kPa. Organic clays with an undrained shear strength as low as 5 kPa have been stabilized successfully with lime/cement and cement.

The type of organic material is generally more important with respect to the increase of the shear strength than the amount. The reaction of the organic material with the hydrated products reduces the pH-value and thus the pozzolanic reactions. Babasaki *et al.* (1996) have reported that the increase of the unconfined compressive strength has been small for soils with a water content larger than 200 per cent even with 35 per cent cement. Sand and gypsum have also been tried.

Lime/cement and cement had to be mixed thoroughly with the soft soil compared with lime and be distributed evenly over the column cross-section. The increase of the shear strength is often 3 to 10 times faster with cement or lime/cement compared with lime 1 to 3 months after the mixing (Åhnberg *et al.*, 1995). It is generally more difficult to mix soft soil with dry cement than with cement slurry. It is therefore preferable to use the wet method to stabilize organic soils, especially when the required cement content is high. However, the shear strength will generally be lower by the wet method at the same cement content compared with the dry method due to the difference of the water content.

The shear strength of samples prepared in the laboratory is often higher than the *in-situ* shear strength. (Kujala and Lahtinen, 1988). The difference has been attributed to poor mixing in the field.

8.5.4 Stabilization of 'dy' and 'gyttja'

'Gyttja' and 'dy', where the plant structure has been destroyed completely, have been stabilized successfully with cement and with cement and blast furnace slag (Holm, 1994; Axelsson *et al.*, 1996). The increase of the shear strength with unslaked lime is usually small for organic soils such as 'dy' and 'gyttja'. The shear strength with 200 kg/m^3 cement has been 2.6 to 6.5 times higher for 'gyttja' than with the same amount of lime/cement (50/50) as reported by Axelsson *et al.* (1996). Mainly the cement has been effective. Even a small amount of humus, 2 to 3 per cent, can reduce substantially the increase of the shear strength. The increase of the shear strength has in general been larger with rapid hardening cement than with standard Portland cement.

The largest increase has been obtained with cement. A 10 to 20-fold increase of the shear strength can often be obtained for 'gyttja'. However, the peak shear strength cannot often be utilized fully in design (Kivelö, 1997, 1998) due to progressive failure caused by the low ductility when the shear strength is high.

Cement or lime/cement is usually required to stabilize organic soils such as peat, 'dy' and 'gyttja' even when the organic content is low (Axelsson *et al.*, 1996). The shear strength increases in general with increasing lime and cement content and with increasing time after the mixing. The shear strength is reduced when the sulphide (FeS_2) content is high (e.g. Bryhn *et al.*, 1983; Kujala, 1984).

8.5.5 Stabilization of sulphide soils

With lime/cement or cement it is often possible to stabilize soils with a high sulphide content mainly iron pyrite (FeS_2) i.e. 'svartmocka'. Such soils are common along the Bothnian coast in northern Sweden and Finland. Several excavations have failed where the sulphite content of the soil has been high. A reduction of the shear strength by 50 per cent has been required in some cases in stability calculations (Schwab, 1976). The volume increase of the soil can be large due to oxidation and the formation of calcium sulphate ($CaSO_4$). However, results from only a few tests are available, which indicate the effectiveness of lime/cement or of cement.

The shear strength of soils with a high sulphide content is generally low, also with cement, especially when the organic content is high. The increase of the shear strength with time is slow. About 2 to 3 months are often required before the columns can be loaded.

8.5.6 *Stabilization of peat*

Peat is common both in Sweden and Finland. About 8 per cent of Sweden is covered by peat, which is usually classified with respect to the decomposition (H1 to H10) of the peat according to a scale proposed by von Post. Fibrous peat (H1 to H4) has a distinct plant structure compared with amorphous peat (H8 to H10) where the plant structure is indistinct. A high cement content is normally required as discussed by Axelsson *et al.* (1996). An undrained shear strength of 30 to 50 kPa can usually be obtained after 14 days with 200 kg/m^3 cement.

The shear strength has been observed to decrease with cement with increasing decomposition of the peat and with increasing content of humic acid and fines (Kujala, 1984). The increase is generally small and slow when the content of humus is high and the pH-value is low. Even a humus content as low as 1.5–2 per cent can reduce the shear strength significantly.

The largest increase of the shear strength has been observed for peat and 'gyttja' stabilized by 50 per cent rapid hardening cement and 50 per cent granulated blast furnace slag. It should be noted that the shear strength increases slowly with blast furnace slag alone especially when the ground temperature is low.

Huttunen and Kujala (1996) have reported an undrained shear strength of 305 kPa after 180 days for a poorly decomposed peat (H2). The cement content was high, 400 kg/m^3, and so was the water content, 1265 per cent. A shear strength of 112 kPa was observed after 180 days for a decomposed peat of grade H3 at a cement content of 400 kg/m^3 and a water content of 981 per cent. Hoikkala *et al.* (1997) found that the shear strength of peat was about 105 kPa after 30 days with 300 kg/m^3 cement and 35 to 75 kPa at a cement content of 150 kg/m^3. A high cement content is thus required when the water content is high.

The shear strength of peat increases in general with increasing cement content and with decreasing water content. Kuno *et al.* (1989) have reported that peat could be stabilized with cement at a water content less than 300 per cent irrespective of the content of humus. Axelsson *et al.* (1996) have investigated the shear strength of peat stabilized by 70 to 400 kg/m^3 rapid hardening cement. The undrained shear strength increased from 80 kPa to over 300 kPa when the cement content was increased from 150 to 250 kg/m^3. However, the shear strength was reduced to 280 kPa when the cement content was increased to 400 kg/m^3. An increase of the shear strength has been observed when water was added during the mixing (Axelsson *et al.*, 1996) when the initial water content of the peat was low.

The shear strength can often be increased with gypsum and cement. The increase with fly ash and lime has been poor (Axelsson *et al.*, 1996). Fine sand and silt has been added to increase the effectiveness of cement and of cement and granulated blast furnace slag. At Kyrkslätt close to Helsingfors in Finland the increase of the shear strength of peat and 'gyttja' was large with rapid hardening cement and sand. The shear strength was over 600 kPa with 275 kg/m^3 rapid hardening cement and with 100 kg/m^3 fine sand.

Several commercial products are available in Sweden and Finland, e.g. Finnstabi and Lohjamix to stabilize peat and 'gyttja'.

8.5.7 Cost

Lime and lime/cement columns have been competitive in Sweden and Finland compared with other soil improvement and soil stabilization methods, when only a marginal improvement of the settlement or of the stability is required in spite of the high cost for finely ground quicklime compared with cement. However, the difference in total cost is often small since less lime than cement is usually required when the organic content is low. Longer time is also required for the installation of lime/cement and cement columns compared with lime columns due to the higher cement content and the lower retrieval rate during the mixing. Unslaked lime is often difficult to store in a humid climate.

It is normally economical to use large-diameter columns since the unit cost decreases rapidly with increasing diameter. However, large-diameter columns can normally be used only in soft clay because of the high torque, which is required during the mixing when the shear strength is high. Also the distribution of stabilizer over the cross-section is frequently uneven for large diameter columns ($\geq$0.8 m) because of the limitation of the maximum air pressure that can be used at the dry method.

A comparison by Nord (1990) and by Carlsten and Ouacha (1993) has shown that the cost to stabilize embankments with lime columns in Sweden has only been one-third of the cost for embankment piles, which was the alternative method. A reduction by 30 per cent of the cost has been reported by Holm (1979).

8.6 Required lime, lime/cement and cement content

8.6.1 Required lime content

The lime content, which is required to stabilize soft inorganic clay is typically 70 to 90 kg/m^3, 6 to 8 per cent with respect to the dry weight of the soil. The shear strength increases rapidly with increasing lime content up to about 10 to 12 per cent. The optimum lime content is 6 to 12 per cent.

8.6.2 Required lime/cement content

A lime/cement content of 15 to 20 per cent with 50 per cent quicklime and 50 per cent cement (50/50) is usually required. Also other proportions are used (i.e. 25 per cent lime and 75 per cent cement (25/75) as well as 75 per cent lime and 25 per cent cement (75/25)). Eriksson and Carlsten (1995) have reported that the shear strength with 25 per cent lime and 75 per cent cement was about the same as with 75 per cent lime and 25 per cent cement. Holm *et al.* (1983a) recommend that 75 per cent lime and 25 per cent cement should be used for long-term stabilization and 50 per cent lime and 50 per cent cement for temporary stabilization.

8.6.3 Required cement content

A cement content of about 4 to 13 per cent with respect to the dry weight is usually required to stabilize clayey silt, 6 to 16 per cent for clay and 16 to 40 per cent for clayey 'gyttja' (Åhnberg *et al.*, 1994). In Japan the cement content is usually 20 to 30 per cent (Okumura, 1996).

8.7 Behaviour of single columns

8.7.1 Transfer length

It is necessary to transfer the load from the embankment to the columns or past weak sections in the columns as shown in Figure 8.6. The estimated transfer length is $2d$, when the load transfer depends entirely on the shaft resistance. When the dry crust is thin or is poorly developed, the transfer length could be up to $10d$. The transfer length decreases with time because of the increase of the shear strength of the soil caused by consolidation. The time required for the consolidation can be large for lime/cement and cement columns due to the low permeability of the stabilized soil. About 10 months will be required to reach 90 per cent consolidation for a 1 m thick clay layer at $c_v = 1\,m^2$/year when the soil is drained on one side only. The transfer

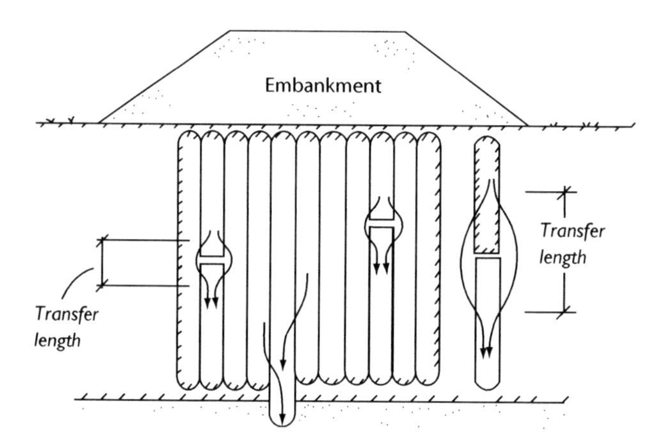

Figure 8.6 Transfer length.

length can also be large at the base of a floating column wall, when the point resistance is low.

The undrained shear strength and the shaft resistance for a normally consolidated clay (OCR = 1.0) increase with increasing effective overburden pressure. The shear strength at a depth of 10 m is estimated to 16 kPa at $c_u/p'_v = 0.22$. The point resistance is 144 kPa (9 × 16) at this depth. The estimated shaft resistance, which corresponds to the undrained shear strength, is 16 kPa.

The settlement within the transfer length L_{tr} is estimated to $0.5q_o\, L_{tr}/M_{soil}$, where M_{soil} is the compression modulus of the unstabilized soil. The transfer length, which depends mainly on the shear strength of the soil around the columns, is 3.2 m (4*d*) for a 0.8 m diameter column at an axial load of 200 kN (400 kPa) and a shaft resistance of 25 kPa, when the point resistance is zero and 4.1 m when the point resistance is 144 kPa.

8.7.2 The stress factor m

The stress factors m_{col} and m_{soil}, the ratio of the stress increase at the same depth in the columns and in the unstabilized soil respectively, and the applied unit load, is low for floating columns because of the low shear strength of the remoulded soil.

$$m_{col} = \frac{1}{\left[\frac{(a + (1 - a)M_{soil}}{E_{col}}\right]} \tag{8.4}$$

where a is the area ratio, M_{soil} is the compression modulus of the unstabilized soil and E_{col} is the modulus of elasticity of the columns. The stress factor m_{col} is equal to 3.7 at $E_{col}/M_{soil} = 10$ and $a = 0.2$. Kivelö (1994a) has estimated the stress factor to 3.0 to 4.5 for lime/cement columns. Liedberg *et al.* (1996a) have proposed similar values.

8.7.3 Weighted average shear strength of lime, lime/cement and cement columns

An average shear strength $\tau_{fu,av}$ is often used to evaluate the stability of embankments, slopes and deep excavations, when the shear strength of the stabilized soil is less than 100 to 150 kPa.

$$\tau_{fu,av} = a\tau_{fd,col} + (1 - a)\tau_{fu,soil} \tag{8.5}$$

where $\tau_{fd,col}$ and $\tau_{fd,soil}$ are the drained and the undrained shear strength of the stabilized and the unstabilized soil, respectively.

The area ratio a is typically 0.10 to 0.25 in Sweden and Finland. A much higher area ratio, 0.5 to 0.9, is often used in Japan, where it is generally assumed that the total weight of an embankment or of a fill is carried by the columns. However, the location of the columns affects the column loads as well as the shear resistance. The shear resistance will be high for the columns, which are located below an embankment or a fill. Conservative values on the shear strength and on the residual angle of internal friction $\phi'_{col,res}$ and on ϕ'_{col} as well as on the axial column loads should be used for the columns, which are assumed to fail along a circular slip surface. The axial load and the shear resistance will be low for the columns, which are located in a slope or next to an excavation or a trench. The bearing capacity will also be low for the columns, which are located in the shear or in the passive zones outside an embankment where the axial column loads are low.

8.7.4 Lateral displacement of single columns

The columns, which are located below an embankment, could be displaced laterally by the high lateral earth pressure in the embankment. However, the lateral displacement of the soft soil and of the columns below an embankment is usually small. A large lateral displacement can be expected, when the stability of the embankment is low and the global factor of safety is less than about 1.5. The lateral displacement of the columns has been up to 0.5 m next to a slope or a deep excavation.

Miyake *et al.* (1991a,b) found from centrifuge tests that the bearing capacity and the lateral resistance of cement columns had been reduced as well as the bearing capacity of the columns at failure. The reduction depends mainly on the ductility of the columns and thus on the failure strain.

The lateral displacements of 1.0 m diameter cement columns have been investigated by Kakihara *et al.* (1996). The lateral displacements were large, up to 400 mm for the columns, which had been installed by the wet method. At the dry method the maximum lateral displacement at the ground surface was 40 mm at a distance of 3 m from the columns.

It has been possible to reduce the lateral displacements by 50 per cent by a 4 m deep trench next to the columns or by 4 m deep, 0.45 m diameter air recovery holes, which are spaced 1.5 m apart. The holes were filled with crushed stone (Uchiyama, 1996). Perforated PVC pipes were placed at the centre of the holes to improve the drainage.

8.7.5 Failure modes

Kitazume *et al.* (1996b) carried out centrifuge tests to determine the failure mechanisms of cement columns. The columns failed one by one when about 10 per cent of the estimated shear resistance of the columns had been mobilized. The unconfined compressive strength was 213 to 750 kPa. Miyake *et al.* (1991a,b) have reported similar results. The columns, which were located at the toe of an embankment, where the axial load was low, failed by bending.

Possible failure modes are shown in Figure 8.7 for single columns. When the slip surface is located close to the ground surface the soil flows past the columns (failure mode a). The moment capacity of the column is sufficient in this case to resist the lateral earth pressure. Failure mode (b) with one plastic hinge is shown in Figure 8.7b. The moment capacity of the column is exceeded in this case.

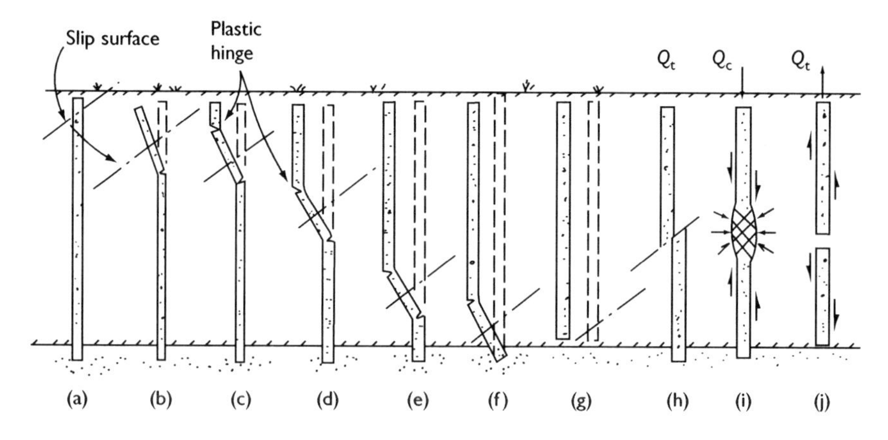

Figure 8.7 Failure modes of single columns.

Only a very small relative displacement is required to mobilize the shear resistance when two plastic hinges develop in the columns at the location of the maximum positive and the maximum negative bending moments as shown in Figures 8.7c,d,e (failure modes c, d and e).

Failure mode (f) is shown in Figure 8.7f, where the column extends down to a firm layer below the soft soil. The slip surface is located in this case close to the bottom of the layer with soft soil. Failure mode (g) is shown in Figure 8.7g where the slip surface is located close to the bottom of the soft soil and the columns move together with the soil.

Failure mode (h), when the shear resistance of the columns governs, is shown in Figure 8.7h. Compression failure of the columns is illustrated in Figure 8.7i (failure mode i). The tensile strength of the columns governs failure mode (j) as shown in Figure 8.7j.

8.7.6 *Bearing capacity of laterally displaced columns (failure mode b)*

The failure modes of the columns, which are located in the active zone, are illustrated in Figure 8.8. The columns fail when the maximum bending moment reaches the moment capacity of the column. The resulting reduction of the bearing capacity has been analyzed below for columns with a circular cross-section.

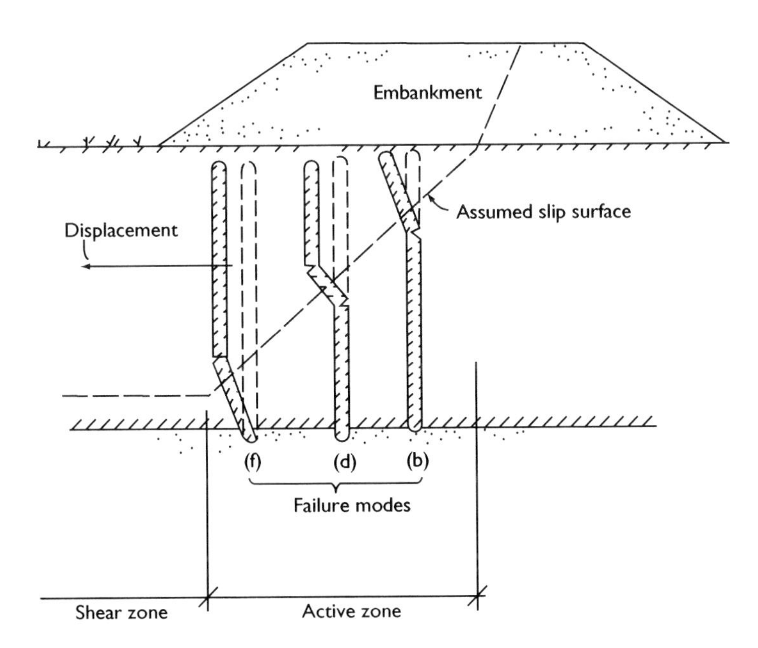

Figure 8.8 Failure mode of columns in the active zone.

The moment resistance when the effective cross-section of the columns has been reduced from A_{col} to ηA_{col} is $q_{u,col}\eta A_{col}e$ as shown in Figure 8.8, where e is the eccentricity of the applied load. It has been assumed that the tensile strength of the stabilized soil is small and can be neglected. The compressive strength $q_{u,col}$ of the columns is assumed to be constant for the column cross-section (ηA_{col}) at the location of the plastic hinge, which is located at a distance $2(f+g)$ below the top of the column and illustrated in Figure 8.9. There, the bending moment in the column reaches a maximum.

For columns with a circular cross-section,

$$0.5Q_{col}\psi\beta f\varepsilon_f d(f+g) + k_1 c_{u,soil} d(f^2 - g^2) = Q_{col}e \qquad (8.6a)$$

where Q_{col} is the axial load in the column and ($Q_{col}e$) is the moment resistance. The coefficient ψ, which depends on the load transfer from the fill to the column is equal to 4.0 when the load is transferred to the column entirely by end bearing and 2.0 when the load is transferred entirely by shaft resistance. The angular rotation of the plastic hinge is $\beta f\varepsilon_f$ and $k_1 c_{u,soil} d$ is the lateral force per unit length of the column. It has been assumed that the coefficient k_1 is equal to 2.0, when the lateral displacement is small and the failure strain corresponds to the peak shear strength of the stabilized soil.

It is likely that the failure strain ε_f could be 2 to 3 times higher than the strain at the peak strength at unconfined compression tests for eccentrically

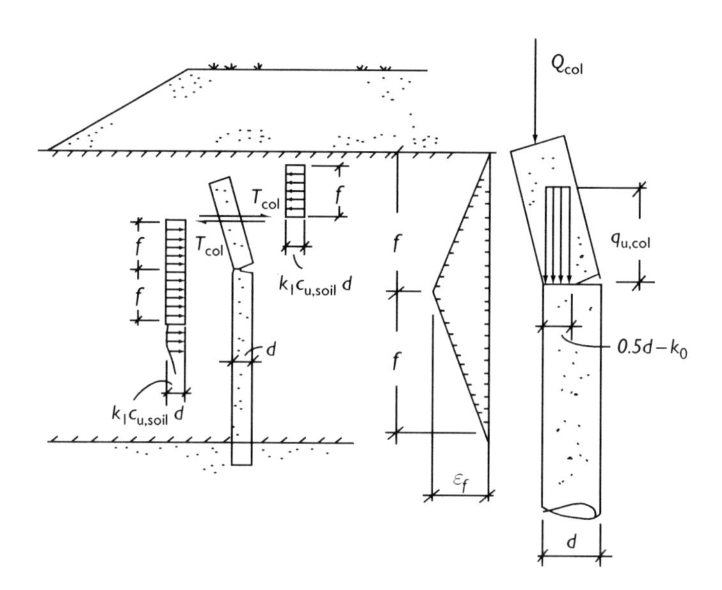

Figure 8.9 Bearing capacity of laterally displaced column (failure mode b).

loaded columns because of the high strain gradient in the column and the high confining pressure by the surrounding soil. The coefficient $\alpha = c_{u,soil}/q_{u,col}$ has been assumed to $^1/_{10}$ and $^1/_{50}$ in the following calculations.

At $Q_{col} = \eta A_{col} q_{u,col}$ and $c_{u,soil}/q_{u,col} = \alpha$, equation (8.6a) can be rewritten as:

$$\frac{(f^2 - g^2)}{d^2} + \frac{0.39\eta\beta f \varepsilon_f \psi(f+g)}{d^2 k_1 \alpha} = \frac{0.79\eta e}{d k_1 \alpha} \tag{8.6b}$$

The strain increase is reduced from ε_f at the plastic hinge to zero at distance f from the hinge, where the bending moment in the column is equal to zero. It has been assumed that the axial strain ε_o before the column has been displaced laterally is small and can be neglected. The angular rotation $\beta f \varepsilon_f$ at the location of the plastic hinge is thus independent of the size of the column. The lateral displacement is $2(f+g)\beta f \varepsilon_f$ at the top of the columns at the distance $2(f+g)$ from the plastic hinge as illustrated in Figure 8.9.

At k_1 $\alpha = 0.02$, $\psi = 4.0$, $g = 0$ and $\varepsilon_f = 0.01$ (1 per cent) the lateral displacement is 80 mm for a 0.6 m diameter column at $Q_{col} = 0.196 Q_{u,col}$ where $Q_{u,col}$ is the ultimate bearing capacity of the concentrically loaded column. The angular rotation at the plastic hinge $\beta f \varepsilon_f$ is 0.059. The tensile strength of the stabilized soil has been neglected as well as the lateral displacement caused by the elastic deformations of the column (Figure 8.10).

The shear resistance of the columns at failure is equal to $2T_{col}/A_{col}q_{u,col}$ as shown in Figure 8.11. It can be seen that the shear resistance is low, when the axial load is high. At $\alpha = 0.02$, $\varepsilon_f = 0.01$, $\psi = 4.0$, $Q_{col} = 0.196 Q_{u,col}$ and $g = 0$, the shear resistance $2T_{col}/A_{col}q_{u,col}$ is only 6.5 per cent of the shear strength of the columns.

Both the failure strain ε_f and the unconfined compressive strength of the column $q_{u,col}$ affect the shear resistance $2T_{col}$, which is less than $A_{col}q_{u,col}$, when the unconfined compressive strength of the columns is much larger than the shear strength of the unstabilized soil around the columns.

The lateral displacement of lime/cement columns is small ($\varepsilon_f < 2$ per cent) at failure compared with the lateral displacement of lime columns ($\varepsilon_f > 2$ per cent) mainly due to the difference in compressive strength. The lateral displacement at failure increases rapidly with increasing failure strain ε_f.

8.7.7 *Bearing capacity of laterally displaced columns, failure mode (c), (d) and (e)*

The shear resistance of lime/cement columns when they function as dowels, has been analyzed by Kivelö (1997, 1998) based on the method proposed by Brinch Hansen (1948) and Broms (1972). This failure mode governs when the shear resistance of the stabilized soil is high and the failure strain is low

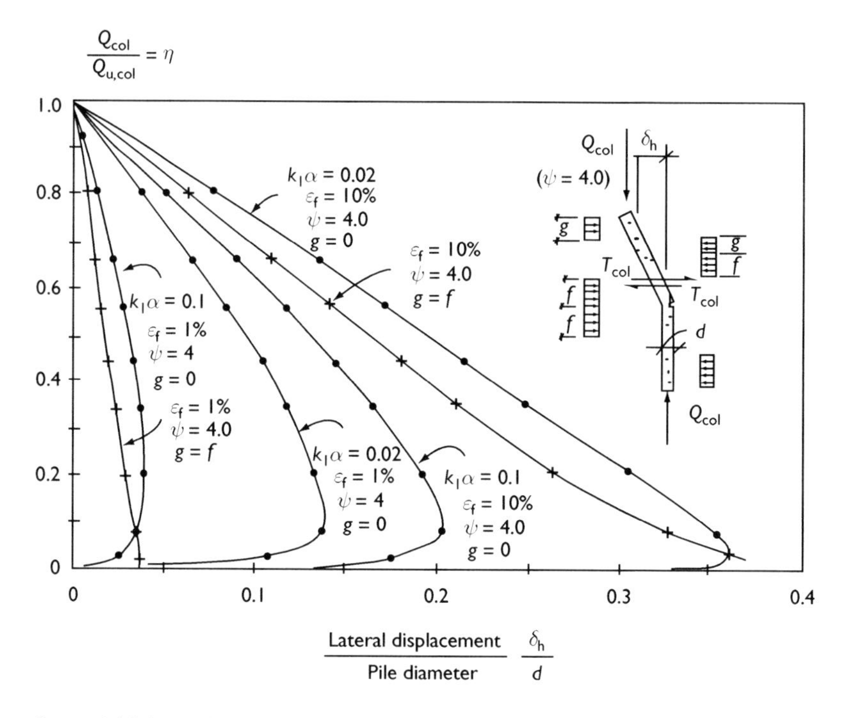

Figure 8.10 Lateral displacement of columns (failure mode b).

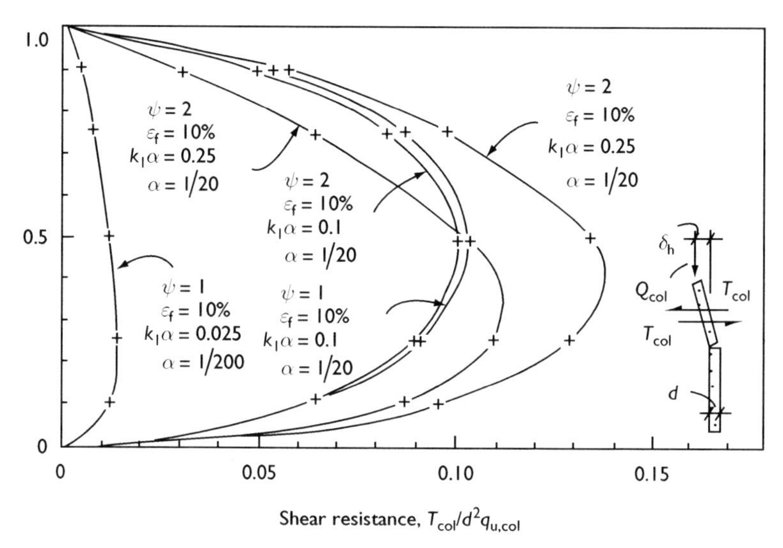

Figure 8.11 Shear resistance of columns (failure mode b).

as shown in Figure 8.12. The length f can be estimated by the following equation:

$$0.5k_1 c_{u,soil} d f^2 + Q_{col} \beta \varepsilon_f f^2/d = Q_{col} e \quad (8.7a)$$

where $c_{u,soil}$ is the shear strength of the unstabilized soil, $\beta f \varepsilon_f$ is the rotation at the plastic hinge, d is the diameter of the column, f is the length of the column when the lateral earth pressure is $dk_1 c_{u,soil}$ and Q_{col} is the axial load.

Equation (8.7a) can be rewritten since $Q_{col} = \eta A_{col} q_{u,col}$ and $c_{u,soil}/q_{u,col} = \alpha$.

$$\left(\frac{f}{d}\right)^2 + \frac{1.57\varepsilon_f \beta f^2 \eta}{d^2 k_1 \alpha} = \frac{1.57 e \eta}{d k_1 \alpha} \quad (8.7b)$$

where ηA_{col} is the loaded area of the column. The lateral displacement at failure of the columns is shown in Figure 8.13.

The shear force T_{col} perpendicular to the column can be estimated by the following equation when the axial load in the column is small and can be neglected as would be the case for a shallow landslide.

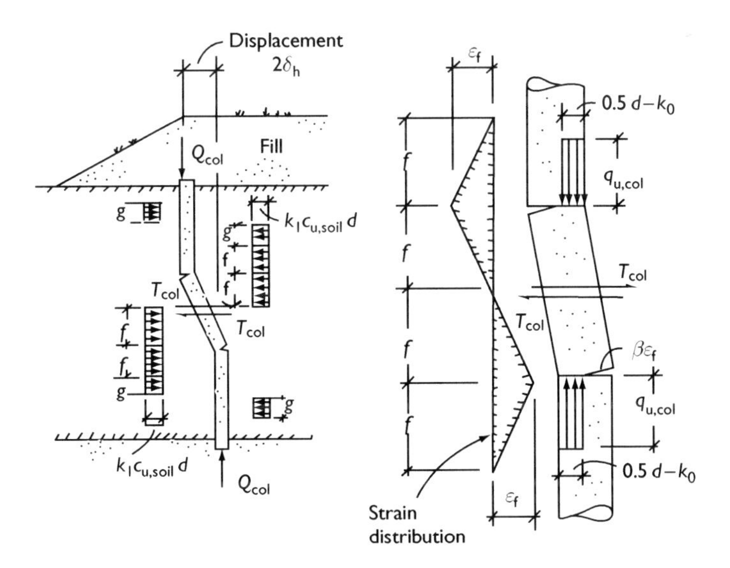

Figure 8.12 Bearing capacity of laterally displaced column (failure modes c, d and e).

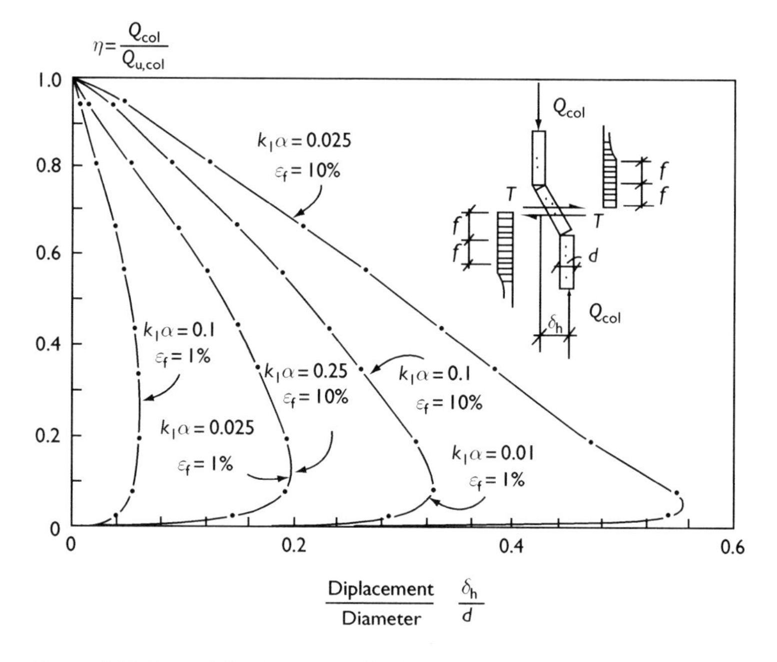

Figure 8.13 Lateral displacement of columns (failure modes c, d and e).

$$\frac{T_{col}}{A_{col}} = \sqrt{\frac{2k_1\, c_{u,soil} d M_{col}}{A_{col}^2}} \tag{8.8}$$

where M_{col} is the moment resistance of the column, $c_{u,soil}$ is the undrained shear strength of the unstabilized soil and d is the diameter of the column. The factor k_1 depends on the relative lateral displacement of the columns with respect to the soil. When the relative displacement is only a few mm, k_1 is 2.0. The moment capacity M_{col} depends on the tensile strength of the stabilized soil, when the axial load is low. The shear resistance of the columns at failure modes c, d and e is shown in Figure 8.14.

It is an advantage to use the wet method with cement as stabilizer to increase the stability of slopes where the axial load in the columns is small. The uniformity of the stabilized soil will improve especially when the shear strength of the unstabilized soil is low. A tensile strength is often high with cement.

The shear resistance $2T_{col}/A_{col}q_{u,col}$ had to be larger than $\tau_{fu,col}$ or $\tau_{fd,col}$ depending on time and the drainage conditions. Otherwise the shear resistance as evaluated by equations (8.1) and (8.2) will govern. The shear resistance $2T_{col}/A_{col}q_{u,col} = 0.054$ at $k_1\alpha = 0.025$, $\varepsilon_f = 0.1$ and $\eta = 0.196$.

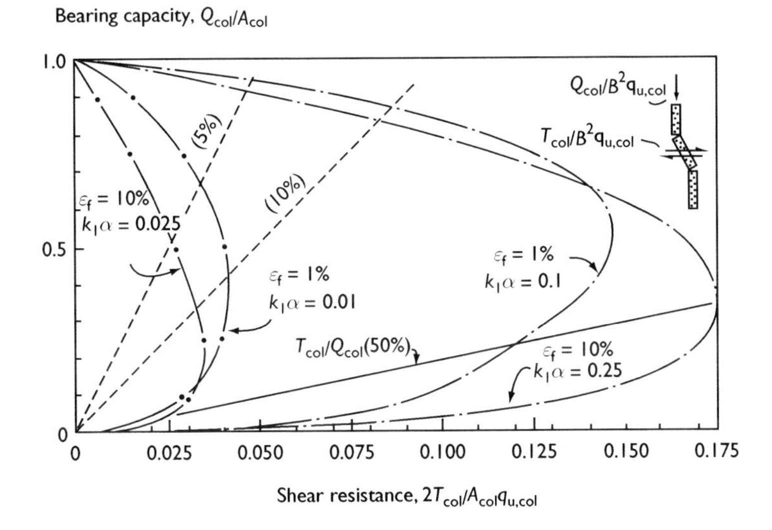

Figure 8.14 Shear resistance of columns (failure modes c, d and e).

Thus the shear resistance at failure of a single column will almost always govern, when two plastic hinges develop in the column.

8.7.8 Bearing capacity of laterally displaced columns, failure mode f

The failure mode of a laterally displaced column when the columns extends into a hard or a firm layer is shown in Figure 8.15. The slip surface is located $(g+f)$ above the hard layer as shown. There is a high concentrated force R_{col} at the tip of the columns due to the high lateral resistance of the hard layer. The force R_{col} is equal to $k_1 c_{u,soil} dg$.

It should be noted that the moment at the plastic hinge is equal to $k_1 c_{u,soil} d(f^2 - 0.5g^2)$ and that

$$Q_{col}\beta f \varepsilon_f(2f - g) + k_1 c_{u,soil} d(f^2 - 0.5g^2) = Q_{col} e \tag{8.9a}$$

This equation can be rewritten since $Q_{col} = \eta A_{col} q_{u,col}$ and $c_{u,soil}/q_{u,col} = \alpha$.

$$\frac{(f^2 - 0.5g^2)}{d^2} + \frac{0.79\eta\beta f \varepsilon_f(2f + g)}{d^2 k_1 \alpha} = \frac{0.79\eta e}{dk_1\alpha} \tag{8.9b}$$

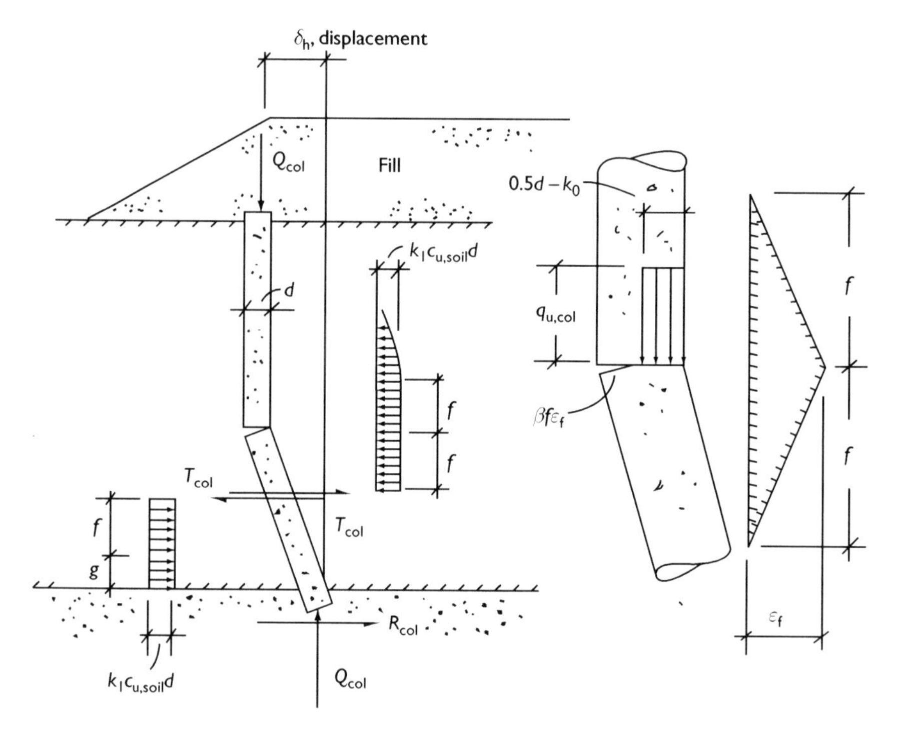

Figure 8.15 Bearing capacity of laterally displaced columns (failure mode f).

The lengths f and g can be calculated from this equation. The lateral displacement of the columns at failure is shown in Figure 8.16. It can be seen that the failure strain ε_f has a large effect on the lateral displacement at failure of the columns.

The shear resistance $2T_{col}/A_{col}\ q_{u,col}$ of the column is less than 0.11 at $k_1\alpha = 0.01$, as can be seen from Figure 8.17 and can be neglected. It can also be seen that $2T_{col}/A_{col}q_{u,col}$ increases with increasing $k_1\alpha$-ratio and thus with decreasing shear strength of the stabilized soil. At $k_1\alpha = 0.1$ and $\varepsilon_f = 1$ per cent and $g = f$, the shear strength of the stabilized soil is at least three times larger than the shear strength of the unstabilized soil.

8.8 Column rows, grids, arches and blocks

8.8.1 *Failure of column rows*

Possible failure modes are shown in Figure 8.18. The unstabilized soil is extruded between the column walls in Figure 8.18a when the slip surface is located close to the top of the column row. A column row fails by overturning

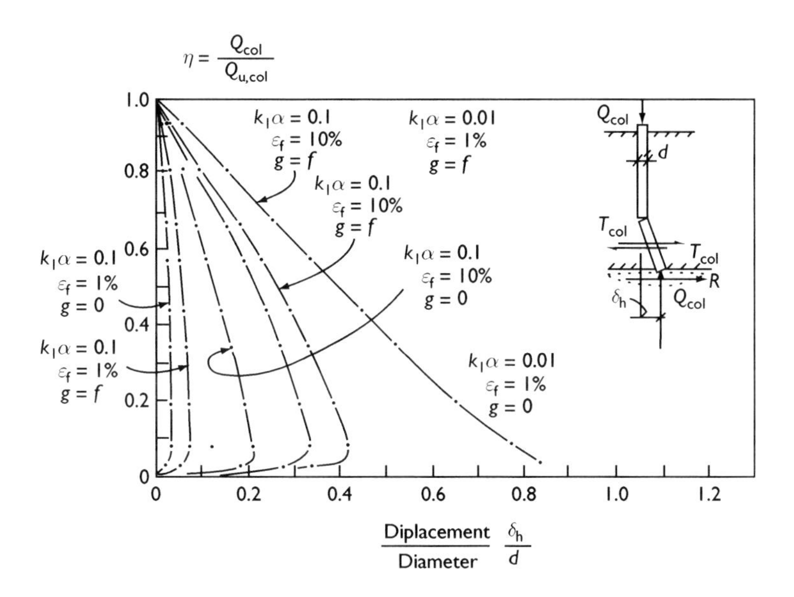

Figure 8.16 Lateral displacement of columns (failure mode f).

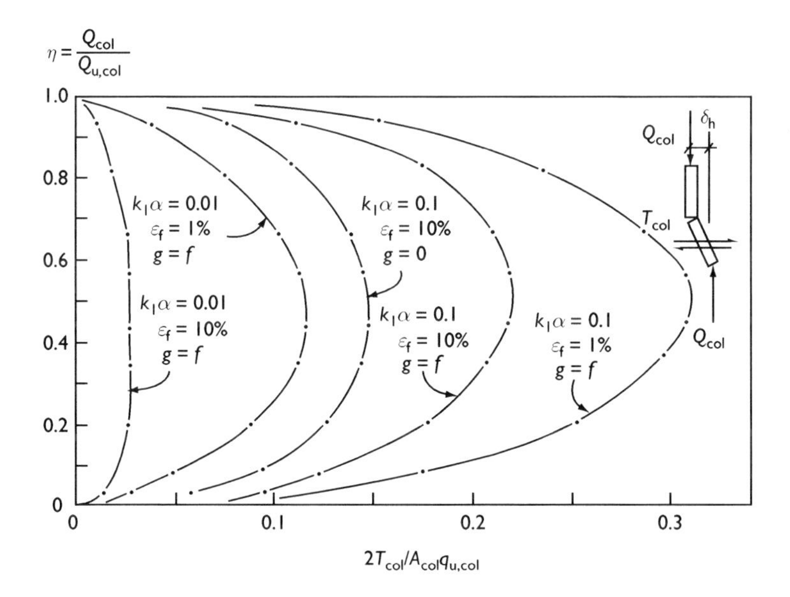

Figure 8.17 Shear resistance of laterally displaced columns (failure mode f).

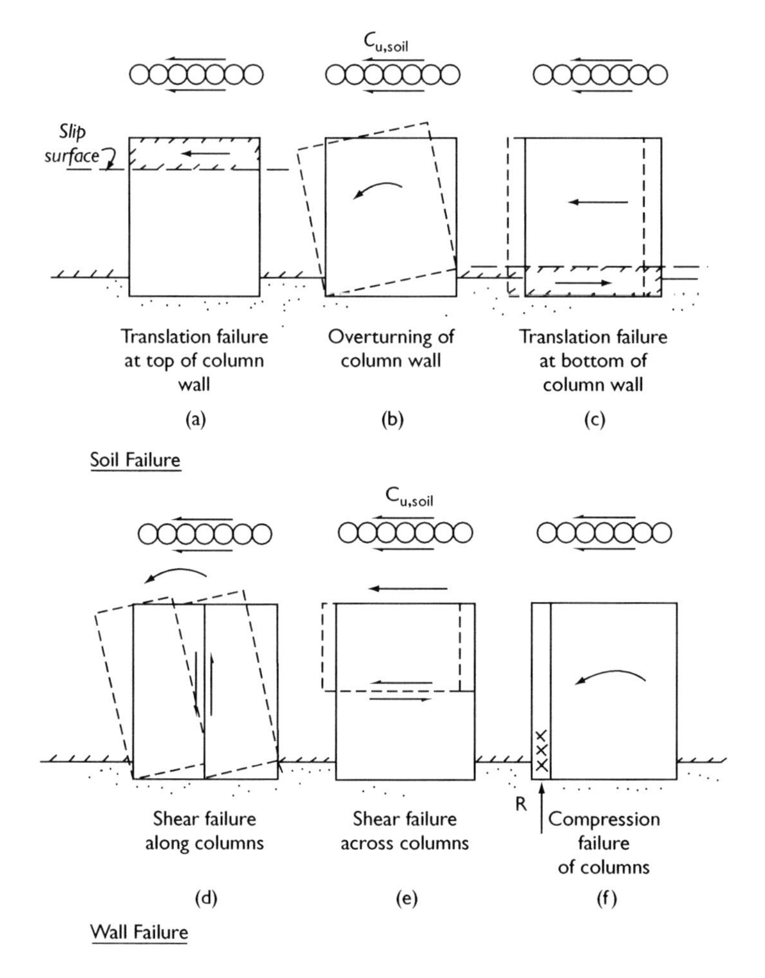

Figure 8.18 Failure modes of column walls.

when the toe resistance is high as shown in Figure 8.18b. The wall is displaced laterally in Figure 8.18c together with the unstabilized soil above the slip surface. The shear resistance depends on the penetration depth of the columns into the hard or stiff layer below the wall as shown in Figures 8.18d,e. The column wall could also fail when the bearing capacity of the columns is exceeded as shown in Figure 8.18f.

It has been assumed in equations (8.6a), (8.7a) and (8.9a) that the peak shear strengths of the columns and of the unstabilized soil between the columns are mobilized at the same time. These equations cannot be used to evaluate the bearing capacity of cement columns since the columns are brittle and the failure strain is low. It should be noted that the efficiency of the columns is low when the axial column load is low.

8.8.2 Extrusion failure for column rows

Terashi *et al.* (1983) have investigated by centrifuge tests the shear resistance at extrusion of the unstabilized soil between the column rows. The shear resistance along the column rows corresponds to the undrained shear strength of the unstabilized soil.

8.8.3 Shear failure of column rows

Shear failure of a column row is shown in Figure 8.19. This failure mode occurs when the vertical shear force in the overlapping zone reaches the shear strength of the columns. The vertical shear force S/b is equal to $c_{u,soil}b$ at $\beta = 0.5$, where β corresponds to the location of the slip below the top of the column row. The vertical shear force S/b, which is equal to $2_{u,soil}b(1-\beta)$ when $\beta > 0.5$, is high at the centre of the column row.

Inclined cone penetration tests (CPT) indicate that the shear resistance of the overlapping zone could be low due to insufficient mixing or insufficient overlap (Larsson and Håkansson, 1998). Yoshida (1996) has reported that the shear resistance of the stabilized soil in the overlapping zone has been only two-thirds of the shear strength of the columns. The shear resistance of

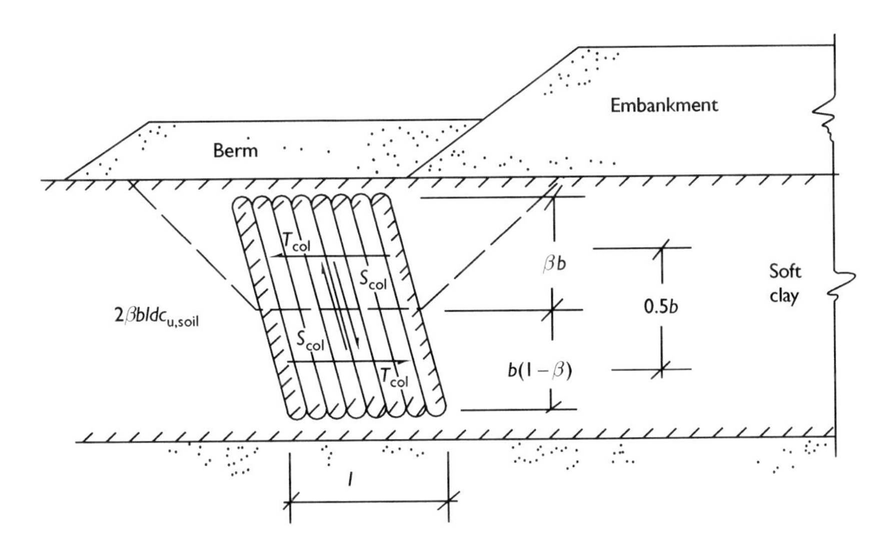

Figure 8.19 Shear failure of column rows.

the overlapping zone could be as low as 40 to 70 per cent of the average shear strength of the stabilized soil in the columns.

The maximum shear force $2S_{col}/b = 2c_{u,soil}b$ occurs at $\beta = 0.5$. However, the shear stress $2T_{col}b/l$ along the overlap with the width $d/2$ had to be less than the allowable shear resistance $q_{u,col}/4$, half the peak shear strength of the stabilized soil. In that case, $2c_{u,soil}b/d < q_{u,col}d/8b$. At $b = 5d$, $q_{u,col} = 200$ kPa and $\beta = 0.5$, the maximum shear strength for the unstabilized soil in 5 kPa. Thus the shear strength of the overlapping some usually governs the stability.

The shear force S_{col}/b, which increases with increasing value of the coefficient β, can be calculated from the following relationship at $\beta = 0.5$.

$$\frac{2S_{col}}{b} = 2\beta c_{u,soil}b/d \tag{8.10}$$

Thus the shear strength of the stabilized soil within the overlapping zones will normally govern the shear resistance of the column rows.

8.8.4 Compression failure of column rows

The maximum shear force when the slip surface is located close to the bottom of a column row with point bearing columns ($\beta = 1.0$) is shown in Figure 8.20. The shear force S_{col}/b should be less than the shear resistance of the overlap.

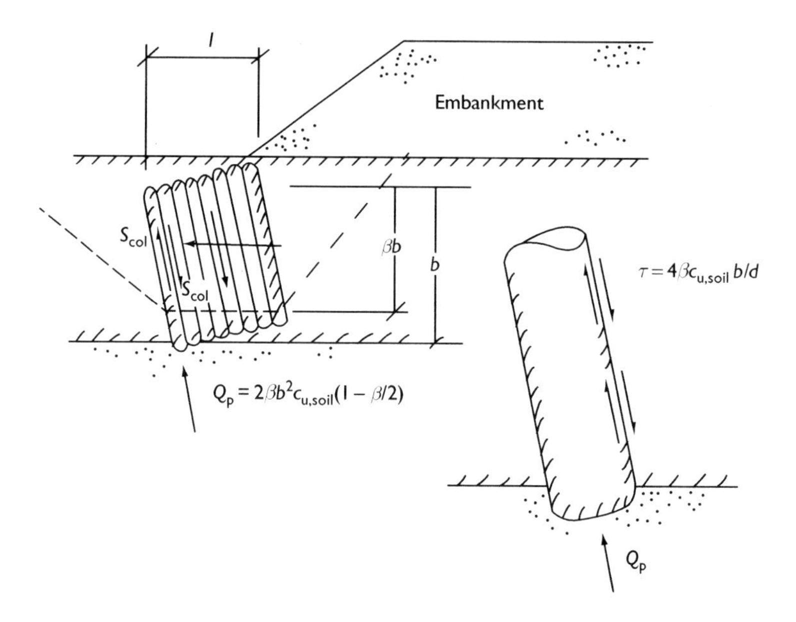

Figure 8.20 Compression failure of column rows.

The axial load is $2c_{u,soil}\beta b^2(1-\beta/2)$ for the column located on the far side of a column row with point bearing columns when the column row rotates in the soil. This axial load should be less than the unconfined compressive strength $q_{u,col}$.

8.8.5 Failure by overturning of column rows

The moment resistance of column rows with respect to overturning is shown in Figure 8.21. The moment resistance increases rapidly with increasing length and thus with increasing number of columns. At $q_{u,col} = 200$ kPa, $\beta = 1.0$ and $b = 5d$ then $c_{u,soil} < 3.14$ kPa. The maximum shear resistance of the unstabilized soil is 1.57 kPa when $q_{u,col} = 400$ kPa and $b/d = 10$.

8.8.6 Failure by separation

Failure by separation of the columns in a column row is shown in Figure 8.22. This type of failure occurs when the slip surface is located close to the top of the column wall. The lateral force at separation can be estimated by assuming that the shear resistance along both sides of the columns is equal to the shear strength of the unstabilized soil.

The columns will separate when the width of the overlap is small and the tensile strength of the overlap is low. Terashi *et al.* (1980) report that the tensile strength is 10 to 20 per cent of the unconfined compressive strength of the stabilized soil.

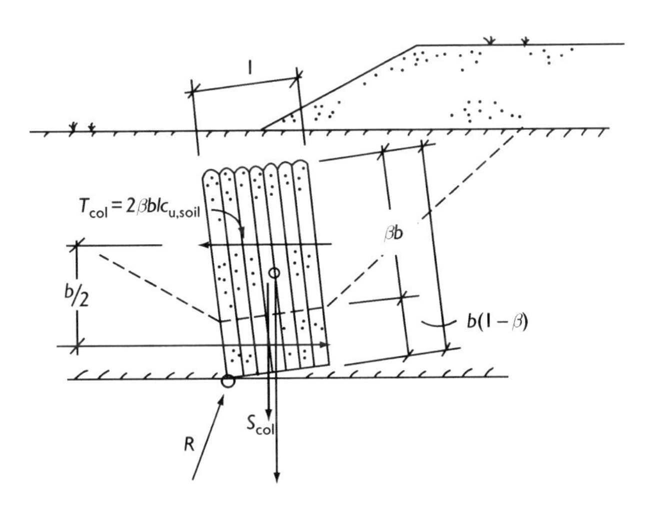

Figure 8.21 Failure of column row by overturning.

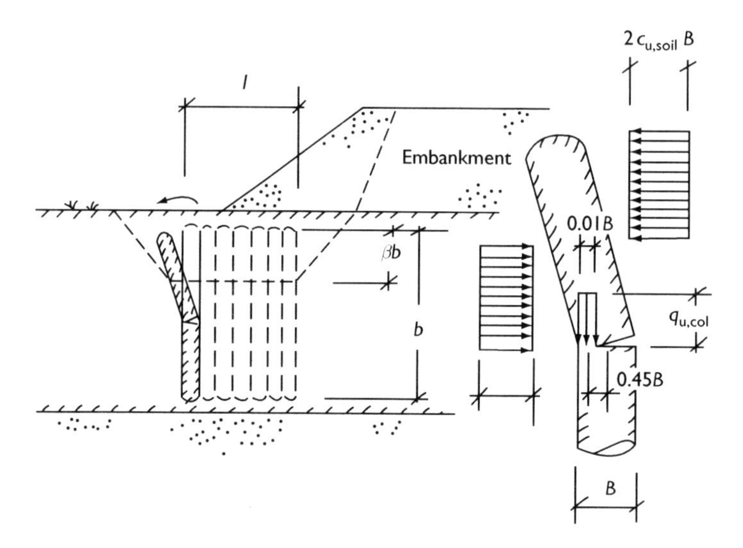

Figure 8.22 Failure by separation of the columns in a column row.

8.8.7 Translation failure

Translation of a column wall is shown in Figure 8.23. The shear resistance of the column wall is sufficient in this case to force the slip surface to the disturbed soil below the wall where the shear resistance has been reduced. Translation of the column wall occurs when the lateral resistance of the part of the wall, which is located above the slip surface is large enough to resist the lateral force below the slip surface. Failure can also occur along a slip surface through the wall.

8.8.8 Column grids, blocks and caissons

Columns are generally installed in single or double rows, as arches or caissons to increase the stability as shown in Figure 8.24. Columns grids and blocks are also used to improve the interaction of the columns with the unstabilized soil.

8.9 Stability of embankments, slopes, trenches and excavations

8.9.1 Stability of embankments

Lime and lime/cement columns are normally used to stabilize road and railroad embankments with a maximum height of 3 to 4 m. However,

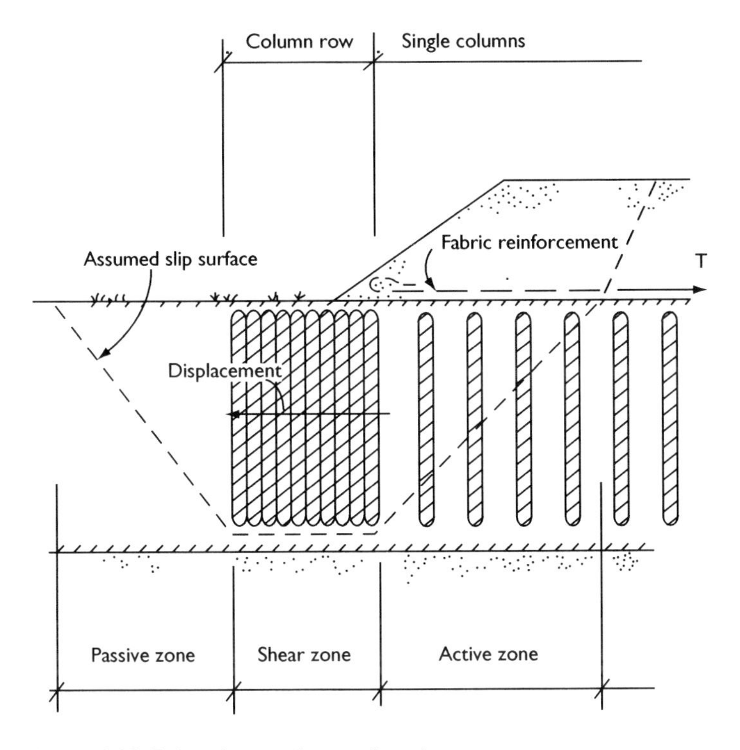

Figure 8.23 Failure by translation of a column row.

embankments with a height of up to 9 m have also been stabilized successfully with lime/cement columns.

The location of the columns affects the shear resistance as indicated by Miyake *et al.* (1991a,b). The columns should be located below an embankment, where the columns are loaded axially and where the creep strength of the columns will be utilized fully. The difference is large when $\phi_{u,col} > 0°$. The columns should overlap to reduce the lateral displacement of the columns. The shear resistance of the columns is then increased as well as the moment capacity.

Berms constructed at the toe of an embankment or a slope can be used to increase the stability. Berms are usually economical, when sufficient space and excess fill are available for the berms. The berms will increase, however, the settlement of the embankment next to the berms.

8.9.2 *Stability of slopes*

Failures have occurred due to separation of the columns, where single columns or column rows have been used to increase the stability of steep

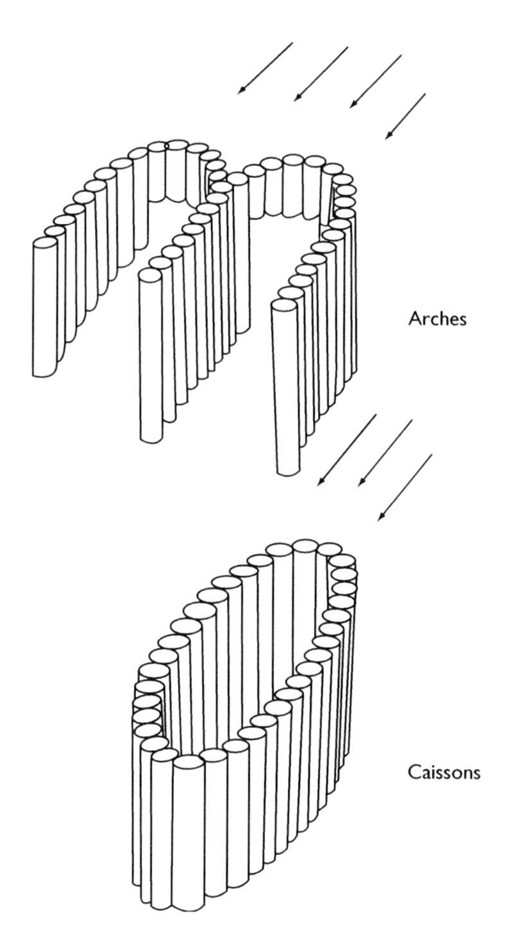

Figure 8.24 Arches and Caissons.

slopes. Failure caused by separation of the columns will occur when the tensile resistance of lime and lime/cement columns is low and the overlap of the columns is small. The stability of the columns located in a slope can be increased by placing the columns in rows, arches or caissons as shown in Figure 8.24. The columns will then function as barriers as the lateral earth pressure in the slope is resisted by the arches in front of the column rows (Blom, 1992).

The column rows transfer the lateral force caused by the lateral earth pressure through the slip surface to an underlying stiff or hard layer. The length of the columns in the arch could be reduced due to the arching.

It is preferable to use double column rows to increase the overlap and the shear resistance. The friction angle ϕ'_{col} is high along the overlap, 35 to 40°, while the effective cohesion c'_{col} is low. The overturning moment caused by the lateral pressure on the arches is high as well as the vertical shear force at the centre of the column row, which could govern the length and the spacing of the columns. The shear strength of the stabilized soil in the overlapping zone could be lower than the shear strength of the stabilized soil in the columns.

The pore water pressure in the unstabilized soil increases during the installation of the columns when the added volume of stabilizer is large. The volume of the column is also increased during the slaking of the lime when water is drawn from the soft soil around the columns. The change of the pore water pressure in the unstabilized soil affects the shear resistance of the columns since the increase of the pore water pressure could approach the pore water pressure in the unstabilized soil around the columns.

The weighted average shear strength of the columns and of the unstabilized soil between the columns is equal to:

$$\begin{aligned}\tau_{fu,av} &= a\tau_{fd,col} + (1-a)\tau_{fu,soil} \\ &= a\left[(\sigma_{f,col} - u_{col})\tan\phi'_{col} + c'_{col}\right] + (1-a)c_{u,soil}\end{aligned} \tag{8.11}$$

where u_{col} is the pore water pressure in the columns and $\sigma_{f,col}$ is the normal pressure on the failure plain. At undrained conditions and a low permeability of the columns,

$$\tau_{fu,av} = a\tau_{fu,col} + (1-a)\tau_{fu,soil} \tag{8.12}$$

In these two equations $\tau_{fu,col}$ and $\tau_{fd,col}$ are the undrained and the drained shear strengths of the stabilized soil, respectively, ϕ'_{col} is the angle of internal friction of the stabilized soil, $c_{u,col}$ and c'_{col} are the undrained and the drained cohesion, respectively, and $\tau_{fu,soil}$ is the undrained shear strength of the unstabilized soil between the columns.

The weighted average shear strength $\tau_{fd,av}$ is used to analyze the long-term stability of embankments and slopes, when failure occurs along a slip surface through thc columns. The shear strengths of the columns $\tau_{fd,col}$ and of the unstabilized soil $\tau_{fd,clay}$ are evaluated by an effective stress analysis.

$$\begin{aligned}\tau_{fd,av} &= a\tau_{fd,col} + (1-a)\tau_{fd,clay} \\ &= a(\sigma'_{f,col}\tan\phi'_{col} + c'_{col}) + (1-a)(\sigma'_{f,soil}\tan\phi'_{soil} + c'_{soil})\end{aligned} \tag{8.13}$$

In this equation ϕ'_{clay} and c'_{clay} are the effective angle of internal friction and the effective cohesion of the unstabilized soil, respectively. Equations (8.11), (8.12) and (8.13) can only be used for columns with a high ductility when the columns fail in shear along a slip surface through the columns, and the shear strength of the stabilized soil is less than 100 to 150 kPa. It is likely that full interaction can be assumed for lime/cement and cement columns at failure if the rotation and the lateral displacement of the columns are large.

In Japan the maximum shear strength, which can be utilized in design, is 100 to 250 kPa. The undrained shear resistance of the unstabilized soil corresponds to the axial strain at the peak shear strength of the stabilized soil (Matsuo *et al.*, 1996; Kitazume *et al.*, 1996b). A reduced shear strength is used for the unstabilized soil between the columns to evaluate the stability of steep slopes and deep excavations.

Column caissons can possibly be used to resist the lateral earth pressure in a slope. The lateral earth pressure is resisted by the columns around the caisson as well as by the unstabilized soil trapped within the caisson. The shear resistance of the caissons is governed by the shear strength of the overlapping zone and by the shear resistance of the unstabilized soil along a vertical plane through the caissons. Caissons may fail by overturning or by sliding along a slip surface below the bottom of the caissons. Failure can also occur along a slip surface through the caisson walls and through the unstabilized soil within the caissons. The lateral displacement of the caissons is governed by the shear modulus of the columns and of the unstabilized soil within the caissons, G_{col} and G_{soil}, respectively. An unreinforced vertical column wall often fails by overturning when a horizontal crack develops just behind the columns in the front row.

Bracing is not required for shallow trenches to facilitate the placement of concrete or steel pipe segments in the trench for sewer lines and water mains.

8.9.3 Stability of trenches

The calculation methods, which have been developed for gravity retaining walls, can also be used to analyze the stability of blocks or rows with lime and lime/cement columns. The stability with respect to sliding and overturning had to be considered. The total stability is analyzed with respect to a circular failure surface through or below the column wall at the bottom of the excavation (Figure 8.25). The lateral pressure on the column wall will increase when cracks behind the wall are filled with water during a heavy rainstorm.

The column block is displaced laterally when the lateral pressure corresponds to the lateral resistance at sliding or overturning. The displacement rate depends on the water pressure in the cracks behind the column wall or block and thus on the water level in the cracks and thus on the intensity of the rainfall and the drainage conditions. Lime columns are preferred due to the high permeability of the columns. By inclining the columns and the sides of the excavation the effectiveness of the lime and the lime/cement columns can be increased as illustrated in Figure 8.26.

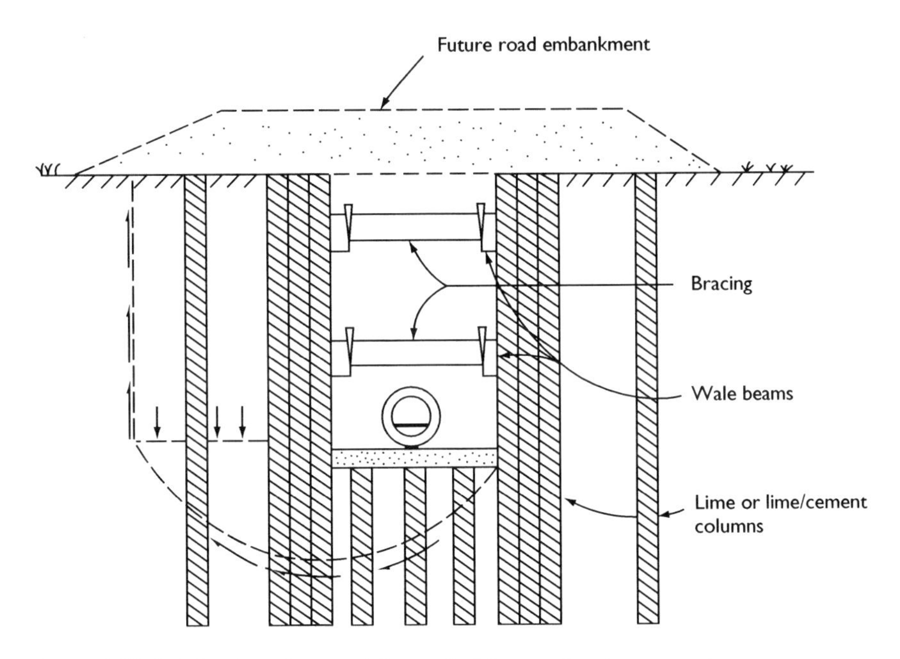

Figure 8.25 Stabilization of a trench by vertical lime or lime/cement columns.

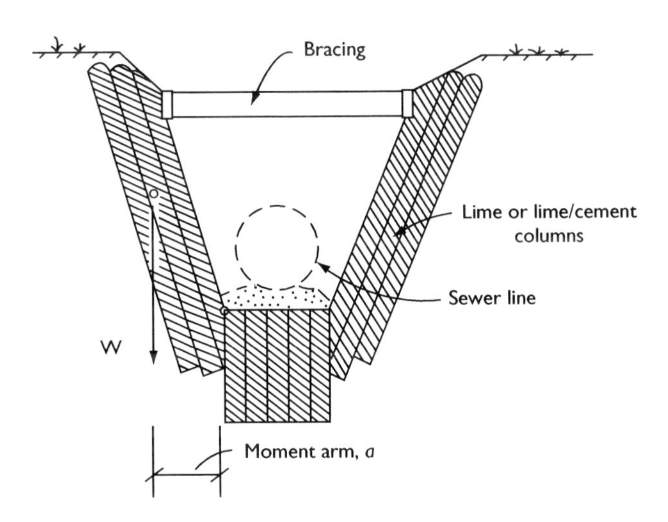

Figure 8.26 Stabilization of a trench by inclined lime or lime/cement columns.

8.9.4 Stability of excavations

The stability of a deep excavation can be improved by single columns, by blocks or rows with overlapping columns placed behind or in front of a sheet pile wall. Lime and lime/cement columns, which are located behind a sheet pile or a retaining wall, will reduce the active earth pressure. Lime and lime/cement columns can also be used to increase the passive earth pressure where the columns are placed in front of a sheet pile or a retaining wall as shown in Figure 8.27. Then the risk of toe failure is reduced. It should be noted that the efficiency of the columns located behind a sheet pile wall is much higher than the efficiency of the columns, which are located in the passive zone in front of the sheet pile wall.

Lime/cement columns have been used in Oslo, Norway to increase the stability of a 7 m deep excavation, which was supported by anchored sheet piles (Long and Bredenberg, 1997). The lime/cement columns were installed as a 2.0 m wide block at the bottom of the excavation in front of the sheet piles and as rows with 4 m long columns in front of the column block.

8.10 Progressive failure

8.10.1 Progressive failure of single columns and of column rows

Terashi *et al.* (1983) have pointed out that the shear resistance of a cement-treated column block can be reduced by progressive failure due to the low failure strain of the columns. The stability is usually calculated by assuming that failure takes place along a cylindrical slip surface through single columns or through column groups without considering the reduction of the bearing capacity caused by progressive failure. Tatsuoka and Kobayashi (1983) have proposed that the residual shear strength of the columns could be used in a slip circle analysis together with the undrained shear strength of the unstabilized soil to take into account the reduction of the bearing capacity caused by progressive failure.

Karastanev *et al.* (1997) observed at centrifuge tests with single columns and with column blocks in soft normally consolidated clay that the columns failed one by one due to the brittle behaviour of the columns. The bearing capacity of the columns located just below the caisson was only 53 per cent of the unconfined compressive strength of the columns when the applied load was vertical. The bearing capacity of the columns located just below the caisson was only 40 to 30 per cent of the unconfined compressive strength, when the lateral load was 23 and 20 per cent of the vertical applied load. The lateral resistance of the columns increased with increasing axial column load until the load was 50 to 75 per cent of the ultimate bearing capacity (Figure 8.28).

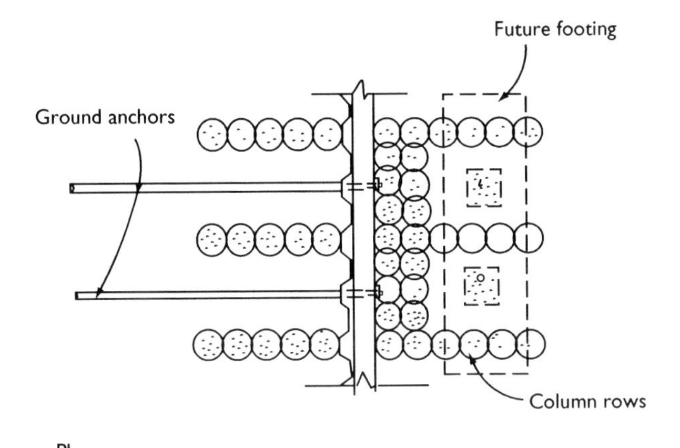

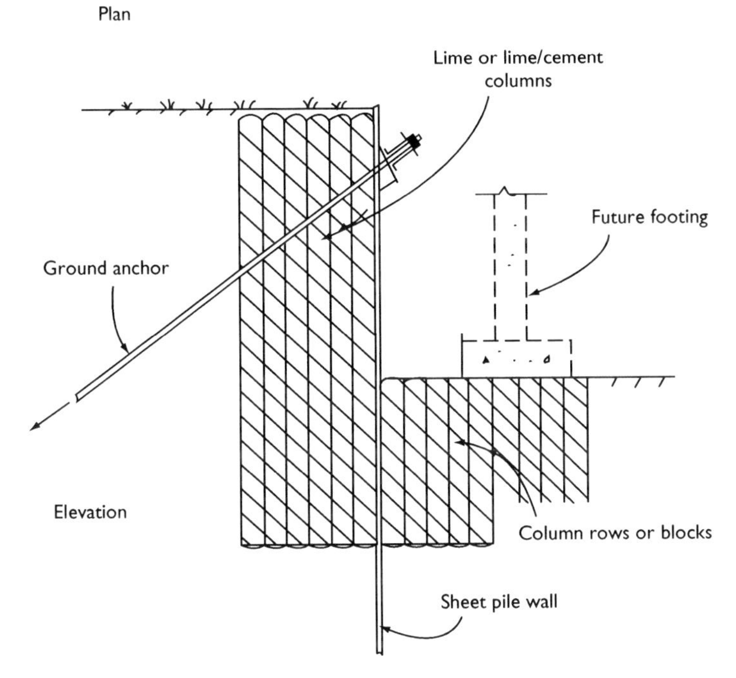

Figure 8.27 Stabilization of a sheet pile wall.

The columns, which are located on both sides of the wedge below the caisson, are displaced laterally. The bearing capacity of the columns is then reduced due to the low failure strain. It was not possible to determine the reduction of the bearing capacity of the columns when a shear strength, which was half of the unconfined compressive strength, was used in the analysis (Karastanev *et al.*, 1997). It was necessary to use a shear strength,

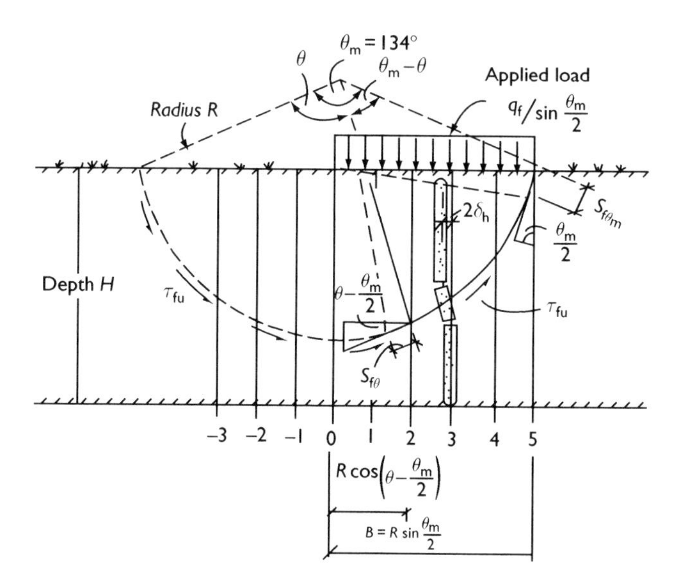

Figure 8.28 Lateral displacement of columns.

which was only 10 per cent of the unconfined compressive strength to obtain a reasonable agreement with the test results.

The interaction of single columns with the surrounding soft clay is uncertain especially for organic soils with a high water content such as peat and 'gyttja' and for soils with a high sulphide content ('svartmocka'). The reduction of the shear resistance can be large when the peak strength of the soil is exceeded and the axial strain at the peak strength is low, 0.1 to 1 per cent. The columns could fail before the peak shear strength of the unstabilized soil has been mobilized.

The reduction of the bearing capacity of single columns and of column rows by progressive failure has been estimated by Broms (1999). It was assumed that the reduction depended on the relative displacement of the soil with respect to the columns along the assumed slip surface and on the compressibility of the unstabilized soil between the columns. The relative displacement, which is large just below an embankment, is assumed to decrease along the assumed slip surface with increasing distance from the embankment as illustrated in Figure 8.29.

The lateral displacement of the unstabilized soil along the slip surface has been calculated in Figure 8.30 at a factor of safety (F_s) of 1.0, 1.2, 1.5 and 2.0. It has been assumed that the center of rotation is located just above the centre of the sloping side of the embankment and that the centre angle of the slip surface is 134° (Figure 8.28). It has thus been assumed that the shear strength of the unstabilized soil is constant with depth and that the ground

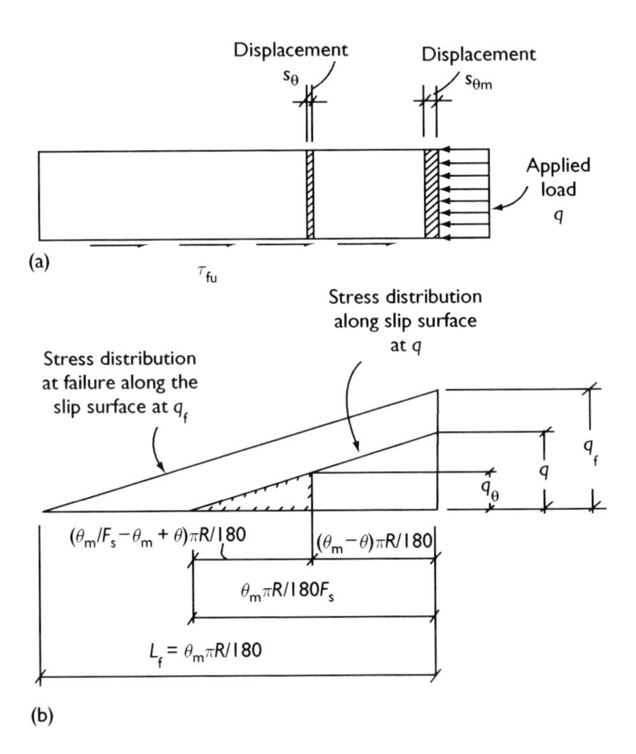

Figure 8.29 Stress distribution along slip surface.

surface is horizontal. The radius R of the slip surface is governed by either the width B of the embankment ($R = 1.09B$) or by the depth H of the soft soil below the embankment ($R = 1.64H$).

It has also been assumed that the relative displacement, which is required to mobilize the peak shear strength of the soil, is small and that the shear strength is fully mobilized along the length of the slip surface, which corresponds to $1/F_s$. At, for example, a factor of safety $F_s = 2.0$ the shear resistance is assumed to be fully mobilized along 50 per cent of the assumed slip surface (Broms, 1999).

The lateral displacement of the columns has a large affect on the bearing capacity of the columns. The reduction depends, besides the lateral displacement of the columns and the factor of safety of the unstabilized soil, also on the failure mode. The largest horizontal displacement will occur of column 4 in Figure 8.30, where the lateral displacement is about 55 per cent of the maximum settlement of the embankment s_{max}. The lateral displacement of column 5 is relatively small since the displacement of the unstabilized soil there is mainly vertical.

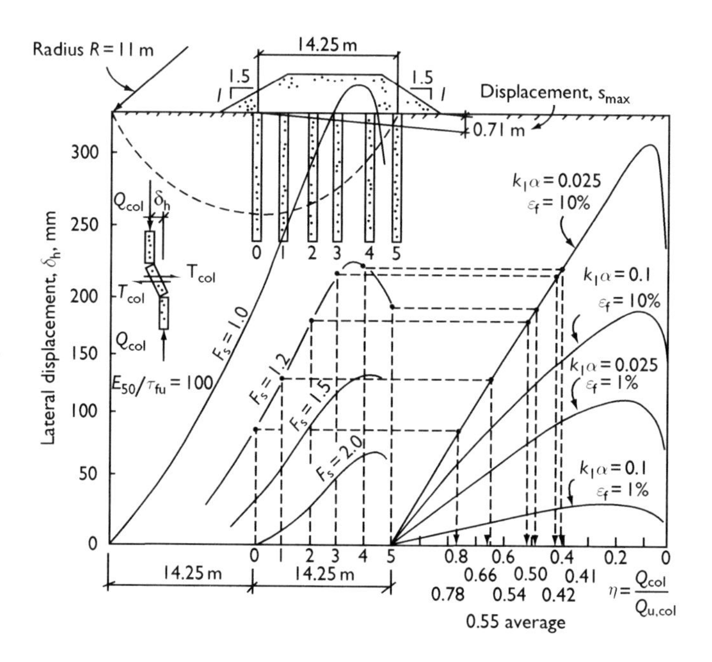

Figure 8.30 Progressive failure of lime and lime/cement columns.

It can be seen from Figure 8.30 that the lateral displacement of the columns decreases rapidly with increasing distance from columns 3 and 4 and with increasing factor of safety. At $F_s = 2.0$ the maximum lateral displacement of columns 3 and 4 is about 25 per cent of the lateral displacement at $F_s = 1.0$.

The relative displacement of the unstabilized soil along a cylindrical failure surface with the radius $R = 11$ m is shown in Figure 8.30 at $\tau_{fu}/E_{col} = 1/100$ and a factor of safety $F_s = 1.2$, when the maximum shear strength has been mobilized along 83 per cent (1/1.2) of the whole slip surface. At $k_1\alpha = 0.025$ and $\varepsilon_f = 10$ per cent the lateral displacement of columns 2, 3, 4 and 5 at $F_s = 1.2$ is large. The bearing capacity of the columns is 78, 66, 54, 42 and 41 per cent of $Q_{u,col}$. The average bearing capacity of the six columns is 55 per cent.

The geotechnical bearing capacity of the unstabilized soil is 34.4 kPa $(0.75 \times 10 \times 5.5/1.2)$ at $F_s = 1.2$ and an area ratio $a = 0.25$ and an undrained shear strength of the unstabilized soil of 10 kPa. The contribution of the columns is 27.5 kPa $(0.55 \times 200 \times 0.25)$ at $a = 0.25$ and $q_{u,col} = 200$ kPa. The total bearing capacity of the columns and of the unstabilized soil is 61.9 kPa (27.5 + 34.4).

8.11 Settlement of buildings and of other structures

8.11.1 Settlement of single columns and of column groups

The main purpose of soil stabilization in the 1960s and 1970s was to reduce the settlements and the angular rotation of relatively light buildings. The bearing capacity of the unstabilized soil without the columns was often sufficient to support up to two-storey buildings without basement when the clay is normally consolidated or is slightly overconsolidated. Most buildings can accommodate relatively large settlements, 0.2 to 0.3 m or more, without structural damage provided the differential settlements are not excessive. Buildings and other structures are often damaged by an angular rotation or by a differential settlement exceeding 1/300 to 1/400 or by a relative deflection exceeding 1/2000 to 1/4000. Embankments can in general accommodate a much larger settlement than buildings, provided the settlements are uniform.

About 75 per cent of the settlements can often be eliminated by 20 m long columns, when the clay is normally consolidated. The length and the spacing of the columns can be varied to take advantage of the increase of the compression modulus and of the undrained shear strength with depth. The load transferred to the columns, which extend down to a stiff or a hard layer, can locally be high. The columns may fail if the bearing capacity of the supporting stiff or hard layer exceeds the peak strength of the columns.

The reduction of the bearing capacity can be large when the peak shear strength of the columns is exceeded. It is therefore often preferable to terminate the columns just above a stiff or a hard layer. It is important that the resistance is less than the bearing capacity of the columns especially when the columns are brittle.

The length of the lime/cement columns supporting the embankments for the 'Svealand' railroad line was varied, 5, 10 and 15 m to take advantage of the increase of the shear strength and of the compression modulus of the soft soil with depth (Rogbeck, 1997).

8.11.2 Settlement calculations

The settlement of embankments and structures, which are supported by single columns or by column groups, is usually checked by assuming that the axial deformations of the columns are the same as the deformations of the unstabilized soil between the columns. The settlement of a column group is governed by the weighted average compression modulus of the columns and of the unstabilized soil between the columns when the shear strength of the columns is low. Test results indicate that the behaviour of the stabilized soil is similar as that of an overconsolidated clay (Pan *et al.*, 1994; Baker *et al.*, 1997). It is expected that the settlements will be small up to the apparent pre-consolidation pressure, which is expected to increase with increasing shear strength of the stabilized soil and thus with increasing cement content. The settlements are also affected by the stress increase in the columns.

8.11.3 Stress distribution

The stress distribution in lime/cement columns has been investigated by Liedberg *et al.* (1996a,b) using the computer program PLAXIS. Also FEM has been used (Kujala, 1984). The analysis shows that the transfer length is small for soft columns, only a few pile diameters. The main difficulty has been to estimate the modulus of elasticity E_{col} for the columns and the compression modulus M_{soil} for the unstabilized soil.

The soil is remoulded during the installation, up to 0.4 to 0.3 m below the columns, which will increase locally the settlement below the columns. There the stress increase depends on the transfer length and on the remoulded shear strength of the unstabilized soil.

The settlement and the settlement rate of the soft soil below the column block is generally calculated from the stress increase as determined by the 2:1 method. However, the stress increase and the settlements could then be overestimated by as much as 27 per cent (Liedberg *et al.*, 1996a,b). It is usually assumed that the total load is transferred to the bottom of the reinforced block without spreading.

The stress increase Δq below the columns is reduced by the shear resistance $c_{u,soil}$ around the perimeter of a column group so that,

$$\Delta q = q_o - 2c_{u,soil}\left(1.0 + \frac{b}{l}\right)lL_{col} \tag{8.14}$$

where b and l are the width and length of the loaded area, respectively, q_o is the applied load and L_{col} is the column length.

It is often not possible to utilize the high bearing capacity of lime/cement and cement columns since the load transferred to the columns is governed by the shear strength and by the thickness of the unstabilized soil above and below the slip surface. The settlements will in that case be governed by the stress increase in the unstabilized soil between the columns.

8.11.4 Settlement of column groups

The settlement of a column group is usually calculated by assuming that the axial deformations of the columns are the same as the deformations of the unstabilized soil and that the behaviour is similar to that of an overcon solidated clay (Pan *et al.*, 1994). Liedberg *et al.* (1996a,b) and Bengtsson and Holm (1984) found that full interaction could be assumed when the spacing of the columns is less than 1.0 to 1.5 m.

The settlement of a column group, s_{group}, is governed by the weighted average modulus of elasticity of the columns and of the compression modulus of the unstabilized soil between the columns, E_{col} and M_{soil}, respectively.

$$s_{group} = \sum \frac{\Delta h q_o}{aE_{col} + (1 - a)M_{soil}} \tag{8.15}$$

where Δh is the thickness of the different layers, q_o is the stress increase from, for example, an embankment and a is the area ratio. It should be noted that M_{soil} decreases with time due to creep.

It is proposed in SGF (2000) that a value of $50c_{u,col}$ can be used on E_{col} for organic soils ('gyttja'), $100c_{u,col}$ for clay and $150c_{u,col}$ for silty clay. At, for example, $E_{col}/M_{soil} = 20$ the stress increase in the columns is 20 times the stress increase in the unstabilized soil between the columns.

Due to the large variation of the compression modulus and of the difference between laboratory prepared samples and the *in situ* compression modulus, it is recommended that the settlements should be calculated for probable maximum and minimum values of the different parameters.

It is also important to monitor the settlements and the settlement rate during the construction to determine if pre-loading will be required to reduce the total and the differential settlements and the time required for the consolidation.

8.11.5 Observed settlements

Holm *et al.* (1983b) and Edstam (1996) have reported that the settlements have been only 40 per cent of the calculated settlements based on a compression modulus determined by oedometer tests. Rogbeck (1997) has reported that the settlement of organic clays has been about half of the estimated settlement down to about 10 m depth. Holm *et al.* (1983b) have reported that the maximum settlement for lime columns was 0.4 m after 2.6 years at an applied load of 50 kPa compared with a maximum settlement of 0.8 m with sand drains. The soil consisted of 7.5 m of soft to very soft clay with an undrained shear strength of 6–9 kPa and a compression modulus of 60 to 175 kPa. The lime content of the 0.5 m diameter columns was 7 to 12 per cent.

Columns with 0.8 m diameter, which are spaced 0.62 m apart, have been used in Finland to stabilize a railroad embankment constructed on peat. The maximum settlement after about 2 months was 175 mm (Hoikkala *et al.*,

1997). Carlsten and Ouacha (1993) have reported that the observed settlement of an embankment in Karlstad, Sweden was larger than the settlement estimated by CRS-tests assuming a compression modulus of $75c_{u,soil}$ since some of the columns could not be installed and that the height of the test embankment was 0.5 to 0.8 m higher than assumed (Nord, 1990).

The settlement of a column group depends mainly on the modulus of elasticity of the columns and on the apparent pre-consolidation pressure. Mainly the modulus of elasticity for the upper part of the columns is important since the compression modulus of the unstabilized soil increases rapidly with increasing depth when the clay is normally consolidated or is slightly overconsolidated.

Carlsten and Tränk (1992) and Ekström *et al.* (1994b) have back-calculated a compression modulus of 8 to 10 MPa for cement columns at a surcharge load of 42 kPa, while CRS-tests indicated a modulus of only 1.5 to 2.5 MPa. The compression modulus as determined by dilatometer tests was 2 to 5 MPa and thus 20 to 50 per cent of the back-calculated compression modulus.

The difference between estimated and observed settlements increases in general with increasing length of the columns. The settlements are in general smaller for organic soils stabilized by lime/cement than by lime mainly due to the higher shear strength with lime/cement.

The height of a 3.5 m high embankment must often be increased 0.9 to 1.0 m to compensate for the settlements caused by the consolidation (Wilhelmsson and Brorsson, 1987). A temporary surcharge load of, for example, 1.2 m can be used to reduce the consolidation time. The factor of safety should be at least 1.2 to 1.3 to prevent failure of the embankment during construction.

Halkola (1984) has reported settlements of 0.3 to 0.4 m for an up to 2.5 m high test embankment up to 2.5 m, where cement columns were used to stabilize the underlying soft soil. The shear strength of the stabilized soil increased from 5 to 10 kPa to 40 to 150 kPa after 30 days. The settlements stopped after the height of the embankment had been reduced from 2.5 to 1.5 m.

At Norrala in Sweden, the predicted settlement was 0.19 m for a test embankment based on the results from triaxial tests. The observed maximum settlement was small, 0.1 m. Triaxial tests by Steensen-Bach *et al.* (1996) indicated an average compression modulus M_{col} of $200c_{u,col}$. The compression modulus varied between $175c_{u,col}$ and $250c_{u,col}$. The settlements close to the ground surface were up to 10 times larger than the estimated settlements due to the low shear strength of the columns, the large transfer length and the weak surface crust. This local settlement accounted for 40 to 70 per cent of the observed settlements (Arnér *et al.*, 1996).

8.11.6 Settlements at Skå-Edeby

The settlements of two loaded areas at the test field of the Swedish Geotechnical Institute (SGI) at Skå-Edeby, located about 30 km west of Stockholm (Boman and Broms, 1975; Broms and Boman, 1979a), where lime columns with a length of 6 m, had been installed below one of the two test fills at the site. The two test fills, 12.6 m × 18.2 m, were loaded to about 10 kPa by a 0.6 m high gravel fill. This applied load corresponds to the weight of a typical one-story family house in Sweden.

The soil below the approximately 1 m thick dry crust consisted of 5 m of soft clay with a relatively high sulphate content and of glacial varved clay down to a maximum depth of 15 m. The clay is normally or slightly overconsolidated with a shear strength, which increases with increasing depth. The minimum shear strength is about 8 kPa at a depth of 4 m. The water content of the soft clay is high, about 100 per cent at a depth of 3 to 4 m.

The settlement below the centre of the two loaded areas increased gradually with time due to consolidation as shown in Figure 8.31. The maximum settlement of the reference area was almost 120 mm, 5 years after the placement of the fill compared with 30 mm for the stabilized area. The settlement of the soft clay below the stabilized block was 18 mm after half a year compared with 40 mm for the reference area. The settlements outside the loaded area were larger for the stabilized area than for the

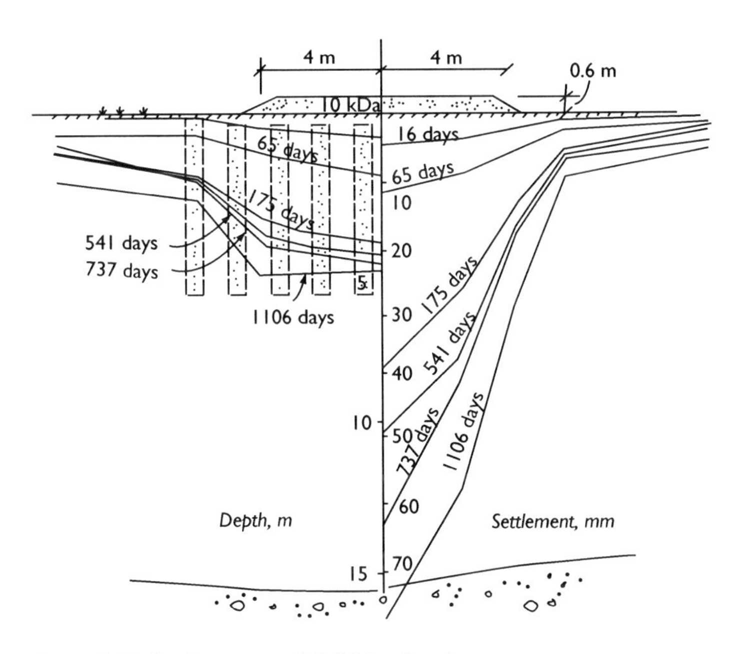

Figure 8.31 Settlements at Skå-Edeby, Sweden.

reference area due to the high average shear stress around the perimeter of the column block.

8.11.7 Differential settlements

Structures are damaged mainly by the angular rotation caused by shear deformations of the unstabilized soil between the columns. Most structures can accommodate an angular rotation of 1:300 to 1:400 without structural damage. The maximum angular rotation α_{max} (Figure 8.32) can be estimated by the following equation (Broms and Boman, 1979a):

$$\alpha_{max} = \frac{\tau_{av}}{G_{soil}} \tag{8.16}$$

where τ_{av} is the average shear stress along the perimeter of the reinforced block and G_{soil} is the shear modulus. The shear modulus, which depends on the Poisson's ratio ν_{soil}, is $0.3M_{soil}$ at $\nu_{soil} = 0.3$. The shear modulus is estimated to be about $100c_{u,soil}$.

The maximum differential settlement occurs usually during the initial loading along the perimeter of the loaded reinforced block. There the total shear force is about 80 per cent of the total weight W of the structure. About 20 per cent of the total weight is transferred directly to the unstabilized soil below the column block. Since a very small relative displacement is required to mobilize the shaft resistance the angular rotation α_{max} is equal to:

$$\alpha_{max} = \frac{\tau_{av}}{G_{soil}} = \frac{0.8W}{2(b+l)L_{col}G_{soil}} < \frac{1}{300} \tag{8.17}$$

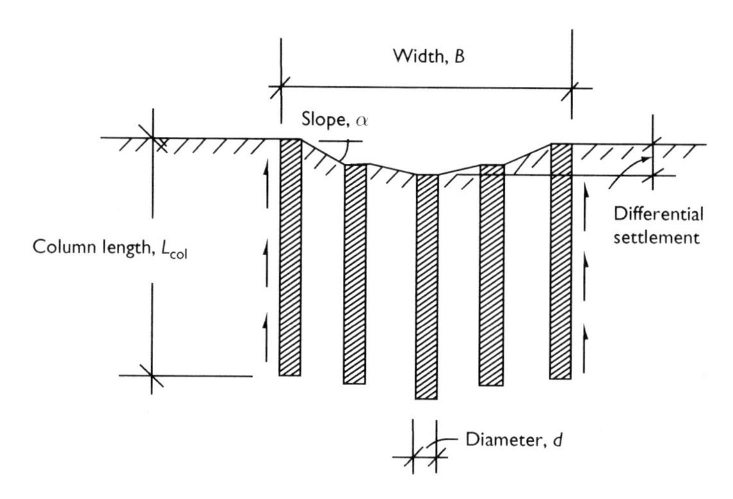

Figure 8.32 Differential settlement.

where b and l are the width and the length of the loaded area, respectively and L_{col} is the length of the columns.

Both the average shear stress τ_{av} along the perimeter and the shear modulus G_{soil} decrease with time. Test results (Broms and Boman, 1977) indicate that the reduction of the average shear stress τ_{av} is often larger than the reduction of the shear modulus G_{soil}. The maximum angular rotation and the maximum differential settlement are therefore expected just after loading of the column group. At $G_{soil} = 100c_{u,soil}$, equation (8.17) can be rewritten as:

$$\tau_{av} = \frac{G_{soil}}{300} = 0.33c_{u,soil} \tag{8.18}$$

At a maximum angular rotation of 1/300 and a partial safety factor of 1.2,

$$\frac{L_{col}}{b} > \frac{1.5q}{c_{u,soil}}\left(\frac{b}{l} + 1\right) \tag{8.19}$$

The length of the columns L_{col} should be at least $2b$ where b is the width of the loaded area. At, for example, $b/l = 0.5$, $q = 2c_{u,clay}$. The differential settlements can be reduced by increasing the length of the columns around the perimeter of the column group.

8.11.8 Pre-loading and surcharging

A surcharge load can reduce the settlements caused by creep and secondary consolidation. The required surcharge load is 1.2 to 1.5 times the weight of the embankment plus any additional dead load. For normally consolidated clays with a liquid limit larger than 150 per cent a pre-load, which is twice the weight of the embankment could be required. A preload of 120 per cent is often sufficient to eliminate the secondary settlements for many clays.

8.12 Consolidaton

8.12.1 Coefficient of consolidation

The settlement rate is governed by the coefficient of consolidation $c_{vh} = M_{av}k_h/\gamma_w$ and thus by the permeability of the soil in the horizontal direction k_h and by the average compression modulus $M_{av} = (aE_{col} + [1\text{–}a]M_{soil})$. It should be noted that k_h is often 3 to 5 times higher than the permeability of the soil in the vertical direction, k_v. It is usually assumed that k_h is $2k_v$ to $3k_v$.

The weighted average compression modulus M_{av} of the column reinforced soil is often 2 to 4 times larger than the compression modulus of the unstabilized soil. The settlement rate could be 5 to 20 times the settlement rate of the unstabilized soil due to the high compression modulus of the columns. The reduction of the permeability by lime/cement is in general compensated at least partly by an increase of the compression modulus due to the high shear strength of the stabilized soil. However, the reduction of the coefficient of consolidation of the stabilized soil is usually less than the reduction of the permeability with lime/cement as pointed out by Baker *et al.* (1997).

The coefficient of consolidation of the stabilized soil below the pre-consolidation pressure is about 10 per cent of the coefficient of consolidation for the remoulded soil. Above the pre-consolidation pressure the coefficient of consolidation decreases rapidly with increasing consolidation pressure. The coefficient of consolidation is usually determined in the laboratory by oedometer tests and in the field by piezocone tests. The coefficient of consolidation could be one order of magnitude lower for organic soils than for inorganic soils.

The coefficient of consolidation c_{vh} with respect to lateral drainage is expected to be 3 to 5 times the coefficient c_{vv} with respect to vertical drainage as determined by oedometer tests. Typical values on c_{vv} are 0.5 to 1.5 m^2/year for normally consolidated soft marine clay and 1 to 3 m^2/year for c_{vh}. Since the coefficient of consolidation of soft clay, which is reinforced by lime columns could be 10 to 20 times higher than the coefficient for the unstabilized soil. The time required for consolidation is then reduced by 90 to 95 per cent.

8.12.2 Consolidation rate

The consolidation of the unstabilized soil between the lime columns occurs rapidly since the columns function as vertical drains as discussed by Broms and Boman (1977), Holm *et al.* (1983b), Halkola (1984) and Åhnberg and Holm (1986). The consolidation is often 90 per cent after about 3 to 4 months when the spacing of the columns is 1.0 m. About 6.5 months is required for the consolidation when the spacing is 1.2 m.

Bergado *et al.* (1996) have reported for the Bangkok clay that the coefficient of consolidation with cement as determined by oedometer tests, decreases almost linearly with increasing consolidation pressure and with increasing cement content. However, the reduction of the permeability is compensated at least partly by an increase of the compression modulus with increasing cement content.

The degree of consolidation of a block stabilized by lime columns can be estimated by the following equation (Hansbo, 1979, 1987).

$$U = U_h + U_v - U_h U_v \tag{8.20}$$

where U_h and U_v are the degree of consolidation caused by drainage in the lateral and in the vertical directions, respectively.

The degree of consolidation U_h caused by the columns is equal to,

$$U_h = 1 - \exp\left[\frac{-2c_{vh}t}{R^2 f(n)}\right] \tag{8.21}$$

where c_{vh} is the coefficient of consolidation with respect to lateral drainage, t is the time required for the consolidation and R is the radius of influence. The factor $f(n)$ in equation (8.21) can be estimated from:

$$f(n) = \frac{n^2}{n^2 - 1}\left[\ln(n) - 0.75 + \frac{1}{n^2}\left(1 - \frac{1}{n^2}\right)\right] + \frac{n^2 - 1}{n^2}\frac{k_{clay}}{k_{col}}\frac{L_{col}^2}{r^2} \tag{8.22}$$

where $n = R/r$, r is the radius of the columns, k_{clay} and k_{col} are the permeability of the unstabilized and the stabilized soil, respectively, L_{col} is the length of the columns when the soil is drained on one side only, and half the column length when the columns are drained top and bottom. The effect of smear around the columns can normally be neglected (Hansbo, 1987). The radius of influence R is equal to $0.56S$ for a square location of the columns and $0.53S$ for a triangular location. The spacing of the columns, S, is usually 0.8 to 1.8 m.

The degree of consolidation due to drainage in the vertical direction U_v depends on the time factor T_v.

$$T_v = \frac{c_{vv}t}{H^2} \tag{8.23}$$

where c_{vv} is the coefficient of consolidation with respect to vertical flow, t is the time and H is the total thickness of the soft clay when the columns are drained on one side only and half the thickness when the columns are drained top and bottom.

The agreement between estimated and measured settlements and settlement rates has in general been satisfactory (Broms and Boman, 1979a,b; Holm, 1979b; Carlsten and Ekström, 1995, 1997). Carlsten and Ouacha (1993) found a good agreement between calculated and measured consolidation rate for an embankment constructed on an 8 m thick clay layer, which had been stabilized by lime columns.

The effectiveness of lime/cement and cement columns as drains is uncertain for long columns. The permeability of the columns had to be at least 300 to 1000 times the permeability of the unstabilized soil for the columns to function as drains without excessive hydraulic lag. Band drains can possibly be combined with lime/cement and cement columns to increase the consolidation rate. Band drains have not been used so far in Sweden together with lime/cement or cement columns.

The time required for the consolidation is short for lime columns, normally a few weeks, when the spacing of the columns is small. Several months or years might be required for the consolidation of lime/cement and cement columns depending on the low permeability and the thickness of the unstabilized soft soil. Silt and sand layers in the soft clay affect the consolidation rate when the distance between the drainage layers is less than the spacing of the columns. It may be economical to vary the length of the columns due to the variation of the permeability and the increase of the compression modulus with depth.

The consolidation rate is affected by the time lag caused by the low permeability of lime/cement and cement columns. For lime columns the consolidation rate at $k_{col} > 1000k_{clay}$ is mainly governed by the permeability of the unstabilized soil between the columns. If the coefficient of consolidation c_{vh} is increased from 3 to $10\,m^2$/year then 94, 75 and 50 per cent consolidation will be reached after 1.0, 0.5 and 0.25 years, respectively at $S = 1.5\,m$, $R = 0.84\,m$, $r = 0.3\,m$, $k_{col}/k_{soil} = 100$ and $L_{col} = 10\,m$. At $k_{col}/k_{soil} = 300$ and $c_{vh} = 10\,m^2$/year 99, 98 and 85 per cent consolidation will be reached after 1.0, 0.5 and 0.25 year, respectively.

A computer program LIMSET has been developed by Carlsten (1989) to estimate the settlement rate, which takes into account the compression modulus of the stabilized and the unstabilized soil, the area ratio, the degree of consolidation and the time between the installation of the columns and the application of load.

8.12.3 Hydraulic lag for lime, lime/cement and cement columns

The last term in equation (8.23) takes into account the time lag caused by the low permeability of the columns. Bengtsson and Holm (1984) indicate that about 1 year will be required to reach 80 per cent consolidation at $k_{col}/k_{soil} = 100$ and about 10 years at $k_{col}/k_{soil} = 1.0$. Thus the permeability ratio k_{col}/k_{soil} has a large effect on the consolidation rate. Hansbo (1994) has pointed out that the influence of the well resistance cannot be neglected for lime/cement columns.

It is required that $k_{col}/k_{soil} > 300$ to 1000 for the columns to function as vertical drains. Otherwise the hydraulic lag in the columns could be excessive especially for long columns, which are drained on one side only. A drainage blanket is usually required below the embankment to improve the drainage.

The diameter of the columns affects the time required for the consolidation since the diameter affects both the coefficient n and the L_{col}/r-ratio. One of the largest uncertainties with presently used design methods is the permeability of the columns and the change of the permeability with time. It is anticipated that the permeability will decrease with time when the soil becomes saturated and the width of cracks and fissures in the stabilized soil is reduced. It is important to monitor the settlement rate in order to verify the effectiveness of the columns as drains.

The time required for the consolidation can be reduced by 50 to 70 per cent by pre-loading the soil just after the installation of the columns, when the columns are still weak. With lime/cement and cement the consolidation rate depends mainly on the increase of the compression modulus with time. The stabilized soil should preferably be loaded as soon as possible after the installation of the columns when the shear strength still is low. Then almost the total weight of the fill used for the pre-loading will be transferred to the soft clay between the columns due to the low bearing capacity of the columns. The load carried by lime or lime/cement columns will gradually increase with time as the bearing capacity of the columns is increased.

8.13 Design of lime and lime/cement columns

8.13.1 *Ultimate and serviceability limit states (ULS and SLS)*

In the design of lime, cement and lime/cement columns it is important to consider,

- that the ultimate bearing capacity and the stability of the stabilized embankments, trenches and slopes are adequate (ULS) as well as the bearing capacity of the columns and of the unstabilized soil between the columns;
- that the total and the differential settlements as well as the lateral deformations are not excessive at the working load (SLS);
- that nearby buildings as well as buried services and other structures are not damaged during and after the installation of the columns.

The environmental impact of the columns should also be considered. The environmental impact is usually small for lime, lime/cement and cement columns compared with many other soil stabilization and soil improvement methods.

8.13.2 *Design of lime and lime/cement columns, $F_{si} > 1.2$*

The bearing capacity and the undrained shear strength of soil stabilized by lime and lime/cement are usually estimated from the peak resistance of the columns and of the unstabilized soil between the columns when the shear strength of the columns is less than 100 to 150 kPa and the required increase of the stability is small. Under favourable conditions a characteristic shear strength of 150 kPa can be used in design, when the global factor of safety without the columns is larger than 1.2 ($F_s > 1.2$).

An effective stress analysis is used to evaluate the long-term bearing capacity with ϕ'_{col} of 30° for lime columns, 35° for lime/cement columns and 40° for cement columns. The pore water pressure in the columns u_{col} could then be as high as the pore water pressure in the surrounding unstabilized soil. A circular slip or failure surface through the columns is normally used to determine the factor of safety.

The load distribution between the columns and the unstabilized soil corresponds to the modulus of elasticity for the columns and to the compression modulus for the unstabilized soil, E_{col} and M_{soil}, respectively. The required global factor of safety is 1.5 according to the Swedish Road Board and the Swedish National Rail Administration.

Single or double rows with overlapping columns are usually required to stabilize slopes, excavations and embankments. The required overlap is 50 mm for columns with 0.6 m diameter. The individual columns should extend into a layer with a high bearing capacity to prevent failure along a slip surface below the columns.

8.13.3 Design of lime and lime/cement columns, $F_{si} < 1.2$

Columns with an unconfined compressive strength exceeding 200–300 kPa are used to increase the stability when the initial factor of safety without the columns is less than 1.2. It is proposed that the stability could be evaluated from the unconfined compressive strength of the columns up to 1.0 MPa, when a large increase of the stability is required,

- by considering the reduction of the stability of the columns and of the unstabilized soil by progressive failure;
- by locating the columns in the active zone below an embankment or a fill;
- by designing the columns to carry the full weight of the embankment or the fill;
- by ensuring that the weight of the embankment or the fill is transferred to and from the columns without excessive settlements;
- by replacing the soft or loose soil above the columns and the weak upper part of the columns by compacted granular material;
- by resisting the lateral earth pressure in the embankment by geofabric or geo-anchors;
- by checking that the columns do not contain weak layers or lenses which could reduce the bearing capacity of the columns.

It is proposed to calculate the settlements with a modulus of elasticity of $150q_{u,col}$ up to an equivalent pre-consolidation pressure $p'_c = 1.7q_{u,col}$ and a maximum creep strength of 1.0 MPa.

8.13.4 Limitations of presently used design methods

Slope and bearing capacity failures have occurred, which indicate that the stability could be overestimated for single columns, when an average shear strength is used to estimate the stability of slopes and of embankments. Even a small lateral displacement could fail the columns when the shear strength of the columns is high and the ductility is low. A 0.6 m diameter lime/cement and cement column could fail when the lateral displacement is only 20 to 30 mm.

The shear resistance of column rows could be less than the peak shear strength. The shear resistance can also be low for floating column rows when the slip surface is located just below a column row with point-bearing columns. The column rows could also fail by overturning, translation, separation or by internal shear along the overlap at the centre of the column row.

8.13.5 Total and effective stress analysis

The bearing capacity and the shear resistance of lime, lime/cement and cement columns are governed by the drained shear strength, depending on the loading rate and the permeability of the columns. The undrained shear strength usually governs the shear strength of the unstabilized soil. The pore water pressure in the unstabilized soil around the columns could govern the shear resistance of lime/cement columns in an effective stress analysis if calculations or measurements don't indicate otherwise since the hydraulic lag can be large in the columns.

The shear strength of lime/cement and cement columns increases with time as the excess pore water pressure in the columns dissipates during the consolidation. The lowest shear strength is expected just below the dry crust where the effective overburden pressures and the shear strength of the unstabilized soil are low.

The undrained shear strength, $\tau_{fu,col}$ of the soil stabilized by lime/cement or cement, which governs the stability of embankments, slopes, trenches and excavations, increases with increasing confining pressure when the stress level is low. The drained shear strength is evaluated by the following equation:

$$\tau_{fd,col} = c_{d,col} + [\sigma_f - u_f]\tan \phi_{d,col} \quad (8.24)$$

where $\phi_{d,col}$ is the drained angle of internal friction of the stabilized soil, which varies with soil type, the confining pressure and with the water content, $c_{d,col}$ is the drained cohesion and σ_f is the total normal pressure on the failure plane through the columns. The shear strength usually increases with increasing clay content and with increasing plasticity index.

The short-term bearing capacity depends on the confining pressure σ_h, which can be estimated by the following equation.

$$\sigma_h = \sigma_{ho} + K_{soil} m_{o,soil} q_o + c_{u,soil}\left(1 + \ln\left[\frac{E_{u,soil}}{2c_{u,soil}(1 - \nu_{soil})}\right]\right) \quad (8.25)$$

where σ_{ho} is the initial total lateral earth pressure, $c_{u,soil}$ is the undrained shear strength of the unstabilized soil, E_{soil}, is the modulus of elasticity and ν_{soil} is Poisson's ratio. The term $K_{soil}\, m_{o,soil}\, q_o$ is the increase of the effective lateral earth pressure caused by the applied unit load q_o, $m_{o,soil}$, is the stress factor for the unstabilized soil and K_{soil} is the coefficient of lateral earth pressure, which is assumed to be at least 1.0 due to the volume increase during the slaking of the lime. At $E_{u,soil} = 200c_{u,soil}$ and $\nu_{soil} = 0.5$

$$\sigma_h = \sigma_{ho} + K_{soil}\, m_{o,soil}\, q_o + 5c_{u,soil} \quad (8.26)$$

8.13.6 Bearing capacity of lime and lime/cement columns

The bearing capacity q_{col} can be estimated by the following equation when the effective confining pressure $(\sigma_h - u_{col})$ is considered.

$$q_{col} = q_{o,col} + K_{col}(\sigma_h - u_{col}) + u_{col} \quad (8.27)$$

where $q_{o,col}$ is the unconfined bearing capacity of the columns when the confining pressure is zero $(\sigma_h = 0)$, K_{col} is the coefficient of lateral earth pressure for the stabilized soil. The bearing capacity of the columns can be estimated by the following equation at $\sigma_{ho} = \sigma_{vo} + 5c_{u,soil} + m_{o,soil} q_o$

$$q_{col} = 2c'_{col}\sqrt{K_{p,col}} + K_{p,col}(\sigma_{vo} + 5c_{u,soil} + m_{o,soil} q_o - u_{col}) + u_{col} \quad (8.28)$$

where c'_{col} is the effective cohesion, σ_{vo} is the initial total overburden pressure, $c_{u,soil}$ is the undrained shear strength of the unstabilized soil and u_{col} is the pore water pressure in the columns. The coefficient of passive earth pressure $K_{p,col}$ is equal to 3.00, 3.69 and 4.60 at $\phi'_{col} = 30°$, 35° and 40°, respectively.

The excess pore water could be high for lime/cement columns and to some extent also for lime columns due to excessive hydraulic lag especially when the length of the columns is large. The pore water pressure in the columns could be as high as the pore water pressure in the unstabilized clay around the columns when the hydraulic lag in the columns is large.

The residual bearing capacity of lime and lime/cement and cement columns at $c'_{col,res} = 0$ can be calculated from:

$$q_{col,res} = K_{p,col}(\sigma_{vo} + 5c_{u,soil} + m_{soil}q_o - u_{col}) + u_{col} \tag{8.29}$$

where u_{col} is the pore water pressure.

8.13.7 Creep strength of columns

The effective cohesion c'_{col} has been reduced by 50 per cent in the following equation due to creep since mainly the effective cohesion c'_{col} is affected.

$$q_{col,creep} = c'_{col}\sqrt{K_{p,col}} + 3(\sigma_{vo} + 5c_{u,soil} - u_{col} + m_{soil}\,q_o) + u_{col} \tag{8.30}$$

It should be noted that creep should only be considered together with the dead load in an effective stress analysis.

8.13.8 Global factor of safety

The variation of the shear strength of the stabilized soil is often large across the diameter and along the length of the columns (Axelsson and Larsson, 1994). It is therefore difficult to determine a characteristic shear strength and a characteristic compression modulus by unconfined compression or by triaxial tests.

A single global factor of safety, $F_s = 2.0$ to 3.0, is generally used in Sweden to determine the allowable load. A minimum factor of safety of 1.2 to 1.5 is usually required for embankments and slopes, respectively, for both short and long-term conditions. The factor of safety should be at least 1.5 according to the Swedish Road Board (Wilhelmsson and Brorsson, 1987). The required factor of safety with respect to bottom heave is 1.3.

The bearing capacity of buildings and of other structures, which are supported by lime or lime/cement columns as well as the stability of embankments, deep excavations and shallow trenches are calculated by an effective stress analysis for the columns and by an undrained analysis for the unstabilized soil between the columns. The bearing capacity of the columns as determined by an effective stress analysis depends on the pore water pressure in the columns. Due to the high hydraulic lag in the columns, the pore water pressure could be as high as the pore water pressure in the surrounding unstabilized soil. The shear resistance of the columns is increased when the excess pore water pressure is reduced.

A high factor of safety is required for column rows due to the often low shear strength of the stabilized soil in the overlapping zone between adjacent columns and the small contact area. The shear force between adjacent columns is high at the centre of a column wall, when the number of overlapping columns in the column rows is large.

It is important that the long-term bearing capacity of the columns is sufficient. The Swedish National Rail Administration requires that the life expectancy of lime and lime/cement columns should be at least 100 years. In sand and silt the shear strength could be reduced with lime when the pH-value of the ground water is low, less than 5.0 and there is a flow of ground water through the columns.

It is expected that the reduction of the long-term bearing capacity will be lower for lime/cement and cement columns than for lime columns due to the low permeability of the stabilized soil with cement or lime/cement. The long-term bearing capacity of lime columns could decrease with time for fibrous peat due to the low pH-value and the relatively high permeability of the stabilized soil. Additional columns could in that case be required.

The lateral displacements of the columns could be excessive for high embankments when the global factor of safety is less than 1.5 to 2.0. An even higher factor of safety might be required for organic soils where the lateral displacement can be large due to creep. Lime/cement and lime columns could be combined with berms at the toe of a slope or of an embankment if the estimated factor of safety is not sufficient.

8.13.9 Characteristic shear strength

The interaction of cement and lime/cement columns with the surrounding unstabilized soil is uncertain when the shear strength of the stabilized soil is high because of the low failure strain. The reduction of the shear resistance and of the bearing capacity can be large when the peak shear strength is exceeded.

The maximum characteristic undrained shear strength is 100 to 150 kPa at full interaction of the columns with the unstabilized soil between the columns. The maximum characteristic shear strength, which is used in design in Sweden, is normally 100 kPa. A characteristic shear strength of 150 kPa can only be applied when the soil and the loading conditions are favourable (SGF, 2000).

The characteristic drained shear strength $\tau_{\mathrm{fd,c}}$ which governs the shear resistance of lime and lime/cement columns, is evaluated by the following equation.

$$\tau_{\mathrm{fd,c}} = c_{\mathrm{d,c}} + \sigma_{\mathrm{f}}' \tan\phi_{\mathrm{d,c}} = c_{\mathrm{d,c}} + [\sigma_{\mathrm{f}} - u_{\mathrm{col}}] \tan\phi_{\mathrm{d,c}} \qquad (8.31)$$

where $c_{d,c}$ is the characteristic drained cohesion, $\phi_{d,c}$ is the characteristic drained angle of internal friction and σ'_f is the normal pressure on the failure plane. The characteristic friction angle $\phi_{d,c}$ is 25 to 35° when the confining pressure is low. A friction angle ϕ_d of 30° is often used in design (Liedberg *et al.*, 1996a,b). It should be noted that the pore water pressure can be negative at failure due to dilatancy of the stabilized soil when the confining pressure is low.

The characteristic cohesion $c_{d,c}$ has been 25 to 50 per cent of the *in situ* undrained shear strength for 0.6 m diameter columns. The average characteristic cohesion $c_{d,c}$ is about 35 per cent of the undrained shear strength. According to Carlsten and Ekström (1995, 1997) the characteristic effective cohesion $c'_c = c_{d,c}$ can be assumed to $0.3c_{u,c}$ for the active zone, $0.1c_{u,c}$ for the shear zone and zero for the passive zone.

8.13.10 Load and partial safety factors

Partial safety factors and load factors are used in Sweden and Finland to take into account the possible variation of the applied load and the shear strength of the stabilized soil across and along the columns. A characteristic shear strength is used as well as a material property coefficient f_d ($f_d = f_k \eta \gamma_m \gamma_n$) where f_k is the characteristic strength of the soil, γ_m is a partial factor of safety, which reflects the uncertainties involved in the evaluation of the shear strength and γ_n is a partial coefficient, which depends on the safety class. The coefficient η takes into account the difference in shear strength between the soil samples and the *in situ* shear strength. The extent of the soil investigation should also be considered as well as the variation of the shear strength and of the compressibility within the site.

Lime and lime/cement columns are in general designed in Sweden according to safety class 2. A partial safety factor γ_m of 1.8 is used with respect to the undrained shear strength τ_{fu} and 1.2 with respect to the effective friction resistance $\tan \phi'_{col}$.

8.13.11 Coefficient of variation

Ekström (1992) observed that the coefficient of variation, the ratio of the standard deviation and the mean shear strength, had a tendency to increase with time. The coefficient of variation was 10 per cent after one week, about 15 per cent after about 2 months and reached 20 to 35 per cent after about 3 months. The increase of the coefficient of variation is mainly caused by the increase of the shear strength with time.

It should also be noted that the variation of the shear strength along the columns is generally less with cement and lime/cement than with lime. The coefficient of variation can for lime columns be as high as 30 to 70 per cent.

The coefficient of variation has also a tendency to increase with increasing water/cement ratio from about 25 per cent at $w/c = 1.0$ to 40 per cent at $w/c = 2.0$ (Matsuo *et al.*, 1996). Yoshizawa *et al.* (1996) have reported that the coefficient of variation decreases with increasing cement content.

8.13.12 Load distribution

The stabilizing effect of the columns, which are located in the active zone, corresponds to the load $(Q_{p,a} + Q_{s,a})$, which is transferred to the columns and through the slip surface. The axial column load $(Q_{p,a} + Q_{s,a})$ corresponds to the sum of the loads transferred to the columns by end-bearing $Q_{p,a}$ and by shaft resistance $Q_{s,a}$ above the slip surface. For floating columns the point and the shaft resistance below the assumed failure surface $(Q_{p,b} + Q_{s,b})$ governs when $(Q_{p,b} + Q_{s,b}) < (Q_{p,a} + Q_{s,a})$.

The load transferred to the columns corresponds to the modulus of elasticity of the stabilized soil E_{col} when the deformations of the columns are the same as the deformations of the unstabilized soil between the columns. For high embankments the creep strength of the columns $Q_{col,creep}$ could be exceeded, which is estimated to 65–80 per cent of the short-term bearing capacity of the columns. The creep strength increases with increasing depth and with increasing time.

For high embankments the column load $Q_{col,creep}$ can be transferred to adjacent columns when the spacing of the columns is small. The settlements will be reduced if the soft or the loose soil above the columns is removed and replaced by compacted granular fill. Geofabric will be required to reduce the lateral displacement of the columns.

The bearing capacity, which can be utilized in design depends mainly on the location of the columns. The bearing capacity will be high for the columns, which are located in the active zone below the embankment. The moment capacity and the shear resistance of the columns, which depend on the axial load, will be low when the surface crust is missing or is poorly developed. The shear strength of the soil located above the columns is usually low since the columns are terminated 0.5 to 1.0 m below the ground surface.

Pre-loading of the columns should be carried out just after the installation of the columns when the bearing capacity is still low, to consolidate the remoulded soil below the columns as well as soft layers or pockets in the columns. It is important to limit the axial load during the pre-loading so that the columns will not be overloaded.

A reduction of the bearing capacity is expected when the peak shear strength of the columns is exceeded. The reduction can be large when the peak shear strength is high and the confining pressure is low. Carlsten and Ekström (1995, 1997) have proposed that the maximum shear strength of the columns to be used in design should be limited to 100 or 150 kPa. Even this reduced shear strength could be too high for the columns, which are located in the shear or in the passive zone of a potential failure surface as indicated by the low shear strength, which has been back-calculated by, for example, Jacklin and Larsson (1994) for lime/cement columns and by Terashi and Tanaka (1983b), and by Kitazume *et al.* (1996b) for cement columns. Also Kivelö (1998) has pointed out the low efficiency of the columns, which are located in the shear and in the passive zones.

The stabilizing effect of the columns in the active zone can be taken into account by reducing the weight of the embankment by the load, which is transferred by the columns through the assumed failure surface. Several methods can be used to calculate the bearing capacity of the columns and of the unstabilized soil between the columns such as:

$$q_{col,res} = K_{p,col}(\sigma_{vo,soil} + 5c_{u,soil} - u_o + m_{soil}q_o) + u_o \tag{8.32}$$

At $\phi'_{res} = 35°$ and $c'_{res} = 0$, the bearing capacity is estimated to be 463 kPa. It is proposed to use a factor of safety of, 1.50 ($q_{col,res}/1.5$) to determine the allowable column load. The allowable column load is 309 kPa (463/1.5), which corresponds to 87.3 kN for a 0.6 m diameter column and 60.6 kN for a 0.5 m diameter column.

Since the bearing capacity of the unstabilized soil cannot be utilized fully, the load caused by the fill, $h_{fill}\gamma_{fill} = 100$ kPa (5×20) had to be carried by the columns. The required spacing of the columns with 0.5 m diameter is 0.78 m ($\sqrt{60.6/100}$) at an allowable load of 60.6 kN. The required spacing of 0.6 m diameter columns is 0.93 m ($\sqrt{87.3/100}$) at an allowable load of 87.3 kN, which corresponds to an area ratio $a = 0.33$.

It is also possible to estimate the required spacing of the columns from the creep strength $q_{col,creep}$ and a reduced undrained shear strength for the unstabilized soil between the columns. The creep strength of the columns is estimated to be 552 kPa ($0.8 \times 200 + 3.67 \times 42.5 + 183 \times 7.5 - 2.67 \times 25 + 3.67 \times 0.27 \times 100$) at an area ratio of 0.2 and a reduced shear strength of 7.5 kPa for the unstabilized soil between the columns. The bearing capacity of the columns $q_{uo,creep}$ is estimated to be 200 kPa. The allowable column load is 54.1 kPa ($0.196 \times 552/2.0$). The required spacing of the columns with 0.5 m diameter is 0.74 m ($\sqrt{54.1/100}$). This spacing is smaller than the spacing required at the residual bearing capacity of the columns.

The reduction of the bearing capacity when the columns are displaced laterally, can be prevented at least partially by placing one or several layers with geofabric just above the columns. The geofabric should be designed to resist the total lateral earth pressure in the embankment or in the fill.

The permeability of lime/cement and cement columns can be low since cement reduces the permeability of the stabilized soil. It is therefore uncertain if lime/cement and cement columns function as effective drains. Band drains might be required to reduce the time required for consolidation of the unstabilized soil between the columns. The permeability and the drainage can be improved by a hole at the centre of the columns.

The lateral resistance T_{col} of the columns is low, when the moment capacity and the axial load in the columns are small. The lateral resistance is reduced when there are soft layers or pockets in the columns. The moment capacity and the shear resistance will also be low for floating columns since lime and cement are injected at 0.3 to 0.4 m above the tip of the mixing tool. The stability of an embankment should therefore be checked for a horizontal slip surface passing through the remoulded soil below the columns.

The shear strength and the load transferred to the columns will increase with time due to reconsolidation of the soil around and below the columns. It is expected that the reconsolidation will be rapid since the remoulded layer below the columns is thin, 0.3 to 0.4 m. The height of the embankment should therefore be increased slowly to allow dissipation of the excess pore water pressure and consolidation of the soil.

The interaction of the lime/cement and cement columns with the unstabilized soil between the columns is uncertain when the ductility of the columns is low. The high shear strength of lime/cement and cement columns can often not be utilized fully in design together with the peak shear strength of the unstabilized soil between the columns because of progressive failure as discussed by Kivelö (1996, 1998).

The shear strength of the unstabilized soil below the columns can also be low due to the disturbance of the soil during the installation of the columns. The shear strength of the remoulded soil below the columns increases with time due to consolidation, which will be rapid due to the limited thickness of the remoulded layer. The time to reach 90 per cent consolidation for a 0.3 m thick layer, which is drained on one side only, is estimated to about 4 weeks at a coefficient of consolidation of 1.0 m^2/year.

Bibliography

Åhnberg, H. (1996) Stress dependent parameters of cement stabilised soil, *Proceedings of the 2nd International Conference on Ground Improvement Geosystems*, IS-Tokyo '96, Tokyo, 14–17 May, Vol. 1, pp. 387–392.

Åhnberg, H. and Holm, G. (1986) Kalkpelarmetoden – Resultat av 10 års forskning och praktisk användning samt framtida utveckling (The lime column method – results from research and practical applications during 10 years and future developments). *Swedish Geotechnical Institute, Linköping, Sweden*, Report No. 31, 122 pp.

Åhnberg, H., Holm, G., Holmqvist, L. and Ljungkrantz, C. (1994) The use of different additives in deep stabilisation of soft soils. *Proceedings of the 13th International Conference on Soil Mechanics and Foundation Engineering*, New Delhi, India, Vol. 3, pp. 1191–1194.

Åhnberg, H., Johansson, S.-E, Retelius, A., Ljungkrantz, C., Holmqvist, L. and Holm, G. (1995) Cement och kalk för djupstabilisering av jord – En kemisk fysikalisk studie av stabiliseringseffekter (Cement and lime for stabilisation of soil at depth – a chemical physical investigation of soil improvement effects). *Swedish Geotechnical Institute*, Linköping, Sweden, Report No. 48, 213 pp.

Ansano, J., Ban, K., Azuma, K. and Takahashi, K. (1996) Deep mixing method of soil stabilization using coal ash. *Proceedings of the 2nd International Conference on Ground Improvement Geosystems*, 14–17 May, IS-Tokyo '96, Vol. 1, pp. 393–398.

Arnér, E., Kivelö, M., Svensson, P.L. and Johnsson, R. (1996) OKB-E4 över Norraladalen. Hög provbank visar på bättre utnyttjande av kalkcementpelarförstärkning (OKB-E4. A high test embankment over the Norrala valley indicates an improved utilization of lime/cement columns), *Bygg & Teknik*, No. 9/96, pp. 19–23.

Axelsson, A. and Larsson, S. (1994) Provningsmetoder på kalkcementpelare – Svealandsbanan. (Test methods for lime/cement columns – the Svealand railway). Final Year Project 94/10, *Royal Institute of Technology, Department of Soil and Rock Mechanics, Stockholm, Sweden*, 62 pp. + 10 Appendixes.

Axelsson, K., Johansson, S.-E. and Andersson, R. (1996) Stabilisering av organisk jord. Förstudie inom Svensk Djupstabilisering (Stabilization of organic soils. A preliminary study within the project Swedish Deep Stabilisation). Report No. 3, 35 pp. + 6 Appendixes.

Babasaki, R. and Suzuki, K. (1996) Open cut excavation of soft ground using the DCM method. *Proceedings of the 2nd International Conference on Ground Improvement Geosystems*, IS-Tokyo '96, 14–17 May, Vol. 1, pp. 469–473.

Babasaki, R., Terashi, M., Suzuki, T., Maekawa, A., Kawamura, M. and Fukazawa, E. (1996) JGS TC Report: Factors influencing the strength of improved soil. Committee Report, Japanese Geotechnical Society, *Proceedings of the International Conference on Soil Improvement Geosystems*, IS-Tokyo '96, Tokyo, Japan, Vol. 2, pp. 913–918.

Baker, S., Liedberg, N.S.D. and Sällfors, G. (1997) Deformation properties of lime cement stabilised soil in the working state, *Proceedings of the 14th International Conference on Soil Mechanics and Foundation Engineering*, Hamburg, Germany, Vol. 3, pp. 1667–1672.

Balasubramaniam, A.S. and Buensuceso, B.R. (1989) On the overconsolidated behavior of lime treated soft clay, *Proceedings of the 12th International Conference on Soil Mechanics and Foundation Engineering*, Rio de Janeiro, Brazil, Vol. 2, pp. 1335–1338.

Bengtsson, P.E. and Holm, G. (1984) Kalkpelare som drän? (Lime columns as drains?). *Nordic Geotechnical Conference, NGM-84*, Linköping, Sweden, Vol. 1, pp. 391–398.

Bergado, D.T., Anderson, L.R., Miura, N. and Balasubramaniam, A.S. (1996) Soft ground improvement in Lowland and other environments. *ASCE Press*, New York, USA, 427 pp.

Björkman, J. and Ryding, J. (1996) Kalkcementpelares mekaniska egenskaper (Mechanical properties of lime/cement columns). Final Year Project 96/1, *Department of Soil and Rock Mechanics, Royal Institute of Technology*, Stockholm, Sweden.

Blom, L. (1992) Kalkpesarforstarkning i ny tappning (New applications of the lime column method) Byggnadsindustrin, No. 3, pp. 16–18.

Boman, P. and Broms, B.B. (1975) Stabilisering av kohesionsjord med kalkpelare. (Stabilization of cohesive soils with lime columns.) *Nordic Geotechnical Conference*, NGM-75, Copenhagen, Denmark, pp. 265–279.

Brandl, H. (1981) Alteration of soil parameters by stabilisation with lime, *Proceedings of the 10th International Conference on Soil Mechanics and Foundation Engineering*, Stockholm, Vol. 3, pp. 587–594.

Brandl, H. (1995) Short and long term behavior of non-treated and lime- or cement-stabilised fly ash, *Bengt B. Broms Symposium on Geotechnical Engineering*, 13–15 December, Nanyang Technological University, Singapore, pp. 39–54.

Bredenberg, H. (1979) Kalkpelarmetoden – Beräkningsmetoder (The lime column method – design methods) Seminar on the practical applications of the lime column method, *Royal Institute of Technology, Department of Soil and Rock Mechanics*, 27 November, 11 pp.

Brinch Hansen, J. (1948) The stabilizing effect of piles in clay. *CN-Post No. 3*, November.

Broms, B.B. (1972) Stabilization of slopes with piles, *Proceedings of the 1st International Symposium on Landslide Control*, Kyoto, Vol. 1, pp. 115–123.

Broms, B.B. (1999) Progressive failure of lime, lime/cement and cement columns, *Proceedings of the Dry Mix Methods Soil Stabilization*, Stockholm, Sweden, 13–15 October, pp. 177–184.

Broms, B.B. and Boman, P. (1977) Lime columns – a new type of vertical drain. *Proceedings of the 9th International Conference on Soil Mechanics and Foundation Engineering*, Tokyo, Japan, Vol. 1, pp. 427–432.

Broms, B.B. and Boman, P. (1979a) Lime columns – a new foundation method. *ASCE, Journal of Geotechnical Engineering Division*, Vol. 105, NoGT4, pp. 539–556.

Broms, B.B. and Boman, P. (1979b) Stabilisation of soil with lime columns, *Ground Engineering*, Vol. 12, No. 4, pp. 23–32.

Brookes, A.H., West G. and Carder (1997) Laboratory trial mixes for lime-stabilised soil columns and lime piles, *Transport Research Laboratory*, TRL Report 306, 16 pp.

Bryhn, O., Løken, T. and Aas, G. (1983) Stabilisation of sensitive clays with hydroxy-aluminum compared with unslaked lime, *Proceedings of the 8th European Conference on Soil Mechanics and Foundation Engineering*, Helsinki, Finland., Vol. 2, pp. 885–896.

Carlsten, P. (1989) Manual to LIMESET. *Swedish Geotechnical Institute*, Varia 248, p. 20.

Carlsten, P. and Ekström, J. (1995 and 1997) Kalk- och kalkcement pelare – Vägledning för projektering, utförande och kontroll (Lime and lime cement columns – guide for project planning, construction and inspection). *Swedish Geotechnical Society*, Report 4:95 and 4:95E, ISSN 1103-7237, Linköping, Sweden, 103 and 111 pp.

Carlsten, P. and Ouacha, M. (1993) Funktionsuppföljning av kalkpelare (Evaluation of the behaviour of lime columns). *Swedish Geotechnical Institute, Linköping, Sweden*, Varia 407, 7 pp.

Carlsten, P. and Tränk, R. (1992) Deep stabilisation with lime and lime/cement columns – a comparison of performance, Nordic Geotechnical Conference NGM-92, Session 1–4, Aalborg, Denmark. *Danish Geotechnical Society*, DGF Bulletin 9, Vol. 1/3, pp. 25–30.

Chida, S., 1981. Development of dry jet mixing methods. Public Works Research Institute, *Ministry of Construction, Japan*, pp. 29–35.

Edstam, T. (1996) Erfarenhetsbank för kc-pelare (Experiences with lime/cement columns). *Swedish Deep Stabilisation Research Center*, Report No. 1, Linköping, Sweden, 154 pp.

Ekström, J.C. (1992) Kalk och kalkcementpelare – Metod under utveckling (Lime and lime/cement columns – a method under development). *Swedish Geotechnical Society* (SGF), Foundation Engineering Day, 15 pp.

Ekström, J.C. (1994a) Kontroll av kalk- och kalkcementpelare (Checking of lime and lime/cement columns) Foundation Engineering Day '94, 9 March. *Swedish Geotechnical Society*, Stockholm, Sweden, 14 pp.

Ekström, J.C. (1994b) Kontroll av kalkcementpelare, Slutrapport med redovisning av fältförsök i Ljungskile (Checking of lime/cement columns, Final Report with Reference to Field Tests at Ljungskile). *Chalmers Technical University*, Gothenburg, Report B 1994:3, pp. 1:1–8:4 + 7 Appendixes.

Eriksson, M. and Carlsten, P. (1995) Tryckförsök på kemisk stabiliserad jord (Unconfined compression tests with chemically stabilized soil). *Swedish Geotechnical Institute*, Varia 435, Linköping, Sweden, 19 pp. + 13 Appendix.

Göransson, M. and Larsson, J. (1994) Kalkpelares deformationsegenskaper (Deformation properties of lime columns). Final Year Project 1994:5. *Chalmers Technological University*, Gothenburg, Sweden.

Green, M. and Smigan, R. (1995) Kalkcementpelare. Materialparametrar och datorsimulering av enskilda pelares function. (Lime/cement columns. Material properties and computer simulation of the behavior of single columns) Final Year Project 95/11, *Royal Institute of Technology*, Stockholm.

Halkola, H.A. (1984) Kontrollmätningar av kalkpelare i Helsingfors (Control of lime columns in Helsinki), *Proceedings of the Nordic Geotechnical Conference*, NGM-84, Vol. 2, pp. 879–886.

Hansbo, S. (1979) Consolidation of clay by bandshaped prefabricated drains, *Ground Engineering*, July, Vol. 12, No. 5.

Hansbo, S. (1987) Design aspects of vertical drains and lime column installations, *Proceedings of the 9th Southeast Asian Geotechnical Conference*, Bangkok, Vol. 2, pp. 8.1–8.12.

Hansbo, S. (1994) Foundation Engineering, *Developments in Geotechnical Engineering*, 75, Elsevier, Amsterdam, 519 pp.

Hoikkala, S., Leppänen, M.S. and Tanska, H. (1997) Blockstabilization of peat in road construction, *Proceedings of the 14th International Conference on Soil Mechanics and Foundation Engineering*, Hamburg, Vol. 3, pp. 1693–1696.

Holeyman, A. Franki, S.A. and Mitchell, J.K. (1983) Assessment of quicklime behavior, *Proceedings of the 8th European Conference on Soil Mechanics and Foundation Engineering*, Improvement of Soils, Helsinki, Finland, Vol. 2, pp. 897–902.

Holm, G. (1979) Praktiska exempel på tillämpning av kalkpelarmetoden (Examples of the practical applications of the lime column method). Seminar on the Application of Lime Columns in Practice, *Royal Institute of Technology, Department of Soil and Rock Mechanics*, Stockholm, 27 November, 7 pp.

Holm, G. (1994) Deep stabilisation by admixtures, *Proceedings of the 13th International Conference on Soil Mechanics and Foundation Engineering*, New Delhi, India, Vol. 5, pp. 161–162.

Holm, G. and Åhnberg, H. (1987) Användning av kalk-flygaska vid djupstabilisering av jord (Application of lime and fly ash at stabilisation of soil at depth). *Swedish Geotechnical Institute*, Report No. 30, pp. 59–91.

Holm, G., Tränk, R. and Ekström, A. (1983a) Improving lime column strength with gypsum, *Proceedings of the 8th European Conference on Soil Mechanics and Foundation Engineering*, Helsinki, Finland, Vol. 2, pp. 903–907.

Holm, G., Tränk, R., Ekström, A. and Torstensson, B.A. (1983b) Lime columns under embankments – a full scale test. Improvement of Ground, *Proceedings of the 8th European Conference on Soil Mechanics and Foundation Engineering*, Helsinki, Finland, Vol. 2, pp. 909–912.

Jacklin, A. and Larsson, U. (1994) Vägbank på kalkpelarförstärkt lera. Utredning av kollaps (Road embankment reinforced by lime/cement columns constructed on clay – analysis of a failure). Final Year Project 1994:3, *Department of Geotechnical Engineering, Chalmers Technological University*, Gothenburg, Sweden.

Kakihara, Y., Hiraide, A. and Baba, K. (1996) Behavior of nearby soil during improvement works by deep mixing method, *Proceedings of the 2nd International Conference on Ground Improvement Geosystems*, 14–17 May, IS-Tokyo '96, Vol. 1, pp. 625–630.

Kamon, M. and Nontananandh, S. (1991) Combining Industrial Waste with Lime for Soil Stabilization, *Journal Geotechnical Engineering Division, ASCE*, Vol. 1, January, pp. 1–17.

Karastanev, D., Kitazume, M., Miyajima, S. and Ikeda, T. (1997) Bearing capacity of shallow foundation on column type DMM improved ground, *Proceedings of the 14th International Conference on Soil Mechanics and Foundation Engineering*, Vol. 3, pp. 1621–1624.

Kitazume, M., Tabata, T., Ishiyama, S. and Ishikawa, Y. (1996a) Model tests on failure pattern of cement treated retaining wall, *Proceedings of the 2nd International Conference on Ground Improvement Geosystems*, Tokyo, 14–17 May, Vol. 1, pp. 509–514.

Kitazume, M., Ikeda, T., Miyajima, S. and Karastanev, D. (1996b) Bearing capacity of improved ground with column type DMM, *Proceedings of the International Conference on Ground Improvement Geosystems*, Tokyo, 14 to 17 May, Vol. 1, pp. 503–508.

Kivelö, M. (1994a) Odränerade provbelastningar av kalkcementpelare i fält (Undrained field load tests of lime/cement columns) Royal Institute of Technology, TRITA – AMI Report 3002, ISSN 1400–1306, Stockholm, Sweden, 62 pp.

Kivelö, M. (1994b) Kalkcementpelare som förstärkt jord eller pålar? (Lime/cement columns as reinforced soil or piles?). Bygg & Teknik, No. 8/94, pp. 42–45.

Kivelö, M. (1996) Spännings-töjningssamband och skjuvhållfasthet hos kalk-cement pelare (Stress–strain relationships and shear strength of lime/cement columns), *Proceedings of the 12th Nordic Geotechnical Conference, Reykjavik, NGM-96*, 26–28 June, Vol. 1, pp. 309–314.

Kivelö, M. (1997) Undrained shear strength of lime/cement columns, *Proceedings of the 14th International Conference on Soil Mechanics and Foundation Engineering*, Hamburg, Vol. 2, pp. 1173–1180.

Kivelö, M. (1998) Stability Analysis of Lime/Cement Column Stabilized Embankments, PhD Thesis, Division of Soil and Rock Mechanics, *Royal Institute of Technology*, Stockholm, Sweden, 170 pp.

Kohata, Y., Maekawa, H., Muramoto, K., Yajima, J. and Babasaki, R. (1996) Deformation and Strength properties of DM cement treated soils, *Proceedings of the International Conference on Ground Improvement Geosystems*, IS-Tokyo '96, Tokyo, Japan, Vol. 2, pp. 905–911.

Kujala, A. (1983a) Andvändningen av tilläggsmedel med kalk i djupstabilisering (Application of lime as additive at deep stabilisation). *Nordic Seminar on Deep Stabilisation*, Esboo, Finland.

Kujala, K. (1983b) The use of gypsum in deep stabilisation, *Proceedings of the 8th European Conference on Soil Mechanics and Foundation Engineering*, Helsinki, Finland, Vol. 2, pp. 925–928.

Kujala, K. (1984) Faktorer som inverkar på djupstabiliserade jordars mekaniska egenskaper (Factors influencing the mechanical properties of soil stabilised at depth), *Nordic Geotechnical Conference, NGM-84*. Linköping, Sweden, Vol. 2, pp. 895–902.

Kujala, K. and Lahtinen, P.O. (1988) The use of cement for deep stabilization, *Proceedings of the Nordic Geotechnical Conference, NGM-88*, Oslo, Norway, pp. 215–218.

Kujala, K. and Nieminen, P. (1983) On the reactions of clays stabilised with gypsum lime, *Proceedings of the 8th European Conference on Soil Mechanics and Foundation Engineering*, Helsinki, Finland, Vol. 2, pp. 929–932.

Kujala, K., Vepsäläinen, P. and Balk, K. (1984) Djupstabiliserade grunders sättningsberäkningsmetoder (Calculation of settlements for foundations stabilised at depth) *Proceedings of the Nordic Geotechnical Conference, NGM-84*, Linköping, Sweden, Vol. 1, pp. 499–510.

Kujala, K., Huttunen, E. and Angelva, P. (1993) Stabilisering av torvmark (Stabilization of peat). Lime Column Day, Stockholm, 12 pp.

Kukko, H. and Ruohomäki, J. (1985) Savien stabilointi eri sideaineilla (Stabilization of clays with various binders), VIT, Research Notes 1682, Espoo, *Technical Research Center of Finland.*

Kuno, G., Kutara, K. and Miki, H. (1989) Chemical stabilisation of soft soils containing humic acid, *Proceedings of the 12th International Conference on Soil Mechanics and Foundation Engineering*, Rio de Janeiro, Brazil, Vol. 2, pp. 1381–1384.

Lahtinen, P.O. and Vepsäläinen, P.E. (1983) Dimensioning deep-stabilisation using the finite element method, *Proceedings of the 8th European Conference on Soil Mechanics and Foundation Engineering*, Helsinki, Finland, Vol. 2, pp. 933–936.

Larsson, S. and Håkansson S. (1998) CPT-sondering i två pelarskivor efter skred. Projekt Hälsingekusten, E4-OKB, Norrala, Söderhamn (CPT soundings in two column rows after a landslide. Project Hälsingekusten, E4-OKB, Norrala, Söderhamn), *Royal Institute of Technology, Dept. of Soil and Rock Mechanics*, Sockholm, Sweden, 1998-04-19, 13 pp.

Liedberg, N.S.D., Baker, S. and Smekal, A. (1996a) Samverkan mellan kalkcementpelare och lera (Interaction of lime/cement columns and clay). *Nordic Geotechnical Conference, NGM-96*, Reykjavik, Iceland, pp. 43–48.

Liedberg, N.S.D., Baker, S., Smekal, A. and Ekström, J. (1996b) Samverkan mellan kalkcementpelare och lera (Interaction between lime/cement columns and clay). Swedish Railroad Authority, Technical Report 1996:2. *Chalmers Technological University*, Report B1996:1, Department of Geotechnical Engineering, p. 107.

Long, P.H. and Bredenberg, H. (1997) KC-förstärkning för schakt inom spont, Filipstad Brygge, Oslo – En numerisk analys med PLAXIS (Stabilization of an excavation with lime/cement columns, Filipstad Brygge, Oslo – A Numerical Analysis with PLAXIS), *Stabilator AB, Stockholm, Sweden*, 19 pp.

Matsuo, T., Nisibayashi, K. and Hosoya, Y. (1996) Studies on soil improvement adjusted at low compressive strength in deep mixing method, *Proceedings of the 2nd International Conference on Ground Improvement Geosystems*, 14–17 May, IS-Tokyo '96, Vol. 1, pp. 421–424.

Mishra, P. and Srivastava, R.K. (1996) Geotechnical aspects of industrial waste utilisation – Indian experience. *Proceedings of the 2nd International Conference on Ground Improvement Geosystems*, 14–17 May, IS-Tokyo '96, Tokyo, Japan, Vol. 1, pp. 425–430.

Mitchell, J.K. (1981) Soil improvement – state-of-the-art report, *Proceedings of the 10th International Conference on Soil Mechanics and Foundation Engineering*, Stockholm, Vol. 4, pp. 510–520.

Mitchell, J.K. (1986) Practical problems from surprising soil behavior, *Journal Geotechnical Engineering Division, ASCE*, Vol. 12, NoSM3, pp. 259–289.

Miyake, M., Wada, M. and Satoh, T. (1991a) Deformation and strength of ground improved by cement treated soil columns, *Proceedings of the International Conference on Geotechnical Engineering Coastal Development, GeoCoast '96*, Yokohama, Japan, Vol. 1, pp. 369–372.

Miyake, M., Akamoto, H. and Wada, M. (1991b) Deformation characteristics of ground improved by a group of treated soil, *Centrifuge 91*, Balkema, Rotterdam, pp. 295–302.

Nagaraj, T.S. Yaligar, P.P., Miura, N. and Yamadera, A. (1996) Predicting strength development by cement admixture based on water content. *Proceedings of the 2nd International Conference on Ground Improvement Geosystems*, 14–17 May, Tokyo, pp. 431–436.

Nieminen, P. (1978) Stabilisation with gypsum and lime. *Proceedings of the 2nd International Conference on Use By-Products Waste Civil Engineering*, Paris, Vol. 1, pp. 229–235.

Nieminen, P. (1979) Use of industrial by-products as binders. *Nordic Geotechnical Conference, NGM-79*, Esbo, Finland, pp. 303–309.

Nord, M. (1990) Comparison between calculated and measured settlements for a road embankment reinforced with lime columns, *Young Geotechnical Engineers Conference*, Delft, Holland, Vol. 1, p. 4.

Okabayashi, K. and Kawamura, M. (1991) Effect of improvement on soft ground by soil cement mixing method, *Proceedings of the International Conference on Geotechnical Engineering Coastal Development, GEO-COAST '91*, 3–6 September, 1991, Vol. 1, pp. 377–380.

Okumura, T. (1996) Deep mixing method in Japan, *Proceedings of the 2nd International Conference on Ground Improvement Geosystems*, 15–17 May, IS-Tokyo '96, Tokyo, Japan, Vol. 2, pp. 879–887.

Okumura, T. and Terashi, M. (1975) Deep lime-mixing method of stabilization for marine clays, *Proceedings of the 5th Asian Regional Conference on Soil Mechanics and Foundation Engineering*, Bangalore, Vol. 1, pp. 69–75.

Pan, Q.Y., Xie, K.H., Liu, Y.L. and Lin, Q. (1994) Some aspects of the soft ground improved with cement columns, *Proceedings of the 13th International Conference on Soil Mechanics and Foundation Engineering*, New Delhi, India, pp. 1223–1226.

Pramborg, B.O. and Albertsson, B. (1992) Undersökning av kalk/cementpelare (Investigation of lime/cement columns), *Proceedings of Nordic Geotechnical Conference, NGM-92*, Danish Geotechnical Society, Bulletin No. 9, Aalborg, Denmark, Vol. 1, pp. 149–156.

Prust, R.E. (1990) A study of the behavior of Swedish soft clays stabilized with lime and cement, *Chalmers Technological University, Department of Geotechnical Engineering*, Gothenburg, Sweden.

Rathmayer, H. (1997) Deep mixing methods for soft subsoil improvement in the Nordic Countries, *Proceedings of the 2nd International Conference on Ground Improvement Geosystems*, 14–17 May, IS-Tokyo '96, Vol. 2, pp. 869–877.

Ravaska, O. and Kujala, K. (1996) Settlement calculation of deep stabilised peat and clay. *Proceedings of the 2nd International Conference on Ground Improvement Geosystems*, 14–17 May, IS-Tokyo '96, Vol. 1, pp. 551–555.

Rogbeck, Y. (1997) Lime-cement columns on the 'Svealand' rail link – Performance observations, *Proceedings of the 14th International Conference on Soil Mechanics and Foundation Engineering*, Vol. 3, pp. 1705–1710.

Rogbeck, Y. and Tränk, R. (1995) Funktionsuppföljning av kalk- och kalkcementpelare, E4, Delen Lövstad – Norrköping, Östergötlands Län (Behaviour of lime and lime/cement columns, E4, Lövstad – Norrköping Östergötland County). *Swedish Geotechnical Institute*, Varia 426, 15 pp. + 24 Appendixes.

Rogers, C.D.F. and Lee, S.J. (1994) Drained shear strength of lime–clay mixes. *Transportation Research*, Record No. 1440, Design and Performance of Stabilised Bases and Lime and Fly Ash Stabilization, Washington, DC, USA, pp. 63–70.

Sandros, C. and Holm, G. (1996) Deep stabilization with the wet cement mixing method, *Proceedings of the 12th Nordic Geotechnical Conference*, Reykjavik, 26–28 June, pp. 129–134.

Schwab, E. (1976) Bearing capacity, strength and deformation behavior of soft organic sulphide soils, *Thesis, Royal Institute of Technology, Department of Soil and Rock Mechanics*, Stockholm, Sweden.

SGF (2000) Kalk och kalkcementpelare (Lime and lime/cement columns). *Swedish Geotechnical Society*. Report 2:2000, 111 pp.

Sherwood, P.T. (1993) Stabilization with cement and lime. State-of-the-Art Review, *Transport Research Laboratory, Department of Transportation*, London, 153 pp.

Steensen-Bach, J.O., Bengtsson, P.-E. and Rogbeck, Y. (1996) Large scale triaxial tests on samples from lime-cement columns, *Proceedings of the 12th Nordic Geotechnical Meeting*, Reykjavik, Iceland, pp. 135–146.

Suzuki, Y. (1982) Deep Chemical Mixing Methods Using Cement as Hardening Agent. *Symposium on Soil and Rock Improvement Technology Including Geotextiles, Reinforced Earth and Modern Piling Methods, Asian Institute of Technology*, Bangkok, 29 November–3 December, pp. B-1-1–B-1-24.

Tatsuoka, F. and Kobayashi, A. (1983) Triaxial strength characteristics of cement treated soft clay, *Proceedings of the 8th European Conference on Soil Mechanics and Foundation Engineering*, Helsinki, Finland, Vol. 1, pp. 421–426.

Tatsuoka, F., Kohata, Y., Uchida, K. and Imai, K. (1996) Deformation and strength characteristics of cement treated soils in Trans-Tokyo Bay Highway project, *Proceedings of the 2nd International Conference on Ground Improvement Geosystems*, 14–17 May, Vol. 1, pp. 453–459.

Terashi, M. and Tanaka, H. (1981) Ground improvement by deep mixing method, *Proceedings of the 10th International Conference on Soil Mechanics and Foundation Engineering*, Stockholm, Sweden, Vol. 3, pp. 777–780.

Terashi, M. and Tanaka, H. (1983a) Settlement analysis for deep mixing method, *Proceedings of the 8th European Conference on Soil Mechanics on Foundation Engineering*, Helsinki, Finland, Vol. 2, pp. 955–960.

Terashi, M. and Tanaka, H. (1983b) Bearing Capacity and Consolidation of Improved Ground of Treated Soil Columns, *Report of the Port and Harbour Research Institute*, Vol. 22, No. 2, pp. 213–266.

Terashi, M., Tanaka, H., Mitsomoto, T., Niidome, Y. and Honma, S. (1980) Foundation Properties of Lime and Cement Treated Soils. *Report of the Port and Harbour Research Institute*, Vol. 19, No.1, pp. 33–62.

Terashi, M., Tanaka, H. and Kitazume, M. (1983) Extrusion failure of the ground improved by the deep mixing method, *Proceedings of the 7th Asian Regional Conference on Soil Mechanics and Foundation Engineering*, Haifa, Israel, Vol. 1, pp. 313–318.

Tielaitos, S. (1993) Deep stabilisation at Veittostensuo. Research Report (in Finnish), *Tielaitoksen selvityksiä 81/1993*, TIEL 3200205, 87 pp.

Tielaitos, S. (1995) Assessment of the quality and functioning of a test structure on the Mire of Veittostensuo. (in Finnish), *Tielaitoksen selvityksiä 54/1995*, TIEL 3200330, 51 pp.

Uchiyama, K. (1996) Prevention of displacements while using the DJM method, *Proceedings of the 2nd International Conference on Ground Improvement Geosystems*, 14–17 May, Vol. 1, pp. 675–680.

Unami, K. and Shima H. (1996) Deep mixing method at Ukioshima Site of the Trans-Tokyo Bay Highway, *Proceedings of the 2nd International Conference on Ground Improvement Geosystems*, 14–17 May, IS-Tokyo '96, Vol. 1, pp. 777–782.

Vepsäläinen, P. and Arkima, O. (1992) Holvautuminen, tiepenger, sementtipilari (The arching of road embankments). *Finnish National Road Administration*, Research Report 4/1992, Helsinki, Finland, 163 pp. + 8 Appendix.

Wilhelmsson, H. and Brorsson, I. (1987) Kalkpelare - grundförstärkning vid väg-byggnad (Lime columns – soil stabilization at road construction). *Swedish Road Administration*, Publication 1986.72, 1987–08, 29 pp. + 3 Appendixes.

Woo, S.M. and Moh, Z.C. (1972) Lime stabilization of selected lateritic soils, *Proceedings of the 3rd Southeast Asian Conference on Soil Engineering*, pp. 369–375.

Wu, D.Q., Broms, B.B. and Choa, V. (1993) Soil improvement with fly ash columns, *Proceedings, 11th Southeast Asian Geotechnical Conference*, 4–8 May, Singapore, Vol. 1, pp. 435–438.

Yoshida, S. (1996) Shear strength of improved soils at lap-joint-face. *Proceedings, 2nd International Conference on Ground Improvement Geosystems*, 14–17 May, ISD-Tokyo '96, Vol. 1, pp. 461–466.

Chapter 9

In situ soil mixing

M. Topolnicki

9.1 Introduction

The use of *in situ* soil mixing (SM) to improve the engineering and environmental properties of soft or contaminated ground has increased widely since its genesis. Use has increased especially in Japan, Scandinavia and the United States, as well as Southeast Asia, China, Poland, France, Germany and UK, and to some extent in other countries. This indicates growing international interest and acceptance of this relatively new and quickly developing technology.

In situ soil mixing is used in diverse marine and land applications, mainly for soil stabilisation and column-type reinforcement of soft soils, construction of excavation-support walls with inserted steel sections and as gravity composite structures, mitigation of liquefaction, environmental remediation and for in-place installation of cut-off barriers. In this method of ground improvement, soils are mixed *in situ* with different stabilising binders, which chemically react with the soil and/or the groundwater. The stabilised soil material that is produced generally has a higher strength, lower permeability and lower compressibility than the native soil. The improvement becomes possible by cation exchange at the surface of clay minerals, bonding of soil particles and/or filling of voids by chemical reaction products. The most important binders are cements and limes. However, blast furnace slag, gypsum, ashes as well as other secondary products and compound materials are also used. For environmental treatment, binders are replaced with chemical oxidation agents or other reactive materials to render pollutants harmless.

SM technology can be subdivided into two general methods: the Deep Mixing Method (DMM) and the Shallow Mixing Method (SMM). Both DMM and SMM include a variety of proprietary systems.

The more frequently used and better developed DMM is applied for stabilisation of the soil to a minimum depth of 3 m (a limit depth proposed by CEN TC 288, 2002) and is currently limited to treatment depth of about 50 m. The binders are injected into the soil in dry or slurry form through hollow rotating mixing shafts tipped with various cutting tools. The mixing shafts are also equipped with discontinuous auger flights, mixing blades or paddles to increase the efficiency of the mixing process. In some methods, the mechanical mixing is enhanced by simultaneously injecting fluid grout at high velocity through nozzles in the mixing or cutting tools.

The complementary SMM has been specially developed to reduce the costs of improving loose or soft superficial soils overlying substantial areas, including land disposed dredged sediments and wet organic soils a few metres thick. It is also a suitable method for *in situ* remediation of contaminated soils and sludges. In such applications, the soils have to be thoroughly mixed *in situ* with an appropriate amount of wet or dry binders to ensure stabilisation of the entire volume of treated soil. Therefore, this type of soil mixing is often referred to as 'mass stabilisation'. Mass stabilisation can be achieved by installing vertical overlapping columns with up and down movements of rotating mixing tools, as in the case of DMM, and is most cost-effective when using large diameter mixing augers or multiple shaft arrangements. With this kind of equipment it is generally possible to stabilise very weak soils to a maximum depth of about 12 m.

More recently, however, another method of mass stabilisation has been implemented, and the mixing process can now be carried out repeatedly in vertical and horizontal directions through the soil mass using various cutting and mixing arrangements that are different from the tools originally developed for DMM. The depth of treatment for this relatively new system is generally limited to about 5 m. Consequently, in the classification scheme used in this chapter the SMM includes both systems of mass stabilisation. It is important to note that the differentiation between SMM and DMM is not solely attributed to the available depth of treatment criterion because, in principle, soil mixing at shallow depth can also be performed with DMM.

In situ soil mixing is a versatile ground improvement method. It can be used to stabilise a wide range of soils, including soft clays, silts and fine-grained sands. Stabilisation of organic soils such as gyttja (sedimentary organic soil), peat and sludges is also possible, but is more difficult and requires carefully tailored binders and execution procedures. However, the engineering properties of the stabilised soil will not only depend on the characteristics of the binder. They will also depend, to a large extent, on the inherent characteristics of each soil and the way it has been deposited, as well as on mixing and curing conditions at a particular worksite. Therefore, a thorough understanding of chemical reactions with the above factors is necessary in order to ensure successful application of this ground improvement technology.

In this chapter, the current status of *in situ* soil mixing is outlined, taking into account recent execution and design practice, international literature and experience. General application areas are identified and discussed, and a few case histories selected from recent international projects are included for illustration. The focus is on civil engineering applications of DMM and, to a lesser extent, of SMM. Some specialised soil mixing issues in relation to environmental projects, such as mass treatment of subsurface hazardous wastes by various processes including solidification, stabilisation and chemical treatments, reactive barriers, etc. are only touched upon, therefore the cited literature should be referred to for more information. Furthermore, overly extensive descriptions of the complicated chemical processes occurring in the stabilised soil when mixed with various binders have been excluded from the contents. This choice, however, should not undermine the importance of this aspect of soil mixing. It may rather reflect the fact that in spite of considerable knowledge about basic reaction mechanisms, identified and described for instance by Babasaki *et al.* (1996) for soils stabilised with lime or cement, it is still not possible to predict the strength of *in situ* mixed soil with a reasonable level of accuracy. As a consequence of this fundamental deficiency, which we are challenged to overcome, it is believed that the development of SM will be continued along a somewhat erratic experimental path, and will be to a large extent dependent on accumulated experiences. Therefore, the scope of this chapter instead concentrates on the characteristics of equipment in current use, execution procedures with reference to selected operational methods, applications, merits and the limitations of the technology. Design aspects as well as quality control and quality assurance issues of DMM are also considered. The design approach outlined herein follows the practice established in Japan, the USA and Europe, assuming that the treated soil is practically an impermeable material. The approach used with respect to deep mixing (DM) columns stabilised with unslaked lime or lime and cement, which may act as vertical drains, has been covered in Chapter 8.

9.2 Historical development and classification

The roots of deep soil mixing go back to the mid-1950s, when the Mixed in Place (MIP) piling technique was developed by Intrusion-Prepakt Inc. (FHWA, 2000), including the US patent of Liver, filed in November 1956 (Jasperse and Ryan, 1992). In this method a mechanical mixer was used to mix cementitious grout into the soil for the purpose of creating foundation elements and retaining walls. The grout was injected from the tip of the mixing tool consisting of a drilling head and separated horizontal blades. Subsequent more intensive use of MIP was observed in Japan for excavation support and groundwater control in the 1960s. Modern deep mixing techniques reflect, however, mainly Japanese and Scandinavian efforts over the last three decades.

The level of research and development activity in Japan in relation to DMM remains the highest in the world. The study on soil mixing was initiated at the Port and Harbour Research Institute (PHRI) in 1967 in the framework of government-sponsored research work. Pilot laboratory investigations with soft marine clays stabilised with granular quicklime or powdered slaked lime were followed by full-scale trials and development of appropriate mixing equipment in cooperation with Toho Chika Koki Co., building the Mark I to III machines, and Kobe Steel Co. Ltd, building the Mark IV machine (CDIT, 2002).

The first commercial application of the established Deep Lime Mixing (DLM) method, utilising a mechanical binder feeding system, was conducted in 1974 by Fudo Construction Co. Ltd., using the Mark IV machine to improve reclaimed soft alluvial clay in Chiba Prefecture in Japan. The first marine use of DLM was in 1975 at Tokyo Port (Terashi, 2002a). In an effort to improve the uniformity of the stabilised soil, a new concept using cement mortar and cement–water slurry as binders was implemented in the mid-1970s, with CMC and DCM methods developed by Kawasaki Steel & Fudo and Takenaka Group, respectively, with a close supervision by PHRI. The first on-land and marine applications of CMC and DCM were conducted in 1976. Also that year, the Seiko Kogyo Co. developed and introduced the Soil Mixed Wall (SMW) method using discontinuous augers and paddles positioned at discrete intervals, usually along three shafts arranged in a row and applied primarily for excavation support and groundwater cut-off walls with the possibility for the installation of reinforcing steel sections within fresh columns. The reinforcing steel sections increase the bending stiffness of the supporting DM elements.

Major marine ground improvement works at Daikoku Pier, beginning in 1977 and continuing for about 10 years, contributed to important developments of the wet-method of deep mixing (e.g. DCM, DECOM, POCOM and others). These developments included the elaboration of design standards and construction control procedures, slowly hardening binders and new positioning systems for offshore applications (Terashi, 2002a). In 1979 the Tenox Corporation introduced in Japan the Soil Cement Column (SCC) method using a single shaft with three pairs of rotating mixing blades and a non-rotating 'shear blade' above the tip to improve the mixing efficiency.

A general method using a variety of stabilising binders in slurry form (wet-method) has been named Cement Deep Mixing (CDM) method. In 1977 the CDM Association was established in Japan to promote and improve the CDM method via a collaboration of general contractors, marine works and foundation works contractors, as well as industrial and research institutes. As a result, new efficient machines were developed, such as CDM-Mega, CDM-LODIC, CDM-Land4 and CDM-Column21 (CDM Association, 2002). In addition, further possibilities of optimisation of the wet-method were investigated; for example, the FGC-DM system (Azuma *et al.*, 2002), which uses a new binder composed of fly ash (F), gypsum (G) and cement (C) to create economical low-strength treated soil.

For marine applications the CDM method has mainly been used to improve the foundations of revetments, as well as quay wall and breakwater foundations. Currently, there are 16 DM barges available in Japan for marine works (Terashi, 2002a). The diameter of the mixing blades ranges from 1.0 to 1.6 m and the maximum depth of improvement is 70 m, below water. For land applications the CDM method has been mainly applied for slide and liquefaction prevention, settlement reduction and to improve the bearing capacity of foundations. The standard CDM machines have two shafts, mixing blades with a diameter of 1 m, and a penetration depth limited to about 50 m (CDM Assoc., 2002). Typical machines for marine and on-land use are shown in Figures 9.1 and 9.2, respectively.

The development of the wet-method in Japan includes successful attempts to combine mechanical mixing with high-velocity injection. In 1984 the Spreadable Wing (SWING) method was introduced. In this unique system a retractable mixing blade mounted on a single drilling shaft allows treatment of specific depths with large diameters (0.6 m with blade retracted and up to 2 m after expansion). Following that, jet grouting was incorporated into SWING and its first application was in 1986. With additional jetting during withdrawal, mechanically mixed and jet mixed concentric zones are produced with a total diameter up to 3.6 m (Kawasaki *et al.*, 1996). Moreover, in 1992 a new method named Jet and Churning System Management (JACSMAN), using combined mechanical mixing and modern cross jet grouting systems, was developed by Fudo Construction Co. and Chemical Grouting Co. (Kawanabe and Nozu, 2002). This versatile method has been in practice since 1994.

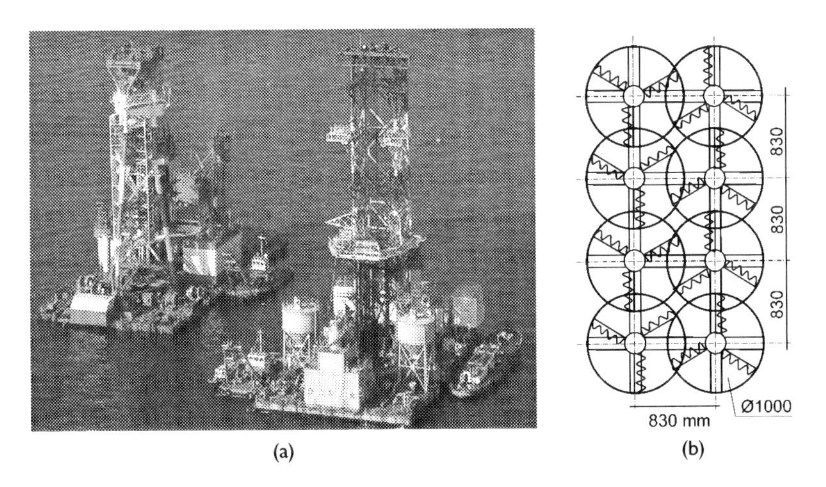

Figure 9.1 (a) CDM barges for marine deep mixing works using the wet-method, Japan (courtesy of CDM Association, Japan); (b) Possible arrangement of eight mixing shafts.

(a) (b) (c)

Figure 9.2 CDM machine for on-land deep mixing works using the wet-method, Japan (note auger screws along the shafts enabling partial extrusion of disaggregated soil).

A variety of wet shallow mixing systems is also available in Japan, which are similar in concept to the mass stabilisation method recently developed in Finland and Sweden, as reported by Terashi (2002b).

The development study of a new Japanese dry-method designated for on-land applications, utilising a pneumatic binder feeding system, and called Dry Jet Mixing (DJM), was initiated in 1978. The work was done at The Public Works Research Institute (PWRI) of the Ministry of Construction, by invited members from PHRI and others who had the most experience with the wet technique. The constructed machine was first applied in 1981. It was subsequently improved, with a landmark project in 1985 on San-yo Motorway, where a 10 m high embankment was constructed on 10 m of sensitive soft clay (Terashi, 2002a). In the early 1980s the DJM Association was established in Japan, with a role similar to that of the CDM Association for the wet-method. The current standard DJM equipment has two mixing shafts, with blades of 1.0 m diameter and a maximum penetration depth of 33 m (DJM Association, 2002), as shown in Figure 9.3. The use of binders in 21 years is as follows: Portland cement 28 per cent, slag cement type B 30 per cent, cement-based agents 33 per cent, quick lime 8 per cent, others 1 per cent (Aoi, 2002).

Because of the variety of the DM equipment developed and built in Japan, as well as due to implementation of appropriate standards for execution and control of the works, the application of deep *in situ* soil mixing has

(a)

(b)

Figure 9.3 (a) DJM machine for on-land deep mixing works using the dry-method, Japan; (b) Two mixing tools diameter 1.0 m.

grown continuously since the mid-1980s. Recently, annual production is around 5 million m^3 of stabilised soil, with roughly $^1/_3$ contribution of CDM on-land, $^1/_3$ of CDM offshore and $^1/_3$ of DJM. At the end of 2001, the cumulative volume of treated soil in Japan amounted to 70 million m^3 (Terashi, 2003).

The development of DMM in Scandinavia was initiated in Sweden in 1967 when laboratory and field research began for a new method of stabilisation of soft clays with unslaked lime. The first light wheel-mounted mixing equipment was manufactured by Linden-Alimak AB in cooperation with Swedish Geotechnical Institute, BPA Byggproduktion (presently LCM AB, Keller Group), and the company Euroc AB. In Finland, research was initiated during that time as well. Commercial use of the Lime Column method started in Sweden in 1975 for support of excavation, embankment stabilisation and shallow foundations near Stockholm. Other types of dry binders, like cement and two-component binders composed of unslaked lime and cement have been subsequently investigated and put into practice. The first commercial project with Lime Cement Column method in Finland was conducted in 1988 and then 1990 both in Sweden and Norway, under Swedish guidance and using Swedish contractors. This type of DMM, generally with two component binders tailored for various soft soils, has been consequently developed over the years. It is now referred to as the Nordic Dry DMM (Holm, 2002a). With an increasing number of proven applications, especially since 1989, this method had become, in the mid-1990s, a predominant technology of ground improvement in Scandinavia, with an estimated current annual production of about 1.2 million m^3 (FHWA, 2000). Relatively light and mobile equipment with one shaft is typically used to produce columns of 0.5–1.0 m in diameter to the maximum depth of about 25 m depending on the soil conditions. The application focus remains on ground improvement to reduce settlement and enhance stability of road and railroad embankments, and soil/column interaction solutions for very soft, highly compressible clayey and/or organic soils. The first application for mitigation of vibrations induced by high-speed trains took place in Sweden in 2000 (Holm, 2002b).

In European countries outside Scandinavia the Nordic method has been mostly used in Poland since 1995, using Swedish contractors, with recent major application for the new city carriageway in Szczecin, involving more than 550 000 lin. m of cement/lime columns (Figure 9.4). Single projects were also conducted in the UK (first in 2001), and field trials in the Netherlands and Germany.

In Finland and Sweden another dry SMM for stabilisation of superficial layers of peat, mud or soft clay to the depth of about 5 m has recently been developed and applied for road and land reclamation projects. This method was invented in Finland, after The Finish Road Administration initiated in 1992 a research project with the objective to develop a suitable and economical method of peat stabilisation, which is much more challenging than stabilisation of clay. The mixing tools of this relatively new method of mass stabilisation have different shapes and are typically attached to the arm of

(a)

(b)

Figure 9.4 Dry-method deep mixing in Szczecin (Poland) using the Nordic method. (a) LCM equipment; (b) Mixing tool diameter 0.6 m.

a conventional or adapted excavator (Figure 9.5a). They can resemble a ship's propeller or may be constructed as mixing/cutting heads equipped with blades rotating about a vertical or a horizontal axis (Figure 9.5b). The mixing process is conducted repeatedly in vertical and horizontal directions through the soil mass in order to obtain a homogeneous soil-binder mixture.

The first commercial project was conducted in Sweden in 1995 in connection with renovation works along Highway 601 Sundsvägen, where about 10 000 m^3 of peat were treated. Interesting applications of this method using rapid cement as a binder include the stabilisation of dredged mud deposited between embankments to create new areas for a container terminal in Port Hamina, and a park at the shoreline of Helsinki where the deposited mud was also contaminated (Andersson *et al.*, 2000). In Finland, other low-cost binders are used more frequently than elsewhere as substitutes for lime and cement. These substitutes include blast furnace slag, ashes, gypsum and other secondary products. These compound binders are blended in a factory, or can be mixed on the worksite.

The application and development of the contemporary DMM in the United States started in the mid-1980s and was comprised initially of the wet-method. In 1986, SMW Seiko Inc. began operations under license from Japanese parent Seiko Kogyo Co. The SMW method was subsequently used in 1987–1989 in a landmark liquefaction mitigation and seepage cut-off project at Jackson Lake Dam, WY, where 130 000 linear metres of column were installed to a maximum depth of 33 m (FHWA, 2000).

Following their cooperation with SMW Seiko on the Jackson Lake Dam project, Geo-Con Inc. developed between 1987 and 1989, the first US soil

(a)

(b)

Figure 9.5 Mass stabilisation of organic soil using the dry-method, Sweden. (a) Equipment; (b) Mass mixing tool diameter 1.0 m (courtesy of LCM).

mixing technologies, i.e. the Deep Soil Mixing (DSM) and the Shallow Soil Mixing (SSM) methods (Figure 9.6). The DSM method currently uses one to six shafts with discontinuous augers of 0.8–1 m in diameter. The SSM method currently uses a large diameter single mixer to economically treat weak superficial soils and contaminated sites to a depth of about 12 m. The SSM has since been extended to accommodate binders in a dry form. This variant using dry-form binders was applied in 1991 to stabilise large lagoons containing contaminated sludge residues from a water-treatment plant at a refinery near Chicago (Jasperse and Ryan, 1992).

The Japanese SCC method was introduced in the United States by SCC Technology Inc. in 1993. The single axis system of Hayward Baker (Keller), with diameters of 0.5–2.5 m, typically 2.1 and 2.4 m, began development in the 1990s and has been applied since 1997 (Burke, 2002). So far, the largest DM works in North America were conducted between 1996 and 1999 for the Central Artery/Tunnel project in Boston. These DM works included application of the SMW method with triple 1.5 m diameter augers. The motor used, similar to those used on the offshore projects in Japan, was one of the world's largest. Depths of up to 40 m were reached. Mass stabilisation with the wet-method (Shallow Soil–Cement Mixing) was also applied, using an excavator equipped with a shallow mixing bucket (Druss, 2002b). The bucket contained mixing blades that rotate about a horizontal axis.

A combined mechanical and hydraulic mixing method called GEOJET (Condon Johnson and Associates, Halliburton) has been developed and modified in the US since the early 1990s. GEOJET equipment includes the

(a)

(b)

Figure 9.6 (a) DSM and (b) SSM equipment for the wet-method, USA (courtesy of Geo-Con Inc.).

soil processor equipped with specially designed cutting blades and multiple jetting nozzles which jet mix at pressures up to 35 MPa. The first commercial application was in 1994, followed by some major retaining wall works and installations of pipe piles in soil. A similar system named HYDRAMECH, capable of creating columns up to 2 m in diameter using mechanical mixing and high-velocity jet grouting, was put into operation by Geo-Con Inc. in 1998.

The first commercial project that used the Nordic method in the US was conducted in 1996 in Queens, NY, by the Stabilator Company (Skanska). Subsequent application for settlement reduction at I-15 in Salt Lake City, UT, took place in 1997. Since 1998, other dry-methods, like, for example, DJM (Raito Inc.) and TREVIMIX (Trevi-ICOS, with Hercules), have been available.

In China the research on DMM started in the late 1970s. The first on-land application for the foundation of an industrial facility in Shanghai took place in 1978, while the first offshore project was conducted between 1987 and 1990 at Tiajin Port by a Japanese contactor. The total volume of Chinese soil treated since with DMM is in excess of 1 million m^3 (Porbaha, 1998). By 1992, the first Chinese CDM equipment for offshore work was built in collaboration with Japan and used at Yantai Port in 1993.

Other applications of wet and dry DM in Southeast Asia include several important projects completed in Taiwan, Singapore, Hong Kong and Thailand, generally in cooperation with Japanese contractors. Very recently, use of the Nordic method has been observed in this region. In 2001, Hercules Grundläggning AB carried out ground improvement works for oil tank foundations in Vietnam (Forsberg, 2002), and LCM-Keller stabilised the soil beneath the realignment of a railway line in Malaysia (Raju *et al.*, 2003). The wet-method was also introduced to New Zealand in late 2002.

In Central Europe, the earliest wet DM activities that took place in the 1980s were oriented towards development of a potentially cheaper alternative to jet grouting. For this purpose, standard hydraulic piling rigs were often equipped with modified drilling shafts, and fitted with simple-angled blades or short shallow-pitch helical-screw mixing tools. More recently, this picture changed somewhat due to development of specialised equipment, including multiple auger and dry mixing capabilities, as well as hybrid mixing (Lebon, 2002). However, the total number of conducted works is still small in comparison to Japan, Scandinavia or the US, and DMM is generally considered a highly specialised technology. Moreover, strict criteria for the mixing energy input are still lacking in most European wet-methods and the execution process is governed by particular features of each system.

In France, Bachy Soletanche developed the COLMIX method in the mid-1980s, in conjunction with the French Railway Authority (SNCF) and the French National Laboratory for Roads and Bridges (LCPC). The COLMIX method appears to be the first development outside Scandinavia. The method features twin, triple or quadruple contra-rotating and interlocking augers, generally 3 to 4 m long and driven via hollow stem rods coupled to a single rotary drive. Blended soil moves from bottom to the top of the hole during penetration, and reverses on withdrawal ensuring very efficient soil mixing and recompaction. Several road and rail embankment stabilisation projects have been completed with this method in France, UK and Italy, as summarised by Lebon (2002).

In Germany, the first application of the Mixed-in-Place (MIP) system developed by Bauer Spezialtiefbau which was based on the Rotary-Auger-Soil-Mixing (RASM) method utilising single shafted crane and wet binder, took place in Nürnberg in 1987. MIP piles were executed to create panels of mixed soil filling-up a 'Berlin'-type temporary retaining wall constructed in sands (Herrmann *et al.*, 1992).

Subsequently, a more advanced triple auger wet mixing system has been developed since the early 1990s. It is comprised of three closely spaced, non-interlocking, full-length augers, arranged in a row and driven as a coupled pair and counter rotating single auger. This system has been in use since 1994, primarily for construction of temporary and permanent panels supporting excavations, cut-off walls, ground improvement and environmental purposes (Außerlechner *et al.*, 2003; Schwarz and Seidel, 2003). Keller Grundbau developed their system based on a single paddle shaft equipped with a short auger and mixing blades above the drill bit. Their commercial ground improvement applications for this system have been ongoing since 1995.

Another high-capacity specialised wet mixing system developed in Germany in 1994 is the FMI method (Fräs-Misch-Injektionsverfahren = cut-mix-injection). It was applied for the first time in 1996 in Giessen (Pampel and Polloczek, 1999). The FMI machine is comprised of a special cutting tree, along which cutting blades are rotated by two chain systems (Figure 9.7a). The cutting tree can be inclined up to 80°, and is dragged through the soil behind the power unit. Due to special blade configuration, the soil is not excavated, but mixed with a binder which is supplied in slurry form through injection pipes and outlets mounted along the cutting tree. With this method it is possible to treat the soil in deep strips, with a mean capacity of 70–100 m^3/h. The width of treatment is 1 m down to a depth of 6 m or 0.5 m down to a depth of 9 m. The applications mainly covered ground improvement works along railways (Figure 9.7b).

In the United Kingdom wet DM for ground improvement was employed in early 1990s by Cementation Piling and Foundations for construction of a few temporary shafts, of approximately 4 m internal diameter and up to 15 m deep (Blackwell, 1994). In this concept two to three concentric unreinforced overlapping rings were created by 75 cm diameter secant columns, which were designed to act together in hoop compression. The columns were installed with a simple auger-type mixing tool, using five

(a)

(b)

Figure 9.7 The FMI (cut-mix-injection) machine, Germany (courtesy of Siedla & Schönberger).

passes of the tool over a 1 m withdrawal length. Until recently, little work was performed with this application.

Around 1995 soil mixing was introduced for geoenvironmental applications, with growing importance since 1997. Currently the UK is probably leading Europe in the research and application of wet mixing to the containment and encapsulation of contaminated soils, including cut-off walls and reactive barriers (Lebon, 2002; Al-Tabbaa and Evans, 2002). In 2001 the UK saw its first use of the dry Nordic Method.

In Italy the Trevi SpA developed in the late 1980s a dry mixing method named TREVIMIX. The equipment has more similarities with the Japanese DJM method than with the Nordic method. In this system one or two (more common) shafts with mixing paddles of 1.0 m (or 0.8 m) in diameter are arranged at variable spacings of 1.5–3.5 m and are used to disintegrate soil structure during penetration with air. Augers are then counter-rotated during withdrawal and dry binders are injected via compressed air through nozzles on shaft below mixing blades. The distinction of this system lies in its ability to operate in dry or semi-dry conditions by adding a controlled amount of water to the soil in order to ensure a hydrating reaction. First applications in Italy have been reported by Pavianni and Pagotto (1991). Another new development is the TURBOJET wet mixing system that uses a tubular Kelly with drilling bit and two mixing blades, and combines mechanical mixing and single-fluid jet grouting technology.

In Poland the wet-method of DM was first introduced in 1999 by Keller Polska, using single axis equipment originally developed in Germany. The first project involved execution of intersecting columns forming a cut-off wall along an old dam of the Vistula River in Kraków. Since then, a considerable number of DM projects designed and conducted by Keller Polska have been completed, including applications such as: improvement of organic soils for a new city road, foundation of several multi-storey buildings on slabs supported by individual columns, stabilisation of soft soils under strip and pad foundations of industrial and municipal buildings, cut-off walls and temporary protection of excavations. The first worldwide applications of the DMM for the foundations of 39 bridges build across and along the new A2 Motorway took place in 2002/2003 (Figure 9.8).

Growing importance of the DMM in Europe has also led to joint research and standardisation activities. A consortium of companies and organisations from England, Finland, Ireland, Italy, the Netherlands and Sweden started in 1997–2000 a research and development program supported by the EU Commission, dubbed EuroSoilStab. As a result, a design guide for stabilisation of soft and especially organic soils was published in 2002 (EuroSoilStab, 2002). The European Standard for Deep Mixing has been also prepared by CEN/TC 288 (provisional EN 14679, currently under approval).

The hitherto development of different technologies and equipment used in SM is difficult to follow without a certain generic classification system. Several similar systems have already been developed for this purpose, e.g. FHWA (2000), CDIT (2002) and CEN TC 288 (2002). The classification format adopted herein is based on three fundamental operational characteristics also identified in the FHWA report. The distinction between wet and dry technologies with respect to the form of binder introduced into the soil is the most straightforward, and hence the most widely used format. In the dry

(a)

(b)

Figure 9.8 (a) Single shaft deep mixing machine at work, bridge WD-82, A2, Poland; (b) Mixing tool diameter 0.8 m (Keller Polska).

mixing methods the medium for binder transportation is typically compressed air, while in the wet mixing methods the medium of transportation is typically water. The second characteristic is related to the method used to mix the binder, i.e. by mechanical action of the mixing tool with the binder injected at relatively low velocity, hydraulic action of the fluid grout injected at high velocity (jet grouting), or by a combination of both aforementioned techniques (so-called hybrid mixing). The third basic characteristic reflects the location, or vertical distance of the drilling shaft over which mixing occurs in the soil. The elaborated classification chart with the allocation of the methods mentioned above (with the exception of jet grouting methods, which are beyond the scope of this Chapter) is shown in Figure 9.9.

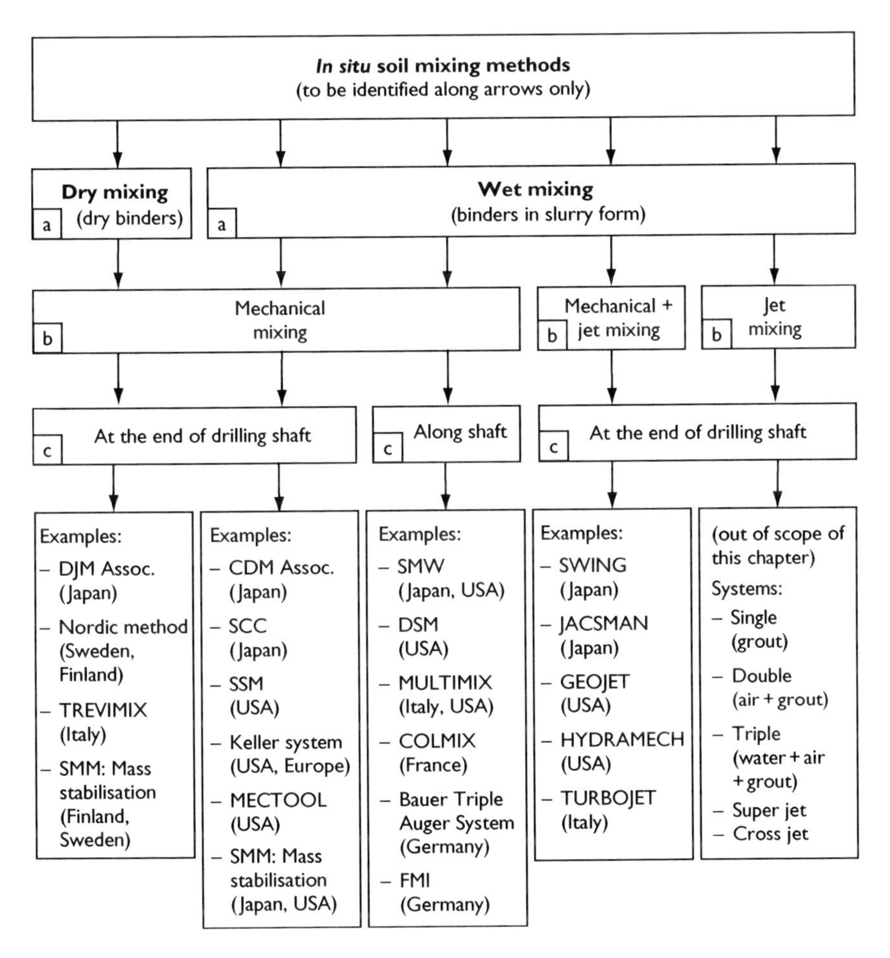

Figure 9.9 General classification of *in situ* soil mixing based on (a) binder form, (b) mixing principle and (c) location of mixing action, with allocation of selected fully operational methods developed in various countries.

When comparing technical features of recently used DMM and SMM machines and operational systems it should be kept in mind that the aforementioned methods have been developed while taking into account various demands and constraints of regional markets, as well as soil conditions prevailing in areas of potential application. Moreover, various operational systems also reflect different objectives of ground improvement and design approaches. Consequently, not all SM methods can be regarded as equivalents, although all are based on the same overall concept of *in situ* soil stabilisation. Despite these variations, the main technical goal of any SM method is to ensure a uniform distribution of binder throughout the treated soil volume, with uniform moisture content, and without significant pockets of native soil or binder.

9.3 Equipment and execution

9.3.1 Dry-method deep mixing

Typical dry-method DM construction equipment consists of a stationary or movable binder storage/pre-mixing and supply unit, and a mixing machine for the injection of binder material and installation of the columns. The binder is delivered to the mixing machine by compressed air. The equipment components generally include: silos with stabilising agents, pressurised tank with binder feeder system, high-capacity air compressor, air dryer, filter unit, generator, control unit and connecting hoses. The two major techniques for dry mixing are the Japanese DJM and the Nordic method.

The DJM mixing machines are equipped with one or two mixing shafts and are able to install columns to a maximum depth of 16–33 m (DJM Association, 2002). A dual mixing shaft is the current standard outfit, while a single shaft may be used in narrow working areas or for sites with headroom restrictions. The driving unit of the mixing shafts is located at the foot of the tower to improve machine stability while the shafts are kept together with a transverse steel bar, allowing for interlocking or tangential positioning of the mixing blades. The bar, and sometimes additional freely rotating (undriven or counteracting) mixing blades, also function to prevent rotation of soil adhering to the driven mixing blades and shaft. The standard mixing tool has a diameter of 1 m and consists of two full-length mixing blades, mounted at the end of the shaft at two different levels, with 90° shift (Figure 9.10). A recently implemented modified version of the tool has 1.3 m diameter (Aoi, 2002). To prevent choking, the injection ports are positioned at the mixing shaft, below and behind angled mixing blades. The lower port is used to inject air during penetration, or air and binder in the case of soils requiring a high amount of stabilising agent. The binder is mostly injected through the upper port into a cavity space created by the mixing blade during withdrawal of the shaft with reversed rotation. Binder quantity is adjusted by changing the rotation speed

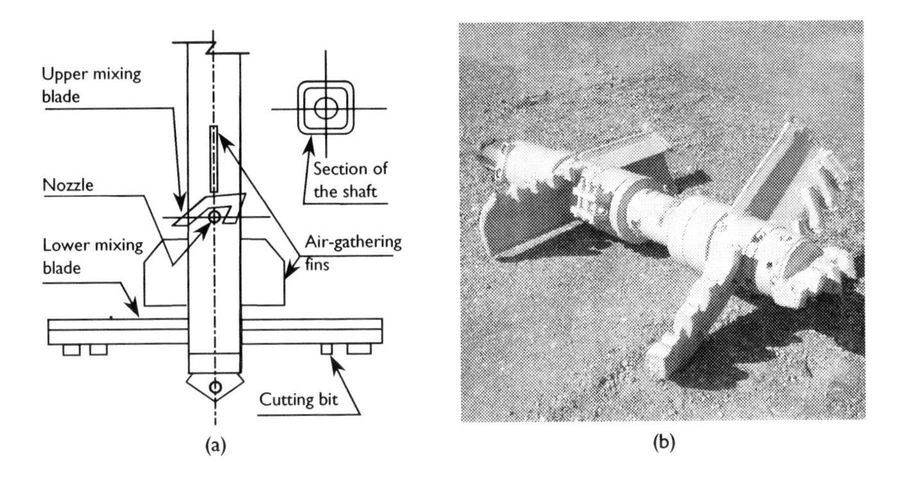

Figure 9.10 Mixing tools of the DJM method. (a) Construction scheme (DJM Association, 2002); (b) Recently used single mixing tool of 1.0 m diameter.

of the feeding wheel. Air pressure and the amount of binder are automatically controlled to supply the specified dosage of binder to the treated zone of soil. A hood covering the mixing tools is lowered to the soil surface to suppress dust emission during work, while a square mixing shaft is generally used to facilitate easier expulsion of injected air from the ground.

With torque capacity in the range of 20–30 kNm, the DJM machines are able to conduct mixing operation in stratified soils with varying resistance (rotation is agitated by hydraulic or electric motors). The limits for execution are 70 kPa maximum shear strength for stiff clays and SPT N-value of 15 in sands (Terashi, 2003). Typical penetration speed in soft soil is 1–1.5 m/min, with 24–32 rpm (electric motors) and an air flow rate of 2 m^3/min to prevent choking of injection ports. During withdrawal (with counter rotation) the speed is typically 0.7 or 0.9 m/min, with 48 rpm or 64 rpm respectively (electric motors) and air flow rate of 5–3 m^3/min at shallow depths. Consumed air volume may vary between 2 and 9 m^3/min, requiring heavy-duty compressors with a capacity of 10.5–17 m^3/min/shaft. The volume of the binder tank is usually 2–3.5 m^3 per mixing shaft (DJM Association, 2002).

The mixing machines developed in Sweden and Finland are lighter than the Japanese rigs, and are equipped with one mixing shaft. They are constructed to work mainly in soft to very soft soils with undrained shear strength below approximately 25 kPa (maximum 50 kPa). The torque capacity at 180 rpm is typically about 7 kNm, and increases to 30–40 kNm at 20–30 rpm (some machines have two engines driving the Kelly rotation). This allows for the installation of 0.6–0.8 m diameter columns to a depth of about 25 m, at a distance 1–5 m from the edge of the base unit. The columns can be also inclined up to about 1:4, maximum 1:1.

The equipment on site usually consists of a drill rig and a separate self-driven mobile shuttle, hosting pressurised binder material tank(s), air dryer and compressor. In addition to the plant items working on the construction site, there is usually a requirement for a pre-mix station including a filter unit, especially when delivery of ready-to-use binder is too expensive. The binder shuttle moves between the pre-mix station and the drill rig, which is normally working several hundred metres away from the pre-mix station. During production, the shuttle is connected to the drill rig by an umbilical through which the binder passes (via compressed air), along with monitoring information on the binder mixing and supply rate. For shallow penetration depths, combined mixing machines with on-board installations are available. The amount of discharged binder is controlled with a cell feeder mechanism, located at the bottom of the supply tank.

The air containing the binder is transported through the hollow Kelly bar to an exchangeable mixing tool, mounted at the end. Typical mixing tools consist of horizontal and curved or angled cutting/mixing blades, as shown in Figure 9.11. The injection outlet is located at the central shaft, close to the upper horizontal mixing blade. After the required depth is reached, the mixing tool is lifted and simultaneously rotated in reverse, while the binder material is horizontally injected to the soil. Typical withdrawal speed is 15–25 mm per rotation, with about 150–180 rpm.

A summary of mixing conditions for selected dry DM methods is presented in Table 9.1.

9.3.2 Dry-method shallow mixing

Shallow dry-method mixing offers a cost-effective solution for ground improvement works or site remediation when dealing with substantial volumes of very weak or contaminated superficial soils with high water content, such as deposits of dredged sediments, wet organic soils or waste sludges. *In situ* mixing of the encountered soil mass with dry reagents to the depth of a few metres can be economically carried out with large diameter single axis augers, or with recently developed mass mixing tools implemented in Finland and Sweden. In such applications it is also quite common for the topsoil to be too weak to provide safe support for heavy mixing machines. Therefore, it is best to use execution methods which employ mixing tools suspended from the crane or mounted on elongated cantilever arms, as they usually offer more flexible operation in the field.

The Shallow Soil Mixing (SSM) method, modified for accommodation of dry binders, utilises a crane-mounted single auger tool, 1.8–3.7 m in diameter (Jasperse and Ryan, 1992). The driver for the tool is a drilling system. It can be a conventional hydraulic drill or a high-torque dual-motor turntable. The auger tool itself is specially designed to break up the soil and/or sludge and

(a) (b) (c) (d)

Figure 9.11 Selected mixing tools of the Nordic method: (a) SD 600 mm; (b) modified SD 600 mm; (c) PB3 600 mm; (d) peat bore 800 mm (courtesy of LCM). Note: changed location of binder outflow hole in relation to the horizontal mixing blade in standard (a) and modified (b) tool.

mix it with dry reagent without bringing the material to the surface. To suppress emissions from the mixing process and/or for environmental applications, the mixing tool can be enclosed in a hood or bottom-opened cylinder to control dust and airborne contaminants (Figure 9.12). Further components of environmental control may also include: a low-pressure blower or vacuum pump to keep negative pressure inside the hood during operation, a dust collector, a fume incinerator or an activated carbon scrubber, depending on site-specific conditions and contaminants (Aldridge and Naguib, 1992).

Treatment reagents are transferred pneumatically to the mixing unit. The delivery system consists of bulk storage tanks, several pneumatic pumps,

Table 9.1 Mixing conditions for selected dry Deep Mixing methods

Technical specification	*Selected dry DM methods*		
	DJM	*Nordic method*	*Trevimix*
Number of mixing shafts	2 (standard), 1	1	2 (more common), 1
Diameter of mixing tool (m)	1.0 (standard) 1.3 (modified version)	0.5–1.0 possible 0.6, 0.8 standard	0.8–1.0 (standard)
Realistic maximum penetration depth (m)	33	25 (30)	30
Penetration/Retrieval velocity (m/min)	0.5–3 (4), 7 (one shaft) typically: P: 1.5, R: 0.7, 0.9 (R: 15 mm/rev.)	P: 2–15 R: 2–6 (R: 15–30 mm/rev.)	P: 0.4 R: 0.6
Penetration/Retrieval rotation speed (rpm)	P: 24, 32 (electr.) R: 48, 64 (electr.) P/R 21–64 (hydr.)	R: 100–220 (150–180 typically)	10–40 P: 20 typically R: 30 typically
Injection during Penetration/Retrieval	R (P used: air/binder)	R (P possible)	R (P used: air/binder)
Max. footprint area of the mixing tool (m^2)	0.78 : 1 × 1.0 m 1.56 : 2 × 1.0 m 2.65 : 2 × 1.3 m	0.28, 0.5 (0.78)	0.78 : 1 × 1.0 m 1.56 : 2 × 1.0 m
Amount of injected dry binder (kg/m^3)	100–400 of cem. in sands, 200–600 of cem. in peat, 50–300 of lime in clay	70–150, 150–250 in organic soils	150–300, 250 typically
Binder supply capacity per shaft (kg/min)	25–120 standard, up to 200 in mod. version	40–230	around 100
Injection pressure (kPa)	P: 100–600 R: 600–100	400–800	600–1000
Productivity (m^3/shift)	300–700	150–300	150–220

Figure 9.12 SSM mixing tool, diameter 3.7 m, for the dry-method (courtesy of Geo-Con Inc.).

portable booster stations and material receivers. The final application of the reagent to the treated soil is made with the hood lowered, by dropping the reagent into the hood through a calculated rotary valve located at the bottom of a material receiver. Various cementitious, chemical or even biological reagents can be added to soil or waste with this method.

The shallow mixing machines currently developed in Finland and Sweden are essentially different from the column stabilisation machines. The mass mixing tools, which are still under development and testing, are typically attached to the arm of a crawler-mounted excavator to enable vertical and horizontal movements of the tool through the soil to complete mixing (Figure 9.13); however, difficulty of horizontal movement increases with depth, eventually leading to the impossibility of movement. The binder is fed from a separate unit which houses the pressurised binder container, compressor, air dryer and supply control unit. The operator injects the binder into the soil in such a manner that the binder is equally distributed

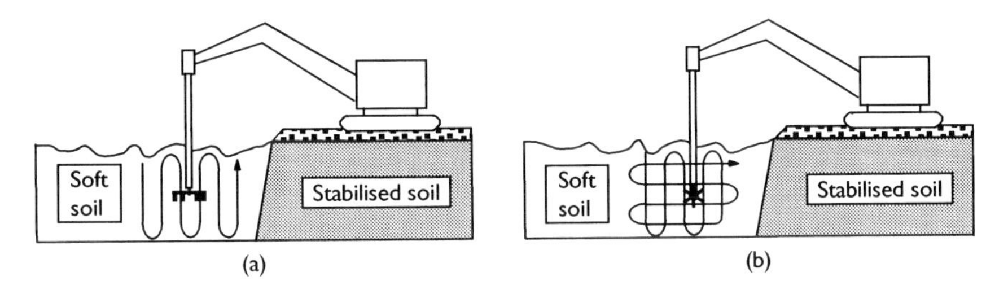

Figure 9.13 Mass stabilisation using dry binders: (a) mainly vertical mixing; (b) vertical and horizontal mixing.

and mixed. Two groups of mixing tools have been used to treat soft organic soils. The tools of the first group are in the form of a propeller and have a binder delivery nozzle at the centre. The second group employs mixing/cutting heads, equipped with multiple blades rotating about a vertical or a horizontal axis. Example tools are shown in Figures 9.14 and 9.5b.

The mixing pattern of mass stabilisation is planned taking into account site-specific conditions and capabilities of the mixing machine and the mixing tool. Usual practice is to stabilise in one sequence a block of soil within the operational range of the machine, typically corresponding to 8–10 m^2 in plan and 1.5–3 m in depth (e.g. 2 m wide × 5 m long × 3 m deep). When the prescribed amount of binder is mixed into the volume treated, remoulding is continued in order to obtain a homogeneous soil-binder

Figure 9.14 Different mass mixing tools for organic soils: (a) after EuroSoilStab (2002); (b) to (d) courtesy of Ramboll.

mixture. The productivity rate is approximately 200–300 m^3/shift of stabilised soil. The amount of binder used in Scandinavia is typically in the range of 150–250 kg/m^3, and the objective for shear strength in peat is 50 kPa (Jelisic and Leppänen, 2003). The diameter of the mixing tool is normally 600–1000 mm, and the rotation speed lies between 80 and 100 rpm. This method can be applied for soft clays and organic soils with shear strength below 25 kPa.

9.3.3 Wet-method mechanical deep mixing

The wet DM methods applied for ground improvement on-land in Japan, USA and Europe are generally developed to produce similar quality columns of stabilised soil, with unconfined compressive strength in the order of 0.5–5 MPa, or even more in granular soils, while the machines, mixing tools, execution procedures and productivity differ considerably.

Typical wet-method DM construction equipment consists of a batch mixing plant to supply proprietary slurry, and of a mixing machine for injection and mixing of slurry into the ground. The plant generally includes silos, water tank, batching system, temporary storage tank, slurry pumps (equipped with flow metres) and power supply unit. The batching system can be varied from manually or computer-controlled colloidal shear mixer, to a very fast inline jet mixing system. The storage tanks have paddle agitators to keep the component materials from settling out of the slurry. Delivery pumps are duplex or triplex reciprocating piston pumps, or variable-speed mono progressive cavity pumps. Pumping rates typically range from 0.08 to 0.25 m^3/min, but can reach up to 1 m^3/min for high-capacity mixing tools. Any changes in the slurry are made by adjusting the weight of each ingredient. Since fluid volume is being introduced into the ground, spoils must come to the surface.

The machines that are used for on-land applications usually have one to four shafts mounted on fixed or hanging leads, and are equipped with specially designed mixing tools. A multi-axis gearbox distributes the torque from a rotary drive unit to each shaft for penetration to the intended depth. The penetration speed is typically in the range of 0.5–1.5 m per minute and is usually increased during withdrawal. The mixing tools are kept in parallel by joint bands mounted at vertical intervals along the drive shafts. With some machines the spacing between individual shafts can be adjusted within prescribed limits to produce overlapped columns, which is beneficial when forming continuous panels or blocks of stabilised soil in a single-stroke operation. Sophisticated single axis systems with double cutting/mixing blades, spaced 30 cm and rotating in opposite directions, have also been developed in Japan (Horpibiulsuk *et al.*, 2002). A summary of mixing conditions for selected wet DM methods used for on-land applications and utilising mechanical mixing is presented in Table 9.2.

Table 9.2 Mixing conditions for selected fully operational wet DM methods used for on-land applications

Technical specification	*CDM (Standard + MEGA)*	*CDM Land4*	*SCC*	*HB-Keller USA/Europe*	*Bauer*	*SMW*	*DSM*	*COLMIX*
Number of mixing shafts	2 1: older system	4	1 possible 2	1	3 possible 1	1–3, 5 usually 3	1–6 usually 4	2, 3, 4
Diameter of mixing tool (m) (shaft spacing)	1.0 : S 1.2/1.3 : M (variable)	1.0/1.2 (variable)	0.6–1.5 1.2 (with 2 shafts)	0.5–2.4 : U 0.6–0.8 : E	3 × 0.37 3 × 0.55 3 × 0.88	0.55–1.5 usually 0.9, (variable)	0.8–1 usually 0.9	0.23–0.75 : 2 0.36–0.50 : 3 0.50–0.75 : 4
Realistic maximum penetration depth (m)	50 (55) 30 : M 1.2 20 : M 1.3	25	20	20 : U 12 : E	0.37 : 10.5 0.55 : 15.7 0.88 : 25	35 (50)	35	20
Penetration/Retrieval velocity (m/min)	P : (0.3) 0.5–1 R : 0.7–1 (2)	P : 0.7–1 R : 1.0	P : 1.0 R : 1.0	P : 0.3–0.5 U P/R : 1.0 U/E	P : 0.2–1 R : 0.7–1 (5)	P : 0.5–1 R : 1.5–2	P : 0.6–1 R : 1–2	P : 0.8 R : 1.0
Penetration/Retrieval rotation speed (rpm)	P : 20 R : 40	P : 20 R : 40	30–60	P : 20–25 R : higher	20–40 (80)	P : 14–20 R : higher	15–25	8–30

Injection during Penetration and/or Retrieval	P and/or R, restroking at the bottom	P and/or R, restroking at the bottom	usually P, restroking at the bottom	P (+R) : U P + R with restroking : E	P and/or R, P (30–50%) restroking	P and R, restroking common	P (+R), restroking at the bottom	P (+R), ev. restroking in clays
Water/Cement ratio	0.6–1.3 av. 1.0	0.6–1.3 av. 1.0	0.6–0.8 clays 1.0–1.2 sands	1–1.5 : U 0.6–1.2 : E	0.6–2.5	0.7–2.5	1.2–1.75 av. 1.5	0.7–2.5
Footprint area of the mixing tool (m^2)	1.5 : 2 shafts 0.8 : 1 shaft 2.17/2.56:M	2.83–3.14 or 4.21–4.52	0.3–1.75 2.25 two shafts	usually 1.1–4.5 : U 0.5 : E	0.44 : 3 × 0.37 0.94 : 3 × 0.55 2.35 : 3 × 0.88	0.7 : 3 × 0.55 1.7 : 3 × 0.9 4.7 : 3 × 1.5	2.5 (4 shafts, tangential)	0.08–0.95 : 2 0.29–0.57 : 3 0.76–1.60 : 4
Amount of added (dry) binder (kg/m^3)	70–300 av. 140–200	70–300 av. 140–200	150–400	150–275 : U 250–450 : E	80–500	200–750	120–400	100–550
Productivity per shift (one rig)	100–200 m^3	500–700 m^3	100 m^2 wall 400 m col.	250–750 m^3 : U 75–120 m^3 : E	30–300 m^3	100–200 m^3	200–300 m^2 wall	100–300 linear m

For marine applications large execution vessels, equipped with the mixing machine, batching plant, storage tanks and a control room, are usually used for rapid treatment of considerable soil volumes. The area of treatment in single-stroke operation with two to eight mixing shafts ranges from 1.5 to 9.5 m^2, and the productivity rates are in excess of 1000 m^3 per day.

The mixing tools for the wet-method are designed for various improvement purposes and are configured to soil type and available turning equipment. Since there is no one tool that can successfully treat all soils, field adjustments are typical. The mixing tools can be broadly classified into blade-based and auger-based constructions (cf. Porbaha *et al.*, 2001).

The tools of the first group have an assortment of flighting and mixing blades of full or near full diameter and different orientation to efficiently break down the soil structure. Steel hard-facing and an arrangement of purposely located teeth serve to aid penetration and reduce maintenance. A small-diameter lead auger/drilling bit usually extends below the cutting blades to centre and control penetration and verticality. The mixing process is mainly conducted at or within a short distance from the tip of the drill shaft(s). Injection nozzles are strategically located on the tool to uniformly distribute the slurry into the soil. They are usually found near the shaft tip, but can also be located along and above the mixing blades. Example single and multiple shaft mixing tools are shown in Figures 9.15 and 9.16.

The second group employs discontinuous or continuous helical augers for drilling and mixing or several levels of inclined paddles located above the cutter head of the mixing shaft. In these systems interlocking or closely

(a)

(b)

Figure 9.15 Single shaft mixing tools: (a) diameter 0.8 m; (b) diameter 2.4 m (Keller-Hayward Baker).

Figure 9.16 Multiple shaft mixing tools: (a) standard CDM 2 × 1.0 m (CDM Association, 2002); (b) CDM Land4 4 × 1.0 m (CDM Association, 2002); (c) SMW mixing paddles 3 × 1.5 m (R. Jakiel) and (d) cutter head (R. Jakiel).

spaced multiple shaft arrangements are typically used and the mixing operation is enhanced by counter rotating action of adjacent shafts. The mixing process occurs along all or a significant portion of the drill shaft(s). The direction of rotation is usually reversed during withdrawal. In most systems the slurry is fed through nozzles located at the bottom of each shaft. Example mixing tools are shown in Figures 9.17 to 9.19.

Figure 9.17 DSM mixing tools: (a) four blade-based mixing shafts; (b) four discontinuous, interlocking augers of diameter 0.9 m (courtesy of Geo-Con Inc.).

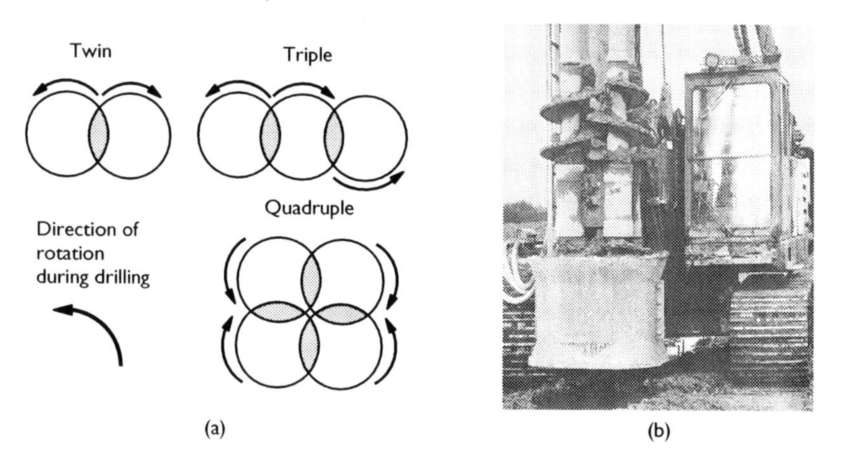

Figure 9.18 COLMIX mixing tools: (a) possible arrangements; (b) four discontinuous, interlocking augers of diameter 0.5 m (courtesy of Bachy-Soletanche).

In addition to the information presented in Table 9.2, a new CDM machine for on-land applications deserves more specific attention. The CDM-Column21 machine uses two shafts with large mixing heads of 1.5 m (1.6 m) diameter, consisting of an upper and lower mixing unit, both of which are equipped with inner and outer mixing blades that rotate in

(a)

(b)

Figure 9.19 Bauer mixing tool with three closely spaced, continuous augers of diameter 0.55 m (courtesy of Bauer Spezialtiefbau).

opposite directions (Figure 9.20). The unique counter rotating action of the blades actualises the shearing mixing effect and ensures uniform mixing of the cement slurry with the soil. The unit is capable of treating harder ground formations sandwiched between softer layers. The area of treatment is $3.5\,m^2$ and the required capacity of slurry supply is up to $1.0\,m^3$/min. The

Figure 9.20 CDM-Column21 mixing tool (courtesy of CDM Association, Japan).

maximum depth of treatment is 40 m. This modern system not only reduces the unit cost of soil treatment due to its very high productivity, but also offers higher-quality soil improvement through increased mixing operation efficiency (CDM Association, 2002).

Except for special situations and projects executed very close to sensitive objects, wet-method deep soil mixing has a very low impact on nearby structures. To avoid net volume increase and corresponding lateral stress in the ground caused by penetration of the mixing tool and injection of cement slurry, a dedicated method called CDM-LODIC (Low Displacement and Control) has been developed and modified since 1985 (Sugiyama, 2002). In this system the upper part of the mixing shafts are equipped with auger screws to forcibly expel equivalent soil volume during penetration and withdrawal stages of the mixing tool. The screws have standardised dimensions (diameter and pitch), and can be changed to best suit the ground conditions. It has been demonstrated in a field test that the execution of standard CDM and CDM-LODIC columns in soft clay, 1.5 m from an inclinometer installed in a vertical borehole, causes at the depth of 17 m a maximum horizontal displacement of 16.11 and 1.01 cm, respectively, confirming efficiency of the LODIC method (Horikiri *et al.*, after CDIT, 2002). In addition to the normal quality control system used for the conventional CDM method, an automatic system has recently been developed to display the volume of extracted soil. Since the cement slurry is normally injected during withdrawal through the nozzles located above the mixing blades, the soil extracted during penetration stage is free of cement and can be, if not contaminated, deposited or reused without any restrictions.

9.3.4 *Wet-method mechanical shallow mixing*

Wet-method mechanical shallow mixing can be used to improve substantial areas of soft or loose superficial soils in ground engineering applications, as well as for stabilisation and fixation of contaminated soils.

The SSM method uses specially designed single augers of 1.8–3.7 m diameter, attached to a hollow-stemmed Kelly rod suspended from a crane. Similar systems offer rigid attachment of the mixing tool to the base unit, and can therefore incorporate down pressure capability. The Kelly transfers the torque and feed pressure to the mixing tool, while the swivel mounted at the top of the rod seals the connection for delivery of the binder during rotation. The binder is usually injected during penetration, in slurry form, through several ports mounted at the bottom of the mixing augers. The pitch on the auger flights and the centrifugal force help to distribute the binder to all parts of the column during rotation. Cycling up and down with reduced binder delivery rates is often performed to improve mixing efficiency. An overlapping pattern of primary and secondary columns is normally used to ensure that the entire volume of treated soil is thoroughly mixed. A high-torque driver in the range of 400–600 kNm and high-capacity batching plant are generally required since the treatment area may reach about 10 m^3 of soil per metre of penetration. Examples of large diameter mixing tools are shown in Figure 9.21.

Wet-method mass stabilisation can also be carried out with specially designed mixing tools that are similar to those presented in Section 9.3.2. Druss (2002b) describes a major project conducted at the Fort Point Channel Site in Boston, where very soft organic, sand and organic silt deposits were shallow mixed prior to the execution of DM in underlying marine blue clay. The works were mostly performed underwater, in areas initially dredged to remove obstructions and timber piles. The objective of stabilisation was to construct a temporary support for a drill bench required for land-based DM operations. Shallow mixing was performed using a sectional barge, excavator with extended reach and a shallow mixing bucket containing blades rotating about a horizontal axis (Figure 9.22). Jet nozzles delivering the fly ash and cement grout were located inside the bucket and were directed towards the mixing blades. The bucket mixed horizontal trenches 1.2 m wide and about 10 m long in 1 m vertical lifts, moving from the surface of soft sediments to the top of clay, or finishing at partial depth.

Similar application has also been mentioned by Terashi (2002b). The original ground was an artificial landmass in Imari city in Japan, reclaimed by dredged sea-bottom clay with undrained shear strength in the order of 1 kPa. A 2 m thick block of treated soil was used to provide a working platform and/or temporary access road floating on the extremely soft soil

(a)

(b)

Figure 9.21 Crane-mounted SSM tools of diameter: (a) 2.4 m and (b) 3.7 m for the wet-method (courtesy of Geo-Con Inc.).

Figure 9.22 Shallow mixing equipment used at Fort Point Channel Site in Boston (R. Jakiel).

deposit. In this particular case, a special floater equipped with four mixing shafts was placed directly on the soft soil and dragged horizontally by winches while the mixing tools were moved up and down vertically. Similar shallow mixing tools as used in Finland and Sweden are also available in Japan.

9.3.5 Wet-method hybrid deep mixing

In addition to mechanical mixing, these methods employ high-velocity jet grouting in order to reduce penetration resistance and improve mixing operation and/or to increase the diameter of the improved ground.

The SWING method, initially developed as a mechanical mixing system, uses a retractable mixing blade mounted at the end of a single drilling shaft. The position of the blade in the ground can be changed from a vertical to horizontal alignment and vice versa, as shown in Figures 9.23a,b. A combination of mechanical and jet mixing with cement slurry enables columns of up to 3 m diameter to be constructed, and the addition of compressed air allows columns greater than 3 m diameter. During penetration of the ground, the soil is broken down by rotation of the blade and jetting action of water. Cement slurry is injected during withdrawal, with the jetting energy supplemented by air pressure. Air is used when the larger diameters are required or when the soils under treatment are too stiff. This method also enables the installation of inclined or even horizontal columns and therefore allows soil mixing in areas of difficult access.

The JACSMAN system consists of two ten-bladed soil mix tools, each combined with a pair of jet grouting nozzles aligned for cross jet (XJET) to ensure that over-cutting does not occur. As compared with the conventional CDM method, JACSMAN offers significant improvements which contribute to a more economical high-quality product. Due to XJET cutting with air-enveloped high-velocity cement slurry during withdrawal, the treatment area of single-stroke operation increases considerably and equals 6.4 m^2 for type A arrangement, with a 75 per cent share of jet mixing, and 7.2 m^2 for type B arrangement, with a 63 per cent share of jet mixing, as shown in Figure 9.24. Moreover, the diameter of the soil–cement column can be controlled and changed over the column's length through stopping and starting XJET action, not affecting the surrounding soil due to the

Figure 9.23 Spreadable Wing (SWING) method: (a) blade position during penetration; (b) blade expanded; (c) demonstration of jetting action (courtesy of SWING Association).

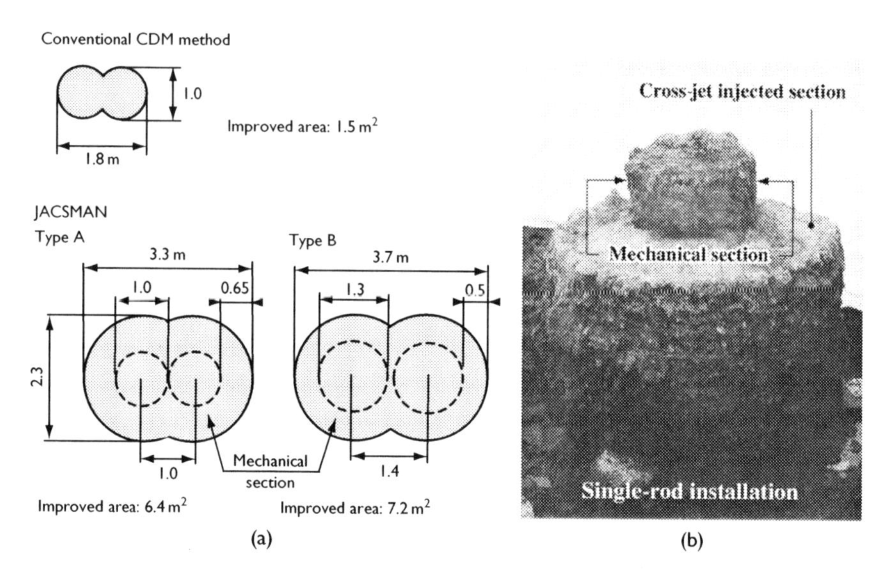

Figure 9.24 The JACSMAN method: (a) comparison of treatment areas; (b) exposed single column (Kawanabe and Nozu, 2002).

dissipation of jet energy at the cross point (Figure 9.25). This allows for the adjustment of the column's diameter to soil stratification, as well as for controlled mixing operations close to structures or excavation walls. The main operating parameters of JACSMAN are as follows: jetting pressure 30 MPa, jetting slurry flow rate 4×150 l/min, air pressure 0.7 MPa, grout

Figure 9.25 JACSMAN mixing tool: (a) twin head assembly (note grout nozzle in front of the tool); (b) XJET demonstration with increased pressure (R. Essler).

flow rate 2 × 200 l/min, grout pressure 5 MPa and withdrawal speed of 0.5 and 1 m/min for type A and B arrangement, respectively (Kawanabe and Nozu, 2002).

HYDRAMECH utilises mechanical mixing with a single shaft, fitted with 1.2 m diameter paddles and a 0.9 m diameter auger, in combination with high-velocity grout injection at 40 MPa through eight 2 mm 'hydra nozzles' on the outer edges of the mixing tool. HYDRAMECH is capable to creating soil–cement columns with diameters of 2 m. Mechanical mixing occurs smoothly in the centre of the column, and chunks of material are forced to the perimeter, where they are disaggregated by the jets. Treatment with jets can be switched on and off throughout the column length to create plugs of treated soil. Realistic maximum penetration depth is 20 m. Penetration/retrieval velocity is 1–3 m/min, with 5–20 rpm during penetration and 10–30 rpm during retrieval (additional mixing). Industrial productivities are in the range of 250–500 m^3/shift. The main objective for developing this method was to improve on current jet grouting technologies that can create subsurface problems with the use of compressed air. HYDRAMECH can create an extended diameter soil–cement column without the injection of compressed air and still provide the continuous overlap that is a very positive aspect of jet grouting systems, particularly when installing horizontal barriers.

TURBOJET (GEOJET in the US) combines mechanical mixing with single fluid jet grouting technology. Jetting is used during insertion of the tool to increase penetration velocity while extraction is carried out solely with mechanical mixing. A specially designed mixing tool (or processor), fitted at the end of a tubular Kelly bar, consists of two levels of inclined blades and is furnished with several 4–8 mm diameter high-pressure nozzles mounted along the shaft and tip (Figure 9.26). The exact nature and composition of the processor can be varied, depending on soil conditions. Grout can be pumped with a discharge rate of 450 l/min at 30 MPa, although lower flow rates and pressure (15 MPa) are the norm. Tool diameters range from 0.6 to 1.5 m, usually 0.9–1.2 m, and the practical available depth of treatment is 25 m (Lebon, 2002). Instantaneous rates include 2–12 m/min (6 m/min typical) during penetration and 15 m/min during withdrawal. Computer control of the equipment during column formation is therefore required. The computer analyses the rate of tool rotation and penetration, slurry pressure, torque, crowd force, and soil mix volume and density as a function of depth. The system also reacts to changing parameters and automatically adjusts to maintain specified soil–cement properties, even in varying subsurface soils. Because of the additional mixing energy supplied, restroking is not required. Industrial production rates in excess of 150 m/h and 1100 m/shift are possible. The system produces low waste volumes (20–30 per cent of ground treated).

(a)

(b)

Figure 9.26 TURBOJET deep soil mixing equipment and processor (M. Siepi).

9.3.6 Installation process

The typical installation process consists of positioning the mixing shaft(s) above the planned location, penetration of the mixing tool, verification and improvement of the bottom soil layer, withdrawal, and movement to a new location if necessary. The details of execution depend on the type of method applied (dry or wet), technical features of the equipment, and the site-specific and functional requirements. Frequently used execution procedures are shown schematically in Figure 9.27.

The position and verticality of the shaft is checked first, and zero adjustments of the logging system are conducted. For on-land application optical survey devices are normally used, whereas for marine operations the use of the Global Positioning System (GPS) has become common. The GPS is also advantageous in the case of large on-land projects, especially those involving treatment of very weak superficial soils.

During penetration the mixing tool is delivered to the required depth. In this phase compressed air (dry-method), or slurry (wet mechanical mixing), or high-velocity jetting with slurry or water and air (hybrid mixing) is used to support mechanical drilling. Mechanical penetration may be difficult when the tool hits a hard layer or when the improvement depth is relatively deep, leading to possible damage or deadlock of the tool in the ground. This danger may be reduced with partial restroking to minimise rotation resistance along the shaft (Figure 9.27b), or by means of pre-boring with an auger machine.

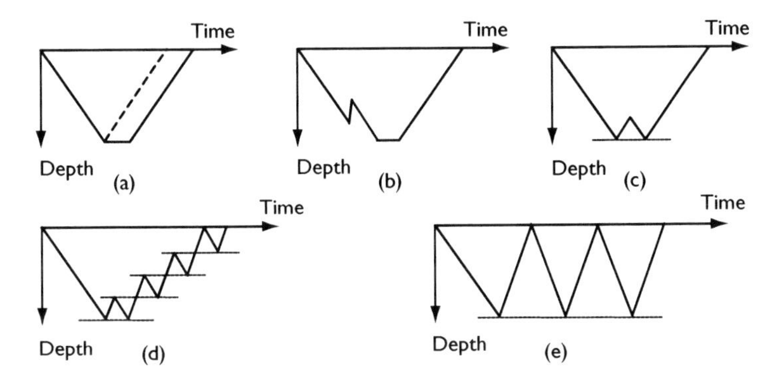

Figure 9.27 Typical execution procedures of deep soil mixing: (a) without or with bottoming; (b) with reversal(s) during penetration; (c) with bottom restroking; (d) with stepped restroking during withdrawal; (e) with full-depth restroking.

After the bottom of treatment depth is reached, the mixing tool remains on the bottom rotating about 0.5–2 min for complete mixing. This phase is often called 'bottoming', and serves to ensure sufficient contact of the column(s) with the bearing subsoil. Penetration into the bearing layer should be confirmed by a rapid change of penetration velocity of the tool, required torque and rotation speed. At this stage the tool can be raised about 0.5 to 1 m and lowered again to treat more effectively the transition zone between soft and bearing soils (Figure 9.27c). Withdrawal may be conducted as a continuous upstroke, but can include stepped or even full restroking if needed (Figures 9.27d,e). Full restroking is beneficial in the case of interchanged soft/stiff layers and stratified soils, leading to more uniform properties of stabilised soil across the depth of treatment.

The accompanying delivery of the stabilising agent to the subsoil is operator/computer controlled and linked to the energy of mixing in the specific layers of treated soil. In general, injection of the stabilising agent can take place during penetration, withdrawal and restroking; however two main injection methods are distinguished:

- the penetration injection method (top–bottom process),
- the withdrawal injection method (bottom–top process).

Penetration injection is typically used for on-land applications of the wet-method because the slurry helps to lubricate the mixing tool and assists in breaking up the soil into smaller pieces. Normally, 100 per cent to about 80 per cent of the total slurry volume is used in this stage. This method is also beneficial to the homogeneity and strength of the manufactured column because the native soil is mixed twice with the binder.

Withdrawal injection is typically used for the dry-method, with usually the whole amount of binder delivered to the soil during this phase. However, if very high binder concentration is needed to reach the design strength, part of the stabiliser may be injected during penetration phase and the rest during withdrawal of the mixing tool. Withdrawal injection also prevails for most marine operations with the CDM method. For on-land applications with the wet-method withdrawal injection is also possible, but usually at a reduced flow rate to minimise the volume of spoil (except of the CDM-LODIC method where withdrawal injection is a standard installation process).

The sequence of mixing operations will need to be adjusted to suit each site's specific conditions, but, in general, the most efficient sequence is to work the stabilisation machine within its radius of operation as much as possible before it is moved.

9.3.7 Details of construction and execution

9.3.7.1 Number of shafts

Models and field observations indicate that multiple shaft arrangements generally provide better homogeneity of DM columns than those produced with single shaft mixing tools furnished with fixed cutting/mixing blades rotating in one direction. This has been especially observed in clays, which may tend to stick to the mixing blades and hence rotate with the mixing shaft, resulting in poor mixing. As a countermeasure, non-rotating vanes have been mounted on single-auger shafts, close to cutting blades. An example is the free blade, developed in Japan (CDIT, 2002), which extends beyond the reach of the mixing blades and is therefore supposed to stay in the ground to provide sharing capability. The 'entrained rotation' phenomenon is significantly reduced with closely spaced augers when the neighbouring shafts rotate in opposite directions, or are eliminated with overlapping augers or interlocking mixing blades providing greater soil shearing and particle milling (the same applies for sophisticated single shaft tools with counter rotating blades). Transverse steel bars used to keep multiple shafts in position have a similar function as the free blade.

Besides contributing to interactive mixing and increased productivity rates, multiple shaft arrangements also minimise the counter-movement against shaft rotation, as compared with single shaft rigs, and improve stability of the machine. This further contributes to more precise control of shaft verticality during penetration. Linear arrangements of mixing shafts, applied for the construction of retaining and cut-off walls, enable easier and safer connection of individual wall panels using the intercut principle. Furthermore, limited adjustments of deviations occurring during penetration can be made by altering the rotation of coupled shafts.

Multiple shaft arrangements are, however, more demanding in terms of constructional requirements, generally leading to more complicated mechanical systems and larger/heavier machines. This may result in reduced flexibility in some applications, as well as increased mobilisation and operational costs, as compared with single shaft machines.

9.3.7.2 Shape and orientation of mixing blades

The function of the mixing tool is to disaggregate the soil during penetration and to facilitate binder injection and immixing with the native soil. In the case of purely mechanical interaction, the mixing tool should also create the appropriate diameter of the column. A wide variety of different mixing tools have been tried so far, ranging from very simple to quite complicated constructions, with the obvious outcome that no single construction method can equally serve all soils. Nevertheless, some general indications can be formulated, keeping in mind that when compared to the wet-method, the dry-method requires more vigorous mixing to achieve the same level of homogeneity of soil-binder mix.

During downward movement of the shaft, the mixing tool has to loosen the soil, while during withdrawal, the soil should be thoroughly mixed with the binder and recompacted as much as possible to reduce excess spoil and to ensure maximum mixed soil density (recompaction does not matter in saturated conditions but is rather important for the dry-method). This generally occurs when the inclination of the blades to the rotating direction produces mixing movements from the outside inwards and from above downwards, opposite to the lifting movement of the tool. The degree of mixing increases when the soil is finely divided into horizontal, inclined and vertical directions during tool rotation. This explains why in the case of single axis shafts the window-type mixing tools, such as shown in Figures 9.11a and 9.11b, may perform slightly better than the tools with several separated horizontal blades, as also corroborated by the investigation conducted by Abe *et al.* (CDIT, 2002). On the other hand, soils like peat require more shearing action to be thoroughly mixed, and this is usually better achieved with multiple horizontal blades. Furthermore, window-type tools cannot overlap when used in multi-shaft arrangements. In the case of continuous or discontinuous augers counter rotation against the direction of auger pitch permits the soil-binder mix to be recompacted during withdrawal. Counter rotation and/or shear bars also generally reduce the mixing energy required.

When designing a mixing tool it is usually necessary to find a balance between a good ability to penetrate stiff or compacted layers of soil, and a good mixing performance in soft soils. The same applies to the speed of rotation and associated higher wear of the mixing blades, as well as the possibility of easy repair and quick replacement of the mixing equipment. Consequently, the goal of designing a mixing device that creates sufficient movement in the soil without a great mixing effort or long mixing time is difficult to achieve.

9.3.7.3 *Position of injection nozzles*

To ensure optimum mixing efficiency, the position of injection nozzle(s) is different for various methods and installation processes. For the penetration injection method the outlet is normally placed close to the bottom end of the mixing tool, while for the withdrawal injection method it is above the mixing blades or at the level of the upper blade. Besides these two standard outfits some mixing tools have nozzles at both levels to facilitate also a combined penetration/withdrawal injection (e.g. CDM and DJM methods). During penetration phase the lower port is used and the upper one is closed. When withdrawing, opposite combination is applied.

With the wet-method and single axis mixing tools there is usually one centre injection nozzle located close to the shaft tip, while large diameter tools have several injection nozzles located along the blades at specified distances from the central shaft. In multiple axis tools the grout is usually fed independently to each shaft, with the outlet port placed at the shaft tip. In some linear arrangements grout can be also supplied through the central shaft, incorporating one or two nozzles at the bottom, depending on the auger diameter, as done for the Bauer Triple method. The direction of grout injection is generally horizontal.

As for the mixing tools of the hybrid method, the high-velocity jet nozzles are purposely located on the outer ends of the mixing blades to increase the range of mixing, but they can also be located at the tip or along the Kelly bar if jetting is primarily used to increase the rate of mixing tool penetration. The direction of the jet stream may be horizontal or inclined, depending on the nozzle orientation.

9.3.7.4 *Degree of mixing*

The efficiency of *in situ* soil mixing with a stabilising agent is one of the key factors affecting column homogeneity and strength. The degree of mixing depends on the mixing time, type of mixer, characteristics of the native soil, and the form of applied binder (slurry or powder) and the energy of injection (low or high output velocity). The overall mixing process is rather complex, especially for the dry-method (cf. Larsson, 1999), and difficult to quantify. Therefore, in an attempt to specify a criterion for the required mixing work, which could be controlled and altered on site during execution, a simplified index named 'blade rotation number' has been introduced in Japan (e.g. CDIT, 2002). The blade rotation number, T, is defined as the total number of mixing blades passing during 1 m of single shaft movement through the soil, and is expressed as follows, considering:

1 complete injection during penetration and outlet located below the blades:

$$T = \Sigma M \times \left(\frac{R_p}{V_p} + \frac{R_w}{V_w} \right) \tag{9.1}$$

2 complete injection during withdrawal and outlet located above the blades:

$$T = \Sigma M \times \left(\frac{R_w}{V_w} \right) \tag{9.2}$$

3 partial injection during penetration and main injection during withdrawal, with the lower outlet active only during penetration and the upper outlet active during withdrawal:

$$T = \Sigma M \times \left(\frac{R_p}{V_p} \times \frac{W_p}{W} + \frac{R_w}{V_w} \right) \tag{9.3}$$

where: T – blade rotation number (rev/min), ΣM – total number of mixing blades, R_p – rotational speed of the mixing tool during penetration (rev/min), V_p – penetration velocity (m/min), R_w – rotational speed of the mixing tool during withdrawal (rev/min), V_w – withdrawal velocity (m/min), W_p – amount of binder injected during penetration (kg/m^3), W – total amount of injected binder (kg/m^3).

The total number of mixing blades, ΣM, is assessed by counting all cutting/mixing blades that are effective in the mixing process, taking into consideration the method of injection and position of the injection outlet(s) in relation to the blades. A full-diameter blade is counted as two blades. For example, when the outlet port is located beneath two levels of blades and when injection is carried out during penetration, as is common for the wet-methods, the total number of mixing blades is $\Sigma M = 4$ and equation (9.1) is used to evaluate T. In case of the withdrawal injection method and the outlet port located above all blades, as is common for the dry-methods, ΣM is also four, but the resulting blade rotation number is lower since only the withdrawal stage is considered (equation (9.2)). This example demonstrates that higher rotational speed is required for the withdrawal injection method to obtain a comparable degree of mixing, while maintaining the same number of blades and withdrawal velocity of the mixing tool.

The blade rotation number is used for mechanical mixing only, and the soil conditions are included indirectly, i.e. through selection of appropriate input values, taking into account accumulated experience and technical specifications of the equipment. Based on field data obtained in loose sands (Mizuno *et al.*) and clays, the blade rotation number of 360 has been recommended in Japan for the wet-method to ensure reasonably low value of the coefficient of variation of the unconfined compressive strength (CDIT, 2002). For the dry mixing methods the blade rotation number is typically 274 or 284 for the DJM and 200–400 for the Nordic method, noting that dry binders are injected mainly only during withdrawal. For special mixing tools using cutting/mixing blades that rotate in opposite directions on a single shaft, like for instance the CDM-Column 21 method, there is a need to conceive a new guideline for the quality of mixing.

9.3.7.5 *Control of binder supply*

The amount of binder injected in a certain soil volume is easier to control for the wet-method than for the dry-method, where the binder is fed into a stream of compressed air.

The wet mixing process blends the materials with water to form a slurry at the design water to binder ratio. The quantity of binder components needed for each batch is weighed and added to the measured water volume in the mixer. Alternatively, ready batched or pre-weighed bagged materials can be used to simplify this process. The binder slurry is then transferred to a temporary storage tank that continually agitates the slurry to ensure that the constituents of the mix do not separate. The slurry is then pumped at a specified flow rate to the mixing tool. In order to obtain the required amount of binder per soil volume, the penetration and withdrawal velocities of the mixing tool and the applied flow rates have to be simultaneously adjusted, taking into account the number of restroking passes with slurry injection. The flow rate of binder slurry is controlled at the delivery pump and monitored with a flow meter.

With the dry-method the weight loss of the binder storage tank, continuously measured by means of load transducers and averaged in such a manner that acceleration components are cancelled out, is used as an indicator of the amount of used binder. This information is combined with the corresponding geometry of stabilised soil to evaluate the binder output in kg/m of column or in kg/m^3 of treated soil. To reach the pre-determined target value it is necessary to control the feeding rate of binder into the air stream until the specified rate of loss of the material is obtained. This is mainly accomplished by adjusting the rotation speed of an impeller provided at the bottom outlet of the binder storage tank (Figure 9.28). The feeding mechanism must be manufactured with a very high precision since the distance between the rotating blades and the cylinder walls is in the

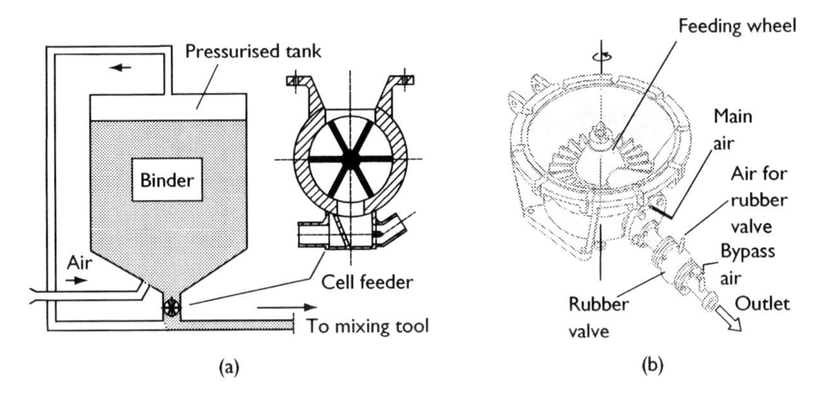

Figure 9.28 Binder feeding systems: (a) cell feeder used in the Nordic method (courtesy of LCM); (b) impeller used in the DJM method (Aoi, 2002).

order of 1/100 mm. However, the throughput of the impeller is not a linear function of the rotating speed and depends also on (Bredenberg, 1999) (1) wear of impeller, blades and wall, (2) pressure and amount of binder in the tank, (3) air and material flow below the impeller, and (4) flow properties of the binder, making binder output control more sophisticated. The downward movement of binder in the tank towards the impeller is facilitated by 'fluidisation' of the stored material, caused by blowing compressed air from the tank bottom. To ensure high productivity with this system it is important that the air blown into the binder storage tank is sufficiently dry and that the binder material is free of particles able to cause blocking or damage to the feeding control system. Moreover, suitable hose diameter for the equipment used must be carefully selected to ensure smooth binder flow to the mixing tool.

9.3.7.6 *Control during construction*

Soil mixing, like other ground improvement technologies, uses indirect control measures to ensure the quality of work and product during execution. The main objective of a control system is to ensure delivery of a correct amount of binder and mixing energy along the installed element. The extent of *in situ* mixing operation monitoring is closely associated with the type of project and the required level of quality control.

For a typical production process of DM the following should be documented: element identification and/or position, mixing tool details, working grade, mixing depth, start time, time at bottom, finish time, mixing duration, agent specification, injection flow rate and pressure, total amount of agent used, tool rpm on penetration, tool rpm on withdrawal and torque of the shaft. From this information the mixing energy and binder content can be calculated to match laboratory data and/or test columns. The standard criteria to ensure quality of tip bearing are the penetration velocity and the applied torque. Centralised control systems are usually available to digitally record all parameters and display information at the control panel to facilitate real-time adjustments (e.g. Yano *et al.*, 1996; Bredenberg, 1999; Burke, 2002; Hioki, 2002). They also simplify the task of preparing daily reports by recording the daily performed activities of soil mixing works.

In special applications requiring automatic and/or more sophisticated control, a variety of measuring systems can be used to control the mixing process, column verticality, or to observe horizontal and vertical ground displacements. A real-time monitoring technique based on an electrical resistance density transducer for control of the amount of binder fed into the ground at any treatment depth is also available (Zheng and Shi, 1996). The computer reacts to changing ground conditions and automatically adjusts injection pressure to ensure specific treated soil parameters are provided for each stratum.

Future advances in computer and transducer technology will be leveraged to promote uniformly high-quality mixing. In particular, the focus is on automatically controlling slurry discharge rates with depth, automatic confirmation of the bearing layer, facilitating the duties of the operators, and tracking *in situ* characteristics of the treated soil (Bruce and Bruce, 2003).

9.4 Applications and limitations

9.4.1 Areas of application

Instead of focusing on a limited number of arbitrarily selected projects involving soil mixing, which can hardly represent numerous interesting international achievements, six areas of SM applications have been identified and reviewed. Similar classification schemes have been presented by Porbaha *et al.* (1998) and Bruce (FHWA, 2000). Other attempts to classify the range of soil mixing applications can be found in CDIT (2002) and CEN TC 288 (2002).

The main areas of SM applications are as follows, with the countries in parentheses indicating their most extensive use so far:

1 Foundation support (Japan, Scandinavia, US, France, Poland)
2 Retention systems (Japan, US, China, Southeast Asia, Germany)
3 Ground treatment (Japan, US, Finland, Sweden, Southeast Asia)
4 Liquefaction mitigation (Japan, US)
5 Hydraulic cut-off walls (Japan, US, Germany, Poland)
6 Environmental remediation (US, UK).

Case histories relating to each application group may be found in the cited bibliography as well as in the proceedings of specialty international conferences on SM, held in Tokyo (1996), Stockholm (1999), Helsinki (2000) and New Orleans (2003). It should be pointed out, however, that in many cases SM works are conducted to fulfil combined functions. Consequently, certain projects fall into more than one general category of application.

9.4.2 Patterns of deep soil mixing installations

Soil mixing can be done to a replacement ratio of 100 per cent wherein all the soil inside a particular block is treated, as is usually the case for shallow mixing applications, or to a selected lower ratio, which is often practised with DM. The chosen ratio reflects, of course, the mechanical capabilities and characteristics of the applied method. Depending on the purpose of DM works, specific conditions of the site, stability calculations and costs of treatment, different patterns of column installations are used to achieve the desired result by utilising spaced or overlapping and single or combined columns. Typical patterns are presented in Figure 9.29.

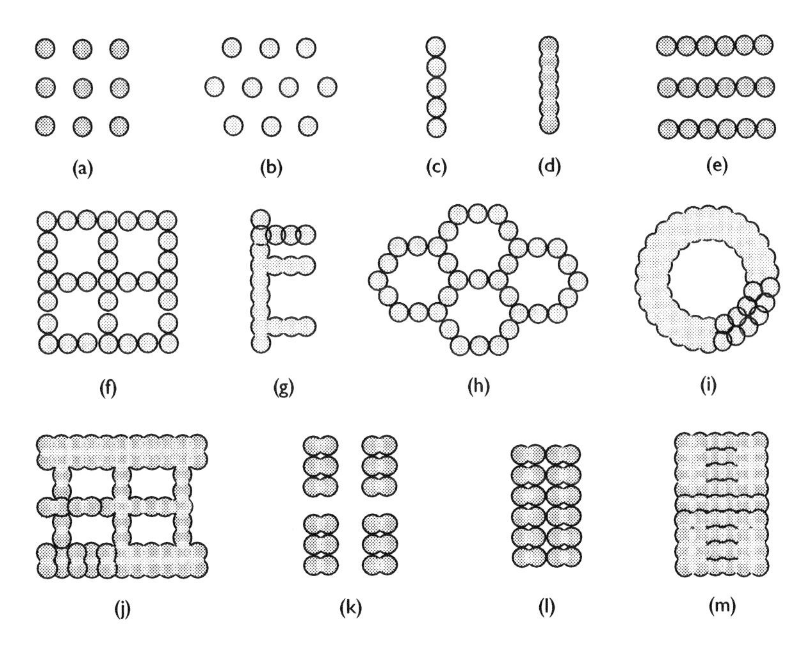

Figure 9.29 Examples of deep soil mixing patterns: (a), (b) column-type (square and triangular arrangement); (c) tangent wall; (d) overlapped wall; (e) tangent walls; (f) tangent grid; (g) overlapped wall with buttresses; (h) tangent cells; (i) ring; (j) lattice; (k) group columns; (l) group columns in-contact; (m) block.

Square or triangular grid patterns of single or combined columns are usually applied when the purpose of DM is reduction of settlement and, in some cases, improvement of stability. Common examples are road and railway embankments. Walls are used for excavation control, to stabilise open cuts and protect structures with shallow foundations surrounding the excavation, and as a measure against seepage. They are also constructed to increase the bearing capacity of improved soil against horizontal or sliding forces, with column rows installed in the direction of horizontal loading or perpendicular to the expected surface of failure. Walls can be constructed with tangential or overlapping elements. Overlapping is particularly important when executing cut-off walls or environmental barriers. In the case of DM machines equipped with linearly arranged multiple shafts, walls are usually executed using intersecting primary and secondary panels, with partial or even full column diameter overlap. Groups of columns can be utilised to support embankments and foundations in order to reduce settlements and/or increase the bearing capacity. Various combinations of columns are also used to build grid, U-formed, cellular or circular installations with tangential or overlapping columns to improve the interaction with the untreated soil. Lattice-type improvements are considered an intermediate, cost-effective system between the wall-type and the block-type improvement. Full blocks are used to create large, highly stable volumes of stabilised soil, which act as gravity structures.

To compare various column patterns in terms of the treatment area, and to evaluate composite properties of the treated elements and the surrounding untreated soil, a purposely defined ratio of area improvement, a_p, is used (cf. Figure 9.30):

$$a_p = \frac{A_t}{A} = \frac{\text{net area of soil mixing}}{\text{respective total area}} \tag{9.4}$$

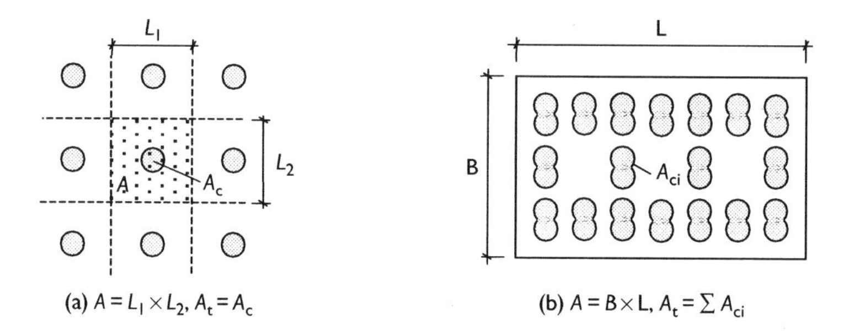

Figure 9.30 Evaluation of the ratio of area improvement for: (a) regular grid of columns; (b) foundation slab.

The upper limit of the ratio of area improvement for a square grid of tangential columns is 78.5 per cent, and for equilateral triangular grid 90.7 per cent. For columns spaced at two diameters a_p is 19.6 and 22.7 per cent, for square and triangular patterns respectively, and for columns spaced at three diameters it is 8.7 and 10.1 per cent respectively. The spacing of three diameters, usually recommended to minimise interaction between piles, can be considered as a practical lower limit of the area improvement ratio. Numerous embankments in Japan have been stabilised with a_p usually between 30 and 50 per cent (due to seismic excitations), while in Scandinavia area ratios between 10 and 30 per cent have been typically applied in case histories.

Column installation patterns may not only vary in plan view but also with respect to the depth of treatment. In the wall-type improvement, short and long walls can be alternately installed in the soft soil to reduce the costs of soil mixing (Figure 9.31a). The long walls transfer the loads exerted by the superstructure and external excitations to the bearing stratum, while the intermediate short walls provide connection between the long walls, increasing the rigidity of the total improved soil mass. In recent times this type of improvement has been commonly applied in port and harbour constructions in Japan (e.g. Kansai Airport man-made island, CDIT, 2002). Another example is the variation of column lengths in transition and/or purposely determined zones of soil treatment, as shown in Figures 9.31b–d. Furthermore, a combination of different soil mixing techniques may be applied to treat specific soil depths, such as with a combined

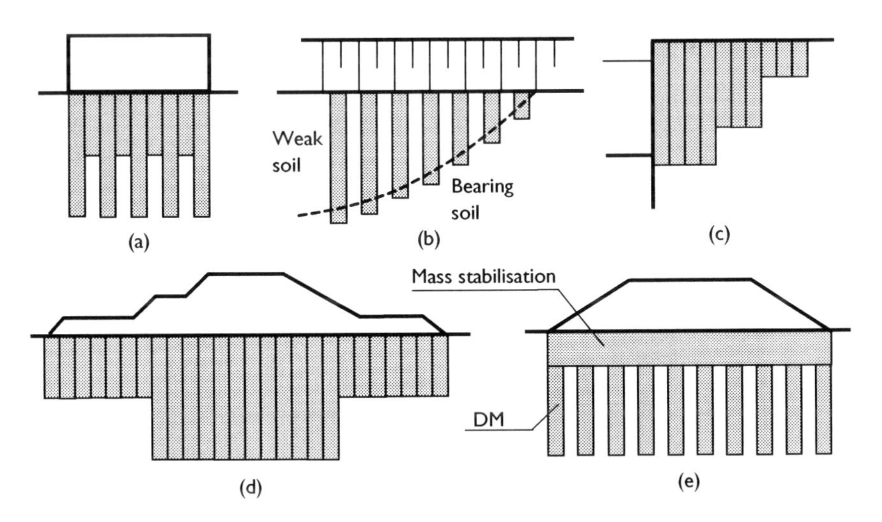

Figure 9.31 Examples of deep soil mixing with varying column lengths (schematic): (a) combined short and long walls; (b) transition zone; (c) stepped columns; (d) embankment with berms; (e) combined mass stabilisation and DM.

application of DM and mass stabilisation resulting in a column-supported raft structure (Figure 9.31e), as practiced in Scandinavia (e.g. Rogbeck *et al.*, 1999; Jelisic and Leppänen, 2003).

9.4.3 Foundation support

The purpose of using DM is mainly the reduction of settlement and the increase of bearing capacity of weak foundation soil, as well as prevention of sliding failure (Figure 9.32). For on-land projects the applications usually comprise road and railway embankments, buildings, industrial halls, tanks, bridge abutments, retaining walls and underground facilities. For waterfront and marine applications they can include quay walls, wharfs, revetments and breakwaters.

The installation patterns typically employ single or combined columns with variable spacing for settlement reduction applications, while combined walls, lattices and blocks are used when dealing with high loads and/or horizontal forces. An increasing tendency to apply economical low values of

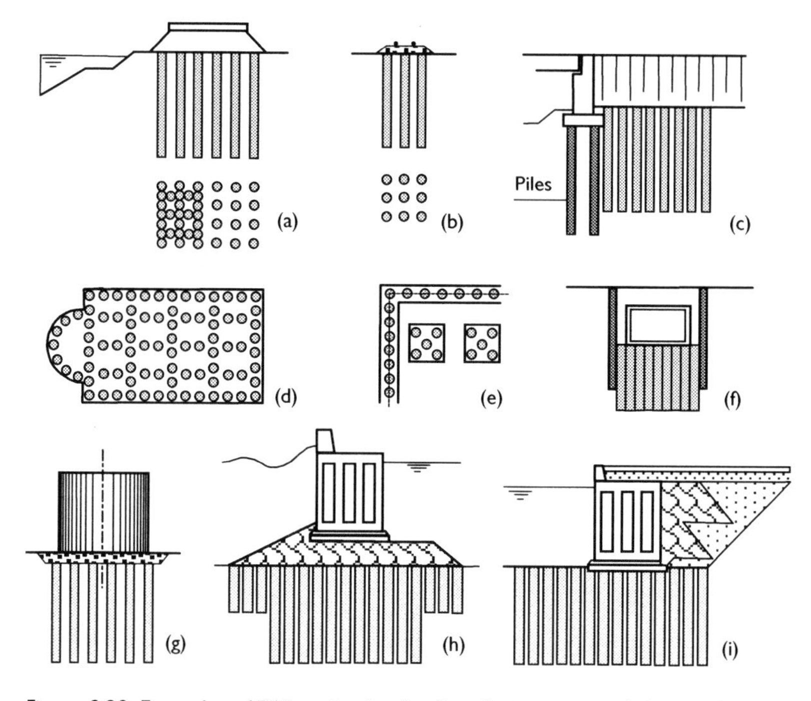

Figure 9.32 Examples of DM application for foundation support (schematic): (a) road embankment; (b) railway embankment; (c) bridge approach zone; (d) slab foundation; (e) strip and pad foundations; (f) culvert; (g) tank; (h) breakwater; (i) quay wall.

the area improvement ratio can be observed in recent times, depending on the adopted DM method and the available column strength. Design of such patterns requires rigorous analysis of the interaction between treated and untreated soil. The strength of DM elements may differ significantly within the range determined by low-capacity lime/cement columns, with say 0.15 MPa shear strength, and high-capacity structural elements having unconfined compressive strength in the order of 5 MPa, which act like piles or caissons. The external loads are usually transferred down to the bearing layer resulting in a fixed type improvement, but can be also partly or wholly transferred to the foundation soil when a more interactive or even a floating type of improvement is desired. The choice of the required strength and of the load transfer system is dictated by the purpose of the DM application, and reflects the mechanical capabilities and characteristics of the particular method used.

When deep soil mixing is applied under embankments or foundation slabs to reduce differential settlement and increase bearing capacity of the foundation soil, it can be noticed that individual column quality is less important, and that it is the overall performance taking into account soil to column interaction that matters most. Such a concept of soil/structure interaction, practised for instance in Scandinavia using the Nordic method and often combined with pre-loading and drainage function of the columns to accelerate settlement, has proved to be efficient and cost-effective compared to other methods. On the other hand, when DM is performed to support high embankments or heavily loaded foundations, and where horizontal loadings or shear forces are significant, the quality of load-bearing columns is essential to prevent progressive failure mechanisms. The same applies for low values of the ratio of area improvement.

In bridge construction the DM columns can be used to act as the pier foundation for the abutment, or to prevent lateral thrust and sliding by reducing the earth pressure behind the abutment. They can also reduce settlement of the bridge approach zone. In the case of buildings, DM is an alternative solution to conventional deep foundation methods, particularly in seismic-prone areas. Since the DM columns can be closely spaced, the foundation dimensions in plan remain relatively small, which contributes to the overall cost-effectiveness of this foundation solution.

Waterfront or marine gravity-type structures are subjected to large horizontal forces caused by earth pressure or wave loading. Therefore, columns patterns typically comprise blocks, lattices or combined walls created by overlapping columns installed by the wet-method.

9.4.4 Retention systems

Retention systems comprise applications associated with restraining the earth pressure mobilised during deep excavations and vertical cuts in soft ground, with protection of structures surrounding excavations, measures against base heave, and prevention of landslides and slope failure (Figure 9.33). In these applications, wall- and grid-type column patterns are mainly used, while the soil-binder mix is typically engineered to have high strength and stiffness. To overcome soil and water lateral pressures the DM columns should have adequate internal shear resistance. Other key requirements for successful construction are a high degree of column homogeneity and maintaining verticality tolerance to achieve the minimum required designed thickness of columns effectively in continuous contact. It is also important that early strength gain is sufficiently retarded to prevent problems when constructing secondary intercut columns.

Steel pipes or H-beams can be installed in DM columns executed with the wet-method to increase the bending resistance and create a structural wall for excavation support (Figure 9.33b). Elongated mixing time and/or full restroking are usually applied to ensure easier installation of soldier elements immediately after mixing. Panels of mixed soil between H-beam reinforcement are designed to work in arching, as in a 'Berlin'-type wall (e.g. Außerlechner *et al.*, 2003). Concrete facing, tieback anchors or stage struts are typically used in combination with the DM walls. Drainage media may be required behind the wall to prevent build-up of excess hydrostatic pressures. Deep circular shafts can be constructed using two to three

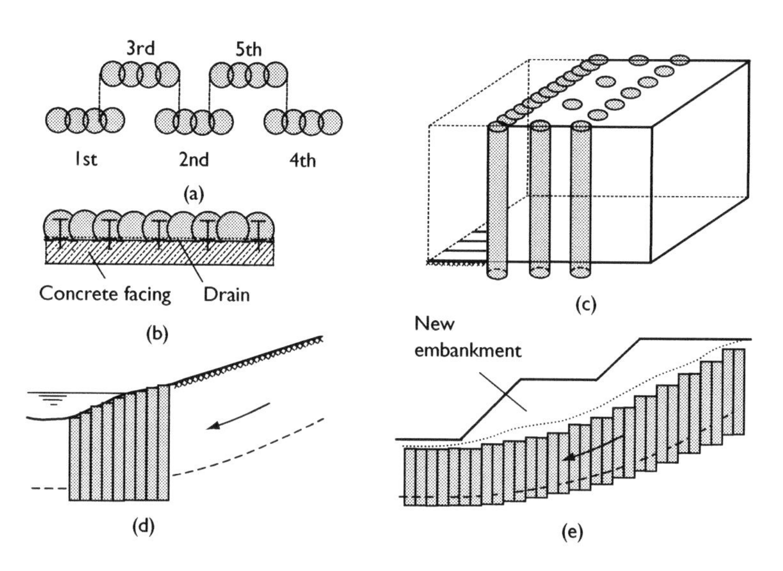

Figure 9.33 Examples of DM applications for retention systems (schematic): (a) typical DM wall; (b) DM wall with concrete facing; (c) composite gravity wall (note: relieving platform not shown); (d) landslide protection; (e) slope stabilisation.

concentric rings of overlapping DM columns, acting together in hoop compression.

Another innovative concept to support vertical excavations is to construct composite gravity structures, which do not require anchors or braces (Andromalos *et al.*, 2000). This vertical earth reinforcement technology, or VERT top-down retaining wall system, typically consists of a continuous front row of DM columns and one or more rows of isolated columns or staggered panels (Figure 9.33c). The back rows of DM columns are sized and spaced to ensure composite action between the wall elements and to provide external stability to the wall in conjunction with the relieving platform constructed from spoil at the top of the wall. The column edge-to-edge spacing should not exceed 1.2–1.5 times a column diameter. The ratio of area treatment is typically between 30 and 40 per cent. The cemented-soil relieving platform is used to tie the DM columns together to transfer the load to the bottom of the vertical columns. This external stability requirement implies the need of a site-specific minimum tensile strength and absolute continuity for the cemented-soil relieving platform (HITEC, 2002). Light steel reinforcement may be used in some of the front face columns to provide anchorage for permanent cast-in-place facing. High-quality DM and rigorous analyses are required for such retaining systems. Limitations may also result from high water level, freeze-thaw durability, surcharges or structures behind the wall, and acceptable horizontal displacements. Terashi (2003) mentioned, for instance, that some cut slopes improved by a group of columns have suffered from excessive horizontal deformation although not documented.

Measures against base heave are comprised of DM columns installed within an excavation site to act like dowels penetrating through potential sliding planes. In some cases the sides of the excavation are stabilised to increase the passive earth pressure and to reduce the penetration length of sheet piling or diaphragm walls.

DM is also applied to stabilise landslides and critical slopes. With suitable column arrangements, typically in the form of walls, grids, cells and blocks which intersect a potential failure surface, the combined shear strength of soil is improved and the factor of safety is increased (Figures 9.33d,e).

There are also novel applications comprising of soil nailing and installation of special anchors using DM. An example is the RADISH (rational dilated short) anchor 40 cm in diameter, which has been used to modify existing embankments to steep slopes (Tateyama *et al.*, 1996). Special anchors can be also installed with the Nordic method.

9.4.5 *Ground treatment*

Ground treatment works usually involve substantial volumes of unobstructed soft soils and fills to be improved on-land, at waterfront areas and offshore with relatively high area improvement ratios. Typical examples are large developing projects including the construction of roads and tunnels on soft soils, stabilisation of reclaimed areas (Figure 9.34a) or river banks, and the strengthening of sea-bottom sediments. The purpose of improvement is mainly the reduction of settlement and an increase of bearing capacity, as well as prevention of sliding failure. Novel applications include the installation of wave-impeding DM blocks of high rigidity beneath or near the foundation to reduce adverse effects caused by vibration on surrounding structures, as well as DM rings around the pile foundation of a vibrating machine (Takemiya *et al.*, 1996). Depending on the project requirements, deep and shallow soil mixing methods can be applied, including mass stabilisation.

Ground treatment works also comprise of dry and wet-method soil stabilisation to a low strength, in the order of 0.2–0.5 MPa UCS, using a reduced amount of cement and cheaper supplementary binders, like fly ash and gypsum. For the wet-method this allows increasing the amount of slurry injected into the soil, hence improving the uniformity of mixing as compared to standard DM applications using cement grout (e.g. Azuma *et al.*, 2002). High initial moisture content of the treated soil may have an adverse effect on the available compressive strength and/or hardening process after treatment, as observed in soft Finnish clay in the Old City Bay area in Helsinki (Vähäaho, 2002). As a consequence, dry mixing may be the better option for very wet soils.

Underground blocks of low-strength DM may be used to increase passive resistance and minimise heave at the bottom of excavation, allowing at the same time easy driving of sheet pilling elements or piles directly into or through the improved ground (Figure 9.34b). Moderate-strength DM can also be used to improve soft soil to allow steady digging by the shield tunnel machine, as applied for instance during construction of the Trans-Tokyo Bay highway project (design UCS 1 MPa, CDIT, 2002).

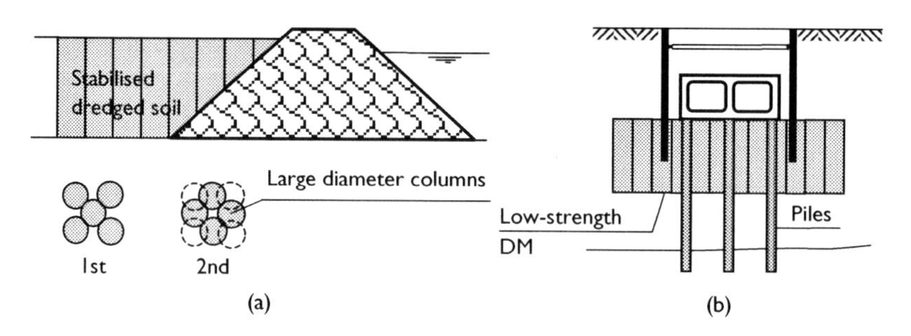

Figure 9.34 Examples of ground treatment with SM (schematic). (a) stabilisation of reclaimed area; (b) low-strength DM.

9.4.6 Liquefaction mitigation

The effectiveness of DMM to prevent liquefaction has been confirmed during the magnitude 7.2 earthquake in Kobe in 1995. A hotel under construction on a reclaimed sand area, supported on drilled piers, actually survived because the piers have been protected with a DM grid against liquefaction and the accompanying lateral flow, while the nearby seawall suffered large lateral movement towards the sea (Kamon, 1996).

Mitigation of the liquefaction potential of a site covered with loose, saturated fine soil can be provided by wall, grid and block DM patterns (Figure 9.35). The use of a grid or lattice patterns is especially effective. The 'cells' reduce shear strain and excessive build-up of pore pressure and contain local liquefied zones during seismic events, preventing lateral spreading. At the same time they can also minimise settlement and/or increase safety against slope failure. DM blocks with low-strength soil-binder mix can also be used to enable further installation of piles and underground facilities in connection with further development of the site. Column groups are generally not recommended because they may suffer from stress concentrations and bending failure.

Seismic prevention by DM to existing structures comprises also perimeter walls installed to isolate and contain liquefiable soils under the structure. The groundwater within the enclosed zone is then permanently lowered to provide non-liquefiable conditions. This solution is used where other more conventional remedial measures are not viable.

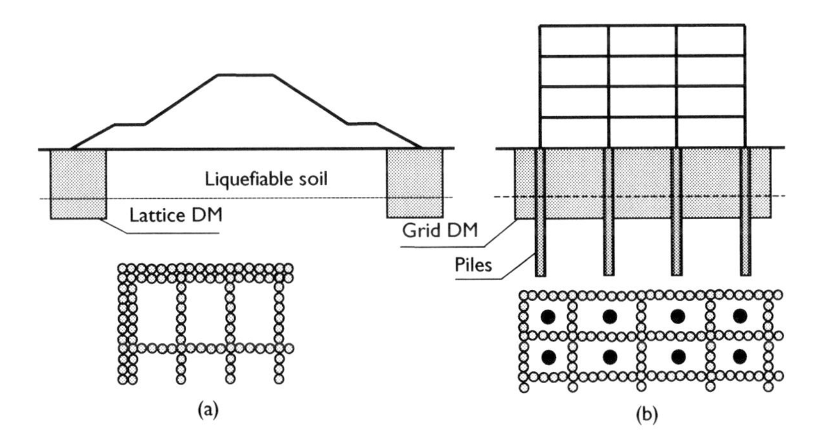

Figure 9.35 Examples of DM application for liquefaction mitigation (schematic): (a) protection of a river dyke; (b) improvement of the lateral resistance of piles.

9.4.7 Hydraulic cut-off walls

Hydraulic cut-offs are constructed by installing DM walls to intercept the seepage flow paths. The columns/panels are typically installed through the permeable strata to some cut-off level, usually penetrating 0.5–1 m into a clay layer or finishing at the top of the bedrock. The soils treated are generally highly permeable coarse deposits, or interbedded strata of fine and coarse-grained soils.

The applications mainly involve rehabilitation and/or upgrading of older water-retaining structures to meet new regulations for safe operation. Typical examples are earth-fill dams, dyke embankments and river banks (Figure 9.36). In the case of excavations, the supporting DM walls may additionally serve to prevent seepage of groundwater towards the pit. When a conventional elevation of a river dyke crest is not possible, steel H-beams can be installed in DM columns to support concrete superstructures or light dismountable protection walls on the crest to prevent overtopping (e.g. Topolnicki, 2003).

Since the hydraulic conductivity and continuity of the cut-off wall are of primary importance, careful design of slurry mixes tailored to soil conditions, and adequate control of overlapping zones and verticality are required, especially when cut-off walls are executed to a large depth with single shaft mixing equipment. For DM walls the unconfined compressive strength is typically in the range of 0.5–3 MPa, and higher if steel reinforcement is installed, while the permeability is normally between 10^{-8} and 10^{-9} m/s. When bentonite and/or clayey stone dust and/or fly ash are added to the slurry mix the permeability can be reduced

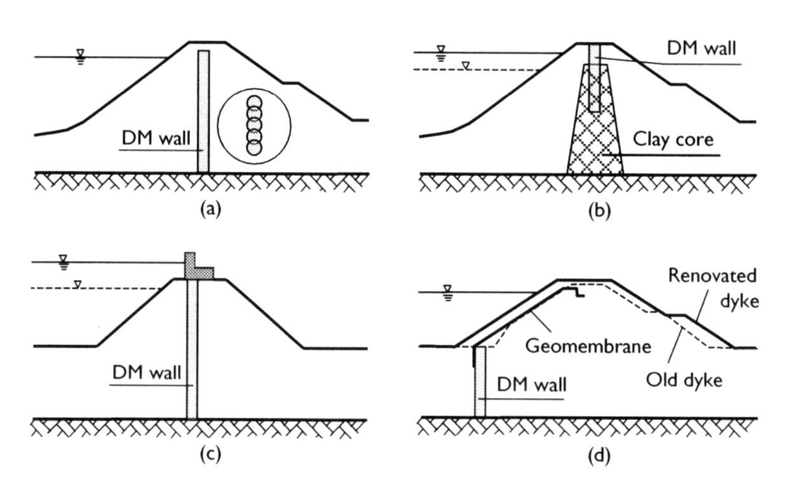

Figure 9.36 Examples of DM applications for cut-off walls (schematic): (a) dam sealing; (b) extension of the clay core; (c) river dyke with superstructure on the crest; (d) seepage protection.

to 10^{-9}–10^{-10} m/s, with associated decrease of the unconfined compressive strength.

Related case histories may be found in: Yang and Takeshima (1994), Walker (1994), Nagata *et al.* (1994), Schwarz and Seidel (2003).

9.4.8 Environmental remediation

Environmental applications emerged in the last 10 years and mainly involve installation of containment barriers and solidification/stabilisation of contaminated soils and sludges. Fixation is much harder to achieve, as it requires contact of the chemical reagent with the contaminant. This is easier in sandy soils but very tough in clayey soils. At an experimental level, soil mixing has also been used to introduce micro-organisms-based grout for bio-remediation purposes, acid/base reagents for neutralisation, and oxidation reagents for chemical reaction.

Soil mixing containment systems include passive and active type barriers constructed around a part or a whole periphery of the contaminated site. Passive barriers resemble hydraulic cut-off walls and are installed to prevent migration of polluted leachates out of the contaminated site. Active barriers have permeability comparable to the native soil. They are typically constructed as 'gates' in passive barriers to reduce significant effects of the containment on the existing groundwater regime (Figure 9.37). With appropriate soil-mixed materials, such as modified alumina silicates, and adsorbance capacities, gates act as microchemical sieves, removing contaminants from groundwater as it passes through and allowing, in principle, only clean water to emerge on the other side. Four recent case histories covering this concept have been reported by Al-Tabbaa and Evans (2002). The DM containment barriers are suitable for existing waste disposal dumps and

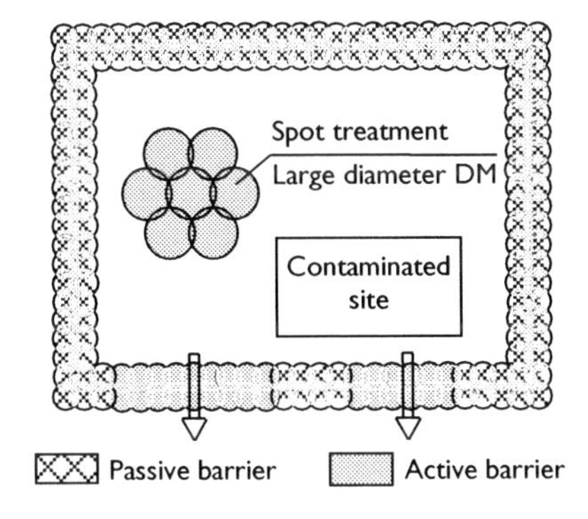

Figure 9.37 Containment system consisting of passive and active DM barriers (schematic).

new landfill facilities. However, grout composition and binder reactions with the contaminants in the short and long-term perspective are key factors in the success of such applications.

Solidification/stabilisation of contaminated soils and sludges containing metals, semi-volatile organic compounds and low-level radioactive materials using wet and dry-method soil mixing started to be recognised as a favoured remediation option because of the advantages over other containment and remediation methods. These include reduced health and safety risks, elimination of off-site disposal, low cost, and speed of implementation. By selecting appropriate equipment and procedures, the reagents can be uniformly injected at depth, and efficiently and reliably mixed with the soil or sludge present. In the case of soil contaminated with volatile compounds, negative pressure is kept under a hood placed over the mixing tool to pull any vapours or dust into the vapour treatment system.

Related case histories can be found in Walker (1992), Aldridge and Naguib (1992), Jasperse and Ryan (1992), Hidetoshi *et al.* (1996), Lebon (2002).

9.4.9 *Advantages and limitations*

When evaluating possible advantages and limitations of the use of soil mixing for each of the six application areas mentioned above, it should be remembered that different soil mixing methods and machines provide different types of treated soil geometries and treated soil parameters. Therefore, one or a few particular soil mixing variants may be practically feasible for consideration under each application, leading to a stimulating competition of solutions also within the general SM technology.

However, in terms of general characterisation of *in situ* soil mixing, as well as relative comparison with other competitive technologies of deep ground improvement, several advantages and limitations of the use of DMM have been recognised and summarised in Table 9.3, based on a similar evaluation conducted by Bruce (FHWA, 2000).

9.5 Design

9.5.1 *General design procedure*

The current design procedures for DM works, mainly developed in Japan for wet and dry mixing and in Sweden for the Nordic method, are to a large extent based on accumulated experience. They are still being modified in accordance with new findings from numerous research studies and developments of the execution methods. In general

Table 9.3 Main advantages and limitations of the use of Deep Mixing Method

Main advantages of DMM	*Main limitations of DMM*
High productivity usually possible, hence economical for large scale projects	Depth limitations (depending on the method applied)
Can be potentially used in all types of soils and fills (without obstructions)	Not applicable in soils that are very dense, very stiff, or that may have boulders
Column's spacing and patterns highly variable, arrangements tailored to specific needs	Limited or no ability to install inclined columns (depending on the equipment applied)
Engineering properties of treated soil can be closely designed	Uniformity and quality of mixed soil may vary considerably in certain conditions
Causes minimal lateral or vertical stress that could potentially damage adjacent structures	Columns cannot be installed in close proximity to existing structures (except hybrid mixing)
No vibration, medium–low noise	Freeze/thaw degradation may occur
Very low spoil (dry-method)	Significant spoil produced with the wet-method
Can be used for on-land, waterfront and marine projects	Weight of the equipment may be problematic for weak soils (depending on the method)
Quality of treatment verifiable during construction	Air pressure or grout injection pressure may cause heave
Minimum environmental impact	Limited ability to treat isolated strata at depth

terms, however, the planning of DM application involves the following main steps:

1 selection of a suitable DM process (i.e. wet or dry) and the construction method,
2 selection of the strength of stabilised soil in specific ground conditions (mix design),
3 selection of the installation pattern and dimensions of improved ground (geotechnical design).

Clearly, all three steps are closely interrelated and iterative procedure is usually needed to achieve the full benefit of the planned DM application (cf. Porbaha, 2000). Laboratory tests, model investigations and even field trials are often conducted to assist this selection process.

In the first step the most suitable treatment process (dry or wet) and binder should be considered taking into account actual soil conditions, site constraints, functional requirements of the structure to be built as well as related economical aspects of the design. At this stage possible advantages and disadvantages of the application of DMM in comparison to other competitive technologies should also be examined.

The expected 'minimum' strength of stabilised soil is selected next in relation to the treatment process, type and amount of binder(s), and working specifications such as the rate of penetration and withdrawal, rotation speed, injection method, mixing tool, water/binder ratio, etc. At this stage a good understanding of the intricacies of soil-binder physics, chemistry and mechanics is required in view of the variability of soils and stabilisation agents. In all cases where the DM columns are much stiffer than the surrounding soil there is a stress concentration on these elements and the internal stability of soil-binder mix starts to be a dominating aspect of the design. The required strength of stabilised soil must therefore be carefully selected, taking into account functional requirements, and inevitable and significant scatter of the field strength. It should also be remembered that the quality of soil-binder mix directly impacts the cost of treatment as it will govern the mixing energy required as well as the binder content and type necessary. Specifying a quality higher than actually needed for acceptable performance can significantly impact the cost of ground improvement.

The geotechnical design determines the installation pattern and the dimensions of the improved ground to satisfy internal and external stability conditions. In the current practice the traditional approach based on the allowable-stress concept is generally used at this stage. Furthermore, displacements of the superstructure are examined. On the other hand, the limit state design (LSD) approach is being progressively adopted in civil and geotechnical engineering in Europe (e.g. Eurocode 7). Also, in Japan a new reliability-based standard for DMM is scheduled for 2006 (Kitazume, 2002). To satisfy ultimate limit states (ULS) requirements, the design of stabilised ground must be such that there is a sufficiently low possibility of collapse of the supported structure with respect to all potential failure modes, including internal and external stability and progressive failure. To satisfy serviceability limit states (SLS) requirements, soil mixing shall be designed in such a way that the total and the differential displacements, including long-term creep movements, are below accepted limit values. Although the failure modes and design formulas remain the same, this new approach cannot be used in practice until many partial safety factors (requiring lengthy calibration work) are established.

In the following part of this section selected issues associated with the analysis and design of DM are considered, complementary to information provided in Chapter 8, especially in relation to possible modes of failure and design of composite ground.

9.5.2 The choice between dry and wet process of soil mixing

As it is with other ground improvement technologies or foundation methods, selection of the most suitable process of deep soil mixing is a matter of somewhat arbitrary engineering judgement. In such an evaluation various technical and economical aspects of both processes should be considered in relation to the type of structure and soil for which DM will be used, including all site-specific conditions and availability of the equipment. Although in practice the decisions in favour of the dry or the wet process usually have a rational background, it is rather difficult to detect distinct application limits and to provide widely valid criteria for selection purposes. Instead, few characteristic features of both processes may be appointed to enable easier selection of the most suitable DM process in the initial stage of design.

From the point of view of engineering properties of soils stabilised with the same type of binder applied in dry or slurry form there is no substantial difference between both mixing processes, as also underlined by Terashi (2003). This has been demonstrated by numerous investigations conducted on laboratory-mixed soils, but is difficult to repeat in field conditions. For example, when a sufficient, but not too high amount of cement in powder or slurry form is thoroughly mixed with a soft soil, higher compressive strength is expected for the dry than for the wet process due to lower value of the resulting water/cement ratio. To obtain a comparable strength with the wet process the amount of immixed cement must be therefore increased, which is usually easily managed at the construction site although more spoil is inevitably produced. However, when still higher strength is requested the amount of cement has to be increased for both methods. In such case the mixing conditions *in situ* start to play a dominant role in the stirring process since it is much easier to dispense and to mix cement slurry with the soil than dry cement, especially when the natural water content of the soil is low and the shear strength is high. Furthermore, problems with complete recovery of the increased amount of air from the ground may also appear in the dry process. Decreasing efficiency of dry mixing will directly affect the homogeneity of DM columns, increasing strength and stiffness variation of stabilised soil. This explains why in the current design practice and for 'normal' soft soils, lower strengths of stabilised soil are adopted for the dry rather than for the wet process. In the case of the wet process, the amount of cement injected is usually between 100 and 500 kg/m^3 of soil to be treated and the strength target is typically between 0.3 and 2 MPa, based on the four-week unconfined compressive strength. For the dry process applied in soils with 40 per cent to more than 200 per cent natural moisture content the amount of cement used is typically 100–300 kg/m^3, with a compressive strength target of about 0.5 MPa for the Japanese DJM method using mainly cement or cement-based binders, and 0.15 MPa shear strength for the Nordic method using mostly combined cement/lime binders. On the other hand, for very soft soils with very high moisture content, reasonable strength gain is easier and more effective using dry binders. This illustrates that the initial moisture content of the soil to be treated, besides the penetration resistance of the ground and the purpose of ground improvement, is a major factor affecting the choice between the dry and the wet process.

A summary of most distinct characteristics of both mixing processes which are relevant for selection purposes is presented in Table 9.4. Related case histories where the applicability of the dry or the wet process was actually investigated by means of field trials are not yet very common. However, three recent case studies are excellent examples of such an

Table 9.4 Characteristic features affecting the choice of dry or wet soil mixing

Item of concern	*Expectations*
Initial water content of the soil to be treated	Cohesive soils with moisture content $w = 60–200\%$ are best suited for the dry process (lower limit $w > 20\%$, note that water content below plastic limit is not fully available for hydration); for soils with very high w dry process is more effective than wet
Quality of mixing	Wet process usually provides better homogeneity of stabilised soil because of longer mixing time, pre-hydration of cement and easier distribution of slurry across the column area
Compressive strength of soil–binder mix	Higher strength is more reliably obtained with the wet process, except for very wet soils
Ability to penetrate through hard soil layers	Much higher for the wet process due to the 'lubrication' effect of the injected slurry and due to higher torque capacity of mixing shafts
Stratified soils	Wet mixing can provide more uniform strength along the column length due to partial soil exchange/movement in the vertical profile; quality control more difficult for the dry process
Spoil	Dry mixing creates very little or no spoil
Use of combined binders and industrial by-products	Quite frequent in dry mixing; slag cement often used in wet mixing, other binders and by-products very rare in wet mixing
Air temperature below 0 °C	Dry process is significantly less affected by low temperatures since compressed air is used to transport the binder
Column reinforcement	Possible in combination with the wet process
Relative unit cost of treatment	Dry mixing is generally less expensive than wet mixing

approach. Huiden (1999) describes field trials for the Botlek Railwaytunnel in The Netherlands, partly embedded in a mass of stabilised soil. The subsoil was very heterogeneous, consisting of peat, clay, silt and sand. Shiells *et al.* (2003) report about trial installations for the ongoing project at the Woodrow Wilson Bridge in Virginia, comprising *c.* 135 000 m^3 of deep soil mixing at the I-95/Route 1 Interchange. The subsoil is built of highly compressible organic silts and clays of 3–12 m thickness. The organic content of the alluvial clay soils commonly varies between 5 and 15 per cent, with maximum organic content approaching 50 per cent at isolated locations. In both cases the wet process was finally chosen. Vähäaho (2002) describes field trials of DM in the Old City Bay area in Helsinki. The area is part of an old sea bed, reclaimed in the 1960s and 1980s. In this case wet and dry mixing was tested in order to stabilise very weak deposit of organic clay and clay, with about 2–3 MPa cone tip resistance. A very high shear strength requirement, set to 750 kPa in clay at 10–19 m depth, was fulfiled only with the dry-method (1300 kPa in average), although a clear asymmetry in the distribution of binder in column cross-section was noticed. The shear strength obtained with the wet-method averaged 230–340 kPa. It should be noted that in all reported cases the final choice was dictated by technical requirements.

9.5.3 *Engineering properties of stabilised soil*

The most common engineering parameters of the stabilised soil that are measured and/or inferred on DM projects and used in the design analyses comprise compressive, shear and tensile strengths, modulus of elasticity, unit density and permeability (hydraulic conductivity). For ground improvement applications with low-strength DM, the coefficient of consolidation and the coefficient of volume compressibility of treated soil also may be important for settlement prediction. In other typical areas of application, however, the working stresses acting on the DM columns are far below the consolidation yield pressure, which for cement-treated soils is typically about 30 per cent higher than the unconfined compressive strength. In a sort of overconsolidated state the coefficient of consolidation of soil–cement is improved at least 5–10 times and the coefficient of volume compressibility is improved to 1/10 or less (CDIT, 2002). Therefore the consolidation settlement of cement-treated ground is small and usually of secondary importance. For dynamic analyses, including seismic, cyclic and repeated loading excitations, assessment of dynamic properties is required, as for instance the shear modulus and damping at different strain levels. More information on dynamic properties can be found in the state-of-the-art paper of Porbaha *et al.* (2000).

Low hydraulic conductivities are of primary concern for cut-off walls and retention systems involving groundwater control, as well as for most environmental applications. Hydraulic conductivity is also important when dealing with long-term behaviour of DM columns installed in aggressive soil–water environments. In other areas of DM application the actual permeability of soil–binder mix should be compared to the permeability of soil before treatment to inspect whether the installed columns are likely to act in the ground as vertical drains or as semi-permeable or (almost) impermeable soldier elements. For instance, significantly increased permeabilities are usually observed after treatment when quick lime is used as stabilising agent in soft clayey soils, with a tendency to increase further in time. In this case, a combined action of the DM columns in the ground must be considered, including strengthening and drainage function. Contrary to that, stabilisation with cement usually effectively reduces initial permeabilities and the treated soils are practically impermeable. Therefore, such columns are not expected to function as drainage elements.

In the case of combined lime/cement treatment, the permeabilities are generally also low, with a tendency to decrease in time and with increasing confining pressure, while the actual permeability of soil–binder mix will depend greatly on the site-specific conditions. It is thus evident that the resulting permeability of stabilised soil will significantly affect the interaction pattern between the treated and untreated ground, and consequently, the design approach.

The initial wet density of the soil may slightly increase or decrease after treatment. Sample Japanese data presented in CDIT (2002) and referring to field investigations indicate that soil density after treatment may change within ±5 per cent in case of stabilisation with quick lime, or increase 3–15 per cent if cement in dry form is used (in peat, an even higher increase was also observed). In the case of the wet process, the density was found to be almost unchanged after treatment, irrespectively of the mass of cement admixed per m^3 of soil within 50–250 kg/m^3 range. Similar observations are known from numerous jet grouting works. However, lower densities after treatment are also occasionally observed, especially on wet grab samples (e.g. O'Rourke *et al.*, 1997, cited after FHWA, 2001). Since the expected changes of soil density are generally small, it is often assumed for design purposes that the wet soil density is not affected by the *in situ* treatment. More critical evaluation is recommended if the soil weight plays a significant role in the design, as for instance in the uplift stability analysis.

Three other basic parameters, i.e.: the shear and tensile strengths and the modulus of elasticity, can be correlated with reasonable accuracy to the compressive strength of stabilised soil. Consequently, the compressive strength, and in practice, the unconfined compressive strength (UCS) due to the simplicity and cost-effectiveness of testing, is the key parameter for the current design practice. In Scandinavia, the shear strength is equivalently used, assuming to be half of the UCS.

At present, the unconfined compressive strength of soil–binder mix cannot be reliably forecasted on the basis of properties of the native soil, and the type and amount of admixed binder(s). Therefore, it is generally recommended that advance appropriate trial tests be conducted on stabilised soils to obtain more adequate data regarding UCS for each project. These pilot investigations typically include testing of laboratory-mixed samples, but may also involve full-scale trials in the cases of more challenging designs. At this stage, it is a common practise to inspect the UCS in relation to the binder factor, $\alpha[\mathrm{kg/m^3}]$, expressed as the weight of injected dry binder divided by the volume of ground to be treated. The weights can refer to the weight of binder used in dry-methods, or the weight of binder used in the slurry in wet-methods. It should be kept in mind, however, that in field situations the injected binder quantities may not necessarily be those that actually remain in place. Therefore the established correlations should be critically evaluated before being used for optimisation purposes. Likely field strength can be also estimated on the basis of accumulated experience from previous projects, and by exercising engineering judgment.

The expected field strengths and permeabilities for ranges of cement factors and different soils are listed in Table 9.5. The corresponding volume ratios, defined as the ratio of the volume of slurry injected to the volume of ground to be treated, vary greatly and reflect the type of DM technique used, but are in the range of 15–50 per cent (25–40 per cent in most cases). Lower volume ratios can be applied when the efficiency of mixing is high, due to higher rotational speed or jet assistance. The data presented in Table 9.5 are useful for an early assessment of technical and economical aspects of the DM design. With additional cement, the strengths generally increase. However, laboratory tests have also indicated that in some clays additional cement dosage will not increase UCS values. In such soils, blast furnace slag has proven to be very effective. Generally, with increased cement dosage,

Table 9.5 Typical field strength and permeability for ranges of cement factors and soil types (data based on soils stabilised with the wet process[1])

Soil type	*Cement factor,* α (kg/m^3)	*UCS 28 days,* q_{uf} (MPa)	*Permeability,* *k* (m/s)
Sludge	250–400	0.1–0.4	1×10^{-8}
Peat, organic silts/clays	150–350	0.2–1.2	5×10^{-9}
Soft clays	150–300	0.5–1.7	5×10^{-9}
Medium/hard clays	120–300	0.7–2.5	5×10^{-9}
Silts and silty sands	120–300	1.0–3.0	1×10^{-8}
Fine-medium sands	120–300	1.5–5.0	5×10^{-8}
Coarse sands and gravels	120–250	3.0–7.0	1×10^{-7}

Note

1 Data compiled from Geo-Con., Inc. (1998), FHWA (2001), Keller Group.

permeabilities decrease, but not to the order of magnitude required for effective cut-off barriers. For this purpose, bentonite or other proprietary reagents should be used to lower the hydraulic conductivity in a more effective way. The amount of spoil increases as the volume ratio increases.

There are numerous factors that can affect the strength increase of *in situ* treated soil, as well as the quality and reliability of data collected and/or reported on soil-binder mix strength. The most important factors are briefly summarised in Table 9.6.

A great deal of valuable published information is available on the relative importance of the factors indicated in Table 9.6, and on the mechanical behaviour of stabilised soils, based on dedicated laboratory and field studies and performance observations (extensive bibliographies can be found in the state-of-the-art reports of Porbaha *et al.*, 2000, FHWA, 2001 and CDIT, 2002). However, the majority of these studies have been conducted on laboratory-mixed soils, actually violating *in situ* mixing and curing conditions except for the amount of binder and the curing time. Field investigations have been mostly carried out to solve site-specific problems and/or to provide quality evidence of the executed works, and therefore often have inherent limitations. Data on long-time performance of DM columns are still insufficient, although generally show promising results (see Section 9.5.4).

Table 9.6 Main factors affecting the observed strength of stabilised soil

Source	*Specific items*
Physical and chemical properties of the soil to be treated	Grain size distribution, mineralogy, natural water content, Atteberg limits, organic matter content, reactivity and pH of pore water
Binder, additives and process water	Type and quality of hardening agent(s), binder composition, quantity of binder and other additives, quality of mixing water
Installation technique and mixing conditions	Tool geometry, installation process, water/binder ratio, energy of mixing (speed and period), time lag between overlaps and working shifts
Curing conditions, time	Curing time, temperature (heat of hydration in relation to treated volume), humidity, wetting/drying and freezing/thawing cycles, long-term strength gain and/or deterioration
Testing and sampling	Choice of testing method, type of test, sampling technique, sample size, testing conditions (stress path and drainage conditions, confining pressure, strain rate, method of strain measurement)

It is therefore rather difficult and challenging to compare extremely detailed experimental data and to assess the real mechanical behaviour of soil–binder mix, especially in view of changing conditions of mixing *in situ* and a variety of native soils. On the other hand it is believed that a good understanding of a generalised behaviour of stabilised soils is needed to meet static and functional requirements of any DM design. Therefore, a synthesised overview of the most pronounced relationships between selected 'input factors' and expected 'responses' of cement-treated soils, as revealed by published experimental evidence, is presented in Table 9.7. Obviously, for any specific project these qualitative relationships must be validated and carefully quantified before being used during design or construction stage.

Bearing in mind that the unconfined compressive strength of stabilised soil is a result of many variables, including construction variability itself, useful relationships and data for practical design have been compiled in Table 9.8. In general, laboratory and field investigations show reasonable correlation between early UCS of 4 or 7 days, and strengths observed at longer cure times. On this basis, relatively quick assessment of the expected strength after 28 or 56 days can be made to confirm initial assumptions. Prediction of soil–binder strength prior to construction is very often based on laboratory-mixed samples and correlations established between the UCS of laboratory and *in situ* mixed soils. Due to inherent limitations of laboratory-prepared samples, discussed above, such strength data must be applied with appropriate safety margins and considered rather as an index of the actual field strength than as a precise prediction.

The information on the secant stiffness modulus and on the axial strain at failure, presented in Table 9.8, gives only a crude engineering estimation of very complex stress–strain behaviour of stabilised soil. For dry mixing with cement the ratio E_{50}/UCS is somewhat lower and in a tighter range than for the wet process, being roughly 25–50 for compressive strength less than 0.3 MPa and 50–200 for strength of 0.3–2 MPa. As the cement factor increases, soil–cement becomes stiffer and more brittle. It should be noted, however, that only in unconfined compression does soil–cement lose nearly all its strength at strains beyond peak strength. In field conditions, when surrounded by untreated soil or a large mass of treated ground, and when loaded axially, it will rather exhibit a ductile behaviour such that sudden breakage or failure does not occur. In undrained triaxial compression, for example, a clear shear band develops and most of the soil–cement strength is actually retained at larger strains, depending on the confining stress (e.g. Yu *et al.*, 1997).

The stiffness modulus of stabilised soil is also highly dependent on the strain-level. When a DM column is stressed in the field only to a low portion of the peak strength, the secant modulus may be unrealistically small, leading to overprediction of deformations. In such cases it may be more appropriate to use the initial stiffness modulus in design analyses.

Table 9.7 A summary of generally observed relationships for cement-treated soils

Factor of influence	*Expected reaction on stabilised soil*
Granular soils	Increase strength and allow reduction of the cement factor, shorten curing time to reach the design strength, simplify distribution of cement throughout the soil, impede very low permeabilities
Clayey soils	Reduce strength and require higher cement factors than sands, slow the rate of initial strength gain compared to sands, involve pozzolanic reaction and strength growth over elongated time (with different rates), impede uniform distribution of agents throughout the soil, enhance low permeabilities of treated soil
Fine soil fractions	The smallest 25% of particle size controls strength, silty sands have noticeably lower strength than clean sands
Natural water content	Compressive strength decreases almost linearly with increasing water content; flow of groundwater may cause cement washout
Organic matter, low pH	Significant negative impact on strength; soils with organic contents over 6% and having pH < 5, difficult to improve
Cement factor, α (typical range 100–400 kg/m^3)	In silts and clays: almost linear strength gain with increasing α, in sands and gravels: over proportional strength increase with increasing α, especially for higher cement factors; higher α improve durability and decrease permeability
Cement type	In clays long-term strength gain is higher for blast furnace slag cement than for Portland cement
Water/Cement ratio, W/C (typical range 0.8–1.2, band extends from 0.5 to 2.5)	Increasing W/C ratios more directly decrease strength than α, higher W/C ratios slow the rate of hydration-related strength gain and lower long-term strength gain beyond 28 days; low W/C ratios minimise extra water, higher promote mixing
Additives, e.g. dispersants, retarders, anti-washout agents, etc.	Change fluid and set properties of slurry mix; prevent binder washout in dynamic water situations
Substitutes for cement (used alone or with cement), e.g.:	Significantly influence all properties, but rare in wet mixing (except for cut-off walls and environmental applications),
– bentonite, clay	– improve stability of high W/C slurries, reduce permeability
– slag	– improve chemical stability and durability, retards strength gain
– kiln dust	– used in environmental applications
– flyash	– increases chemical durability, reduces heat of hydration
– lime, gypsum	– used in low-strength DM
– silicates, polymers, etc.	– used in special environmental applications

Air entrainment (in jet mixing)	Lowers strength, may increase freeze-thaw resistance
Mixing operation	Mixing efficiency improves with higher rotary speeds, is usually easier with thinner blades; UCS improves and variation of strength decreases with increasing blade rotation number
Installation process	High volume ratios cause high volume of spoil; weaker strength observed in overlapping zones, when separated in construction by considerable time
Sampling	Good-quality core samples have often higher strength than wet grab samples; small samples usually yield higher strength than bigger samples

Table 9.8 Typical correlations and data for cement-treated soils using the wet process[1]

Selected parameters	*Expected values/ratios or relationships*
UCS – rate of strength gain	28 days UCS = *c.* 2 × 4 days strength 28 days UCS = 1.4–1.5 × 7 days strength (silts, clays) 28 days UCS = 1.5–2 × 7 days strength (sands) 56 days UCS = 1.4–1.5 × 28 days strength (clays, silts) long-term strength increase typically observed
UCS – coefficient of variation, ν (COV)	0.2–0.6 (typically 0.35–0.5), COV is lower for laboratory-mixed samples than for field samples
UCS – relative strength ratios:	
– core samples to laboratory-mixed samples, λ,	0.5–1, lower values for clays, higher for sands (1.0 – for offshore works in Japan)
– core samples to wet grab samples	1–1.5
Shear strength (direct shear, no normal stress)	0.4–0.5 × UCS, for UCS < 1 MPa 0.3–0.35 × UCS, for 1 < UCS < 4 MPa 0.2 × UCS, for UCS > 4 MPa
Tensile strength	0.08–0.15 × UCS, but not higher than 200 kPa, indirect splitting tests yield lower values than direct uniaxial tests
Secant stiffness modulus, E_{50}, at 50% peak strength	50–300 × UCS, for UCS < 2 MPa 300–1000 × UCS, for UCS > 2 MPa (ratio increases with increasing UCS)

Table 9.8 (Continued)

Selected parameters	*Expected values/ratios or relationships*
Axial strain at failure, ε_f:	
– unconfined compression test (crushing failure)	0.5–1.0% for UCS > 1 MPa 1–3% for UCS < 1 MPa,
– confined compression tests (plastic shear failure)	2–5% (undrained triaxial test)
Poisson's ratio	0.25–0.45, typically 0.3–0.4
Unit density	No noticeable relationship with UCS

Note

1 Data compiled from CDM (1996), BCJ (1997), Porbaha *et al.* (2000), FHWA (2001), CDIT (2002), Matuso (2002).

9.5.4 Long-term strength gain and deterioration of stabilised soil

For a rigorous design of DM it is necessary to collect data on long-term behaviour of stabilised soils. Terashi (2002b) has recently summarised Japanese field investigations concerning two aspects of long-term characteristics of stabilised ground with respect to lime and cement-treated soils using both wet and dry processes.

One aspect is the strength gain in the long-term. The data presented in Table 9.9 reveals that stabilised soils exhibit significant strength increases, although the rate of strength increase strongly varies with different case records. Roughly, a strength increase of 2–3 times may be expected during 10–20 years after treatment. Stronger increase is observed for the dry process, while increasing W/C ratios may impede long-term strength gain. No substantial change with time of the unit density and moisture content of treated soil was noticed. In the current design practice, possible long-term gain of the compressive strength, beyond a 90-day period, has not been yet accounted for.

The second aspect is the long-term deterioration of stabilised soil exposed to different environmental conditions, like untreated soil, fresh water, salt water, polluted ground, etc. The possibility of deterioration was first addressed by Terashi *et al.* in 1983. They found that a slow deterioration process starts from the exposed boundary of treated soil and progresses inwards. Strength reduction is associated with leaching of Ca dissolved from hydrated cement component into pore solution of treated soil, and then migrating to the surrounding environment. Saitoh (1988) observed that the rate of deterioration depends mostly on the strength of treated soil and partly on the type of soil and binder. Existing experimental evidence from laboratory and field observations suggests that the rate of deterioration is almost linear with logarithm time. The depth of deterioration is smaller for soil with a higher cement factor. For the treated soil at Daikoku Wharf, the depth of deterioration is 30–50 mm over the past 20 years (Terashi, 2002b).

Table 9.9 Long-term strength gain based on Japanese field data[1]

Mixing process	*Soil type*	*Binder*	*UCS and standard deviation, s_d, short-term*	*UCS and standard deviation, s_d, long-term*	*UCS ratio*
Dry deep mixing	Reclaimed clay, 60% clay, 39% silt, 1% sand	Quick lime, 12.5% dry weight	64 days: 1.02 MPa mean, $s_d = 0.2$ MPa	11 years: 3.5 MPa mean, $s_d = 0.78$ MPa	3.4
Dry deep mixing	Upper volcanic ash, underlined by peat, clay, silty fine sand and silt	Blast furnace slag cement type B, 290 kg/m^3 upper 3 m, 130 kg/m^3 lower 5 m	28 days: 0.2–0.5 MPa at 0–1 m depth,	17 years: 1.5 MPa mean $s_d = 0.98$ MPa	>3
			0.2–0.5 MPa at 4–6 m depth	0.53 MPa mean, $s_d = 0.35$ MPa	1.5
Dry deep mixing	4.5 m peat layer $w = 300$–500%, 5.5 m organic clay $w = 150$–200%	Cement-type agent, 265 kg/m^3	28 days: 0.58 MPa mean	14 years: 3.5 MPa mean	6.0
Wet shallow mixing	Reclaimed clay, 30% clay, 70% silt	Cement-type agent W/C = 1.5, 5% wet weight	21 days: 74 kPa mean	15 years: 220 kPa mean $s_d = 139$ kPa	3.0
Wet deep mixing	Deep marine clay (offshore)	Cement	93 days: 6.1 MPa mean, $s_d = 2.0$ MPa	20 years: 13.2 MPa mean, $s_d = 5.19$ MPa	2.2

Note

1 Data compiled from Terashi (2002b), based on the investigations of Terashi and Kitazume (1992), Yoshida *et al.* (1992), Hayashi *et al.* (2003), Inagaki *et al.* (2002), Ikegami *et al.* (2002).

In view of the above investigations it may be tacitly assumed for current design practice and unpolluted soils, as long as new information are available, that long-term strength gain and deterioration compensate.

9.5.5 *Design strength of stabilised soil*

A common feature of all DM applications is a large scatter of the field strength of stabilised soil. A rational selection of the design compressive strength of soil–binder mix should be therefore based on a statistical approach. According to accumulated experimental evidence, it can be assumed, with reasonable accuracy, that the UCS data fits a normal distribution curve (e.g. BCJ, 1997; Matuso, 2002). Bearing in mind the limitations of the unconfined compression test to represent the actual field strength of stabilised soil, mentioned in Section 9.5.3, the design compressive strength, f_c, may be related to the mean, $\overline{q}_{uf}$, and the standard deviation, s_d, of the strength of field samples using the following equation (see Figure 9.38):

$$f_c = \overline{q}_{uf} \quad ms_d \tag{9.5}$$

The m-value determines the confidence that any measured $q_{uf} \geq f_c$ (f_c is sometimes referred to as guaranteed strength). For a relatively high confidence level of 95 per cent, as is usually applied in the case of a structural concrete, m is equal to 1.64. For wet soil mixing using cement $m = 1.3$ has been recommended in Japan by BCJ (1997), corresponding to 90 per cent of confidence.

Introducing the coefficient of variation (COV), $v = s_d/\overline{q}_{uf}$, equation (9.5) can be rewritten as:

$$f_c = (1 - mv)\overline{q}_{uf} = \eta_1 \overline{q}_{uf} \tag{9.6}$$

where η_1 is a measure of the scattered strength. Based on BCJ (1997) data, Taki (2003) reported that the average ratio between compressive strengths of DM columns, evaluated from 26 case histories of strength tests performed on full-scale columns of 0.6–1 m diameter, and the mean UCS of the cores obtained from these columns was $\eta_1 = 0.62$ for cohesive soils and

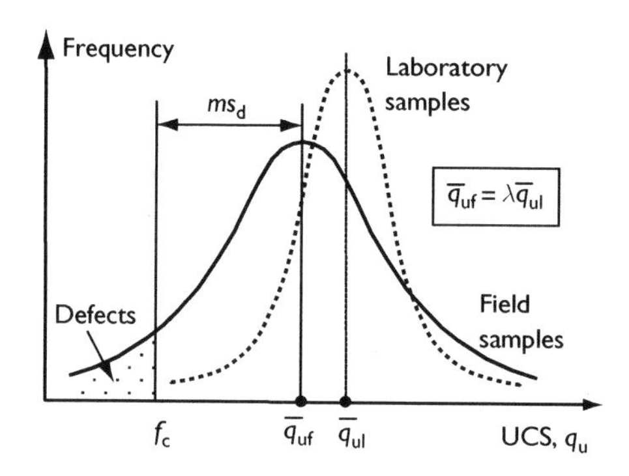

Figure 9.38 Normal distribution curves for field and laboratory strength data and assessment of the design strength f_c.

$\eta_1 = 0.80$ for cohesionless soils (mean for all soils was 0.69; CDIT, 2002). Both values are relatively high due to rather low COV, equal to 0.26 and 0.18 respectively. For a more conservative estimation Taki proposes to use $\eta_1 = 0.5$ for cohesive soils and $\eta_1 = 0.64$ for cohesionless soils, which corresponds well with the design recommendation of CDIT (2002) quoting $\eta_1 = 0.5$–0.6. Applying the latter range of η_1 and assuming that $m = 1.3$ is a reasonable choice for the acceptance criterion, it can be back-calculated that the corresponding coefficients of variation should not exceed $\nu = 0.38$ for cohesive soils and $\nu = 0.31$ for cohesionless soils. Although both variation coefficients fall into the observed range specified in Table 9.8, a relatively high degree of mixing must be assured to fulfil these criteria. It should be noted that the reported field data of BCJ are average values of many different mixing methods. For a specific project, the level of confidence, and especially the expected coefficient of variation, should be carefully assigned in relation to the type of support and the selected method of treatment to obtain a reliable estimate of the field design strength.

Equation (9.6) can be used directly in the case of non-overlapping execution of DM columns. If intersecting columns are installed, the strength in the overlapped area may be lower than in other parts of the columns, depending on the time interval until overlapping, execution capacity of the DM machine and the type of binder used. The corresponding reliability coefficient of overlapping, η_2, is typically set in Japan in the range of 0.8–0.9 (CDIT, 2002). Furthermore, when the improved ground is composed of serial overlapping columns, small areas of untreated soil remain enclosed between the joint columns (Figure 9.39). In this case a correction factor for the effective width of treated column, η_3, is used to compensate for the untreated part. Typical values of η_3 are within 0.7–0.9 (CDIT, 2002), depending on the applied installation pattern.

With the additional correction factors, equation (9.6) reads:

$$f_c = \eta \bar{q}_{uf}, \quad \text{where } \eta = \eta_1 \eta_2 \eta_3 \tag{9.7}$$

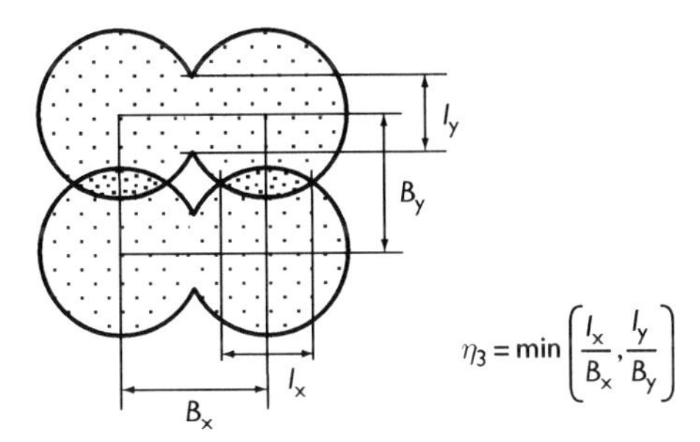

Figure 9.39 Correction factor for the effective width according to CDIT (2002).

If the mean UCS of laboratory-mixed specimens, $\overline{q}_{ul}$, is used in equation (9.7) instead of $\overline{q}_{uf}$, then the following practical relationship for an early estimation of the design field strength is obtained (cf. Figure 9.38):

$$f_c = \eta\lambda\overline{q}_{ul} \tag{9.8}$$

where λ is the overall correction factor representing strength discrepancy of the field and laboratory-mixed soils. A large scatter of λ has been observed in a number of investigations, with values ranging roughly from 0.3 to 1.5 for the wet-method and about 0.5–2 for the dry-method (based on CDM and DJM experience). For design purposes λ is usually assumed in the range of 0.5–1 because lower mixing efficiency *in situ* is generally expected (see Table 9.8).

The allowable compressive strength, f_{ca}, of stabilised soil is subsequently calculated as follows:

$$f_{ca} = \frac{f_c}{F_s} \tag{9.9}$$

where F_s is the adopted safety factor. Since f_{ca} is based on UCS in which no account for creep and cyclic loading is incorporated, relatively high global safety factors of 2.5–3 for static conditions and 2 for dynamic (earthquake) conditions are typically used. They also incorporate the importance of the structure, the type of loads and the design method (CDIT, 2002). The corresponding allowable shear and tensile strengths of stabilised soil can be calculated from the allowable compressive strength f_{ca} using empirical relationships specified in Table 9.8. If more advanced analyses with respect to shear deformation are required, direct data from drained or undrained triaxial compression tests may be used to account for actual confining pressure and drainage conditions in the field, while the residual shear strength, rather than the peak shear strength, should be used in the design.

If the limit state design philosophy is followed, then F_s can be formally replaced with appropriate partial safety factors applied to various elements of the design according to the reliability with which they are known or can be calculated. The obtained factored design compressive strength, $f_{cd} < f_c$, is then compared with the maximum factored design normal stress acting on the column. The same applies for the shear and tensile strengths.

In general, the overall procedure aiming for selection of the design strength of stabilised soil should be more rigorous for all cases where failure of a single column may be critical to the performance of stabilised ground. For DM applications with a sufficiently high ratio of area improvement and appropriate installation patterns, individual column quality is usually less important and lower safety margins can also be accepted. Evidently, the final judgement regarding the field design strength should be based on local experience in terms of the improvement effect on the soils found in the region, the properties of the stabiliser, data from pre-construction trial tests, the sensitivity of the project, the experience of the contractor (i.e. the COV), and the expected level of quality control and quality assurance, as pointed out by Porbaha *et al.* (2000).

9.5.6 Geotechnical design

The purpose of geotechnical design is to determine the final installation pattern and dimensions of improved ground on the basis of appropriate stability analyses to satisfy functional requirements of the supported structure. Depending primarily on the adopted arrangement of DM columns and on the selected design strength of stabilised soil, which in general may represent hard to semi-hard material, the improved ground can be considered as a rigid body or as a geocomposite system.

If a stabilised soil is likely to behave as a rigid structural member embedded in the ground, its external stability can be evaluated under modes of failure typical for gravity-type structures, including horizontal sliding, overturning, bearing capacity and rotational sliding. Related DM patterns which can be analysed with this approach comprise mainly block-type improvement and, with certain simplifications, also 'blocks' composed of long and short walls (cf. Figure 9.31a), as it has been practised in Japan for various port facilities (cf. CDIT, 2002). In the latter case, however, it is also necessary to examine the extrusion failure mode of untreated soil remaining between the long walls of stabilised soil when subjected to unbalanced active and passive earth pressure, generated for instance by an earthquake (Figure 9.40). One method to inspect this failure mode is to assume that the soft soil in between the long walls moves as a rigid rectangular prism, while the height of the prism is varied (cf. Terashi *et al.*, 1983). Any safe design requires also that the stresses inside the stabilised soil body do not exceed the capacity of soil–binder material. The earth pressures applied to the

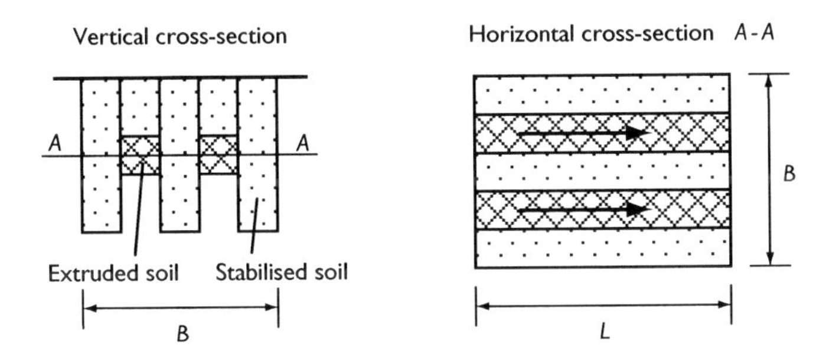

Figure 9.40 Extrusion failure of combined wall-type DM.

internal stability analysis should be carefully assigned in relation to the margin of safety adopted for the external stability to maintain compatibility of displacements. Sequence and method of construction may also affect the internal stability of the soil–binder product as cold joints may form at the intersections of primary and secondary panels or overlapping columns. After the final pattern of treatment is determined by the above analyses, the immediate and long-term displacements of the stabilised ground should be calculated. Since the deformation of the stabilised soil itself is usually small, the displacement of the improved ground results from the deformation of the soft layers surrounding or underlying the treated soil mass.

For a majority of on-land applications there is a general tendency to use more economical installation patterns than block treatment, especially for settlement reduction and improvement of stability of embankments, and for foundation of structures. Consequently, column-, group column-, lattice- and grid-type solutions are frequently designed with area improvement ratios between 15 and 50 per cent, depending on application. A common feature of these solutions is that untreated soil is surrounding an individual column or a column group, or is left within enclosed spaces formed by stabilised soil (cf. Figure 9.29). As a result, a geocomposite system is created and the interaction between the stabilised and the native soil must be carefully considered and understood at the stage of design. The mode of deformation and/or the mode of failure of composite ground are primarily dependent on the relative stiffness of stabilised soil, on the selected installation pattern and on external loading conditions.

Corresponding stability analyses of composite ground usually begin with initial assumption of the installation pattern (i.e. with selection of area improvement ratio, a_p) and initial evaluation of necessary compressive strength of stabilised soil with respect to vertical loading (internal stability). Examination of sliding stability is carried out considering equilibrium of horizontal forces acting on the boundary of improved ground assumed to behave as a unit block. Rotational sliding is usually checked by the slip circle analysis, taking into account the average shear strength of composite ground, $\bar{c}$, calculated as a weighted mean of the strength of the columns and the strength of the unstabilised soil, also noting the difference in the strain levels, i.e.,

$$\bar{c} = f_t a_p + r c_0 (1 - a_p) \qquad (9.10)$$

where f_t is shear strength of stabilised soil, c_0 is shear strength of untreated soil and r represents reduction factor for soil strength due to limited strain level. Centrifuge tests conducted by Kitazume *et al.* (2000) revealed, however, that the group of columns might fail by several failure modes governed not only by shear failure but also by bending failure of columns, as shown in Figure 9.41. Consequently, the design procedure based only on simple slip

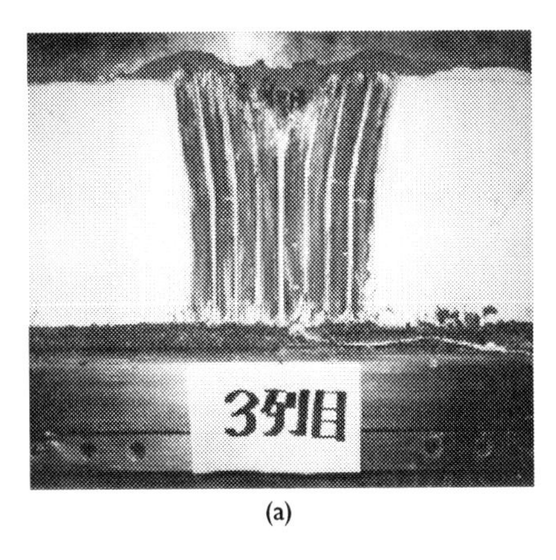

(a)

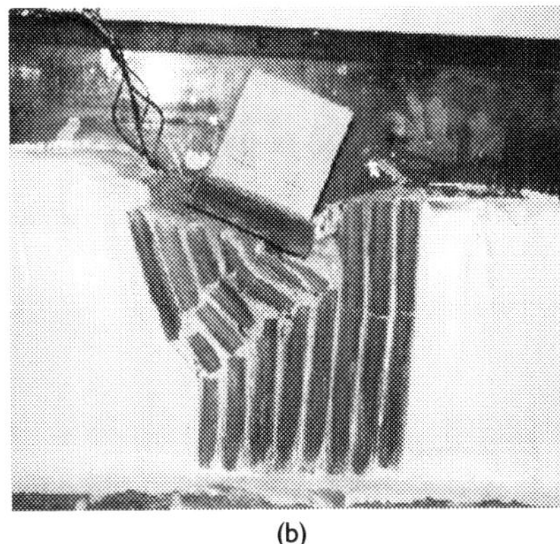

(b)

Figure 9.41 Typical modes of failure observed in centrifuge tests: (a) vertical loading – shear failure of columns just beneath the foundation, and bending failure of the outer columns; (b) inclined loading – all columns collapsed by bending failure (Kitazume *et al.*, 2000).

circle analysis with averaged shear strength may overestimate the external stability of group columns, especially if horizontal forces in excess are encountered, leading to progressive bending failure, which actually happened in a couple of unreported cases as mentioned by Terashi (2003). Unexpected deformations of column-stabilised embankments have also occurred, despite the fact that the undrained shear strength of the columns was much higher than the design shear strength adopted for stability calculations (Kivelö, 1998). To avoid too risky designs with low a_p values and high-strength elements, it has been generally recommended that the width of the improved ground should be larger than half of the thickness of the soft soil (CDIT, 2002).

Deformation and stability analyses for composite ground are generally very complex, except for vertical or nearly vertical loading (horizontal forces less than 3 per cent) and regular patterns of separated columns. In such cases, classical calculation methods based on stress concentration and uniform settlement of the stabilised and untreated soil may be applied. For more complicated installation patterns, complex loadings and difficult soil conditions, sophisticated 2-D or 3-D FEM calculations as well as model or even field tests are indispensable. In general, numerical analyses and parametric studies are exceptionally good means of assessing how alternative column patterns and combinations of soil and column stiffness will affect the behaviour of the structure, especially when the properties of stabilised soil are not well known.

When considering the geometry of the *in situ* treatment volume, one should also consider the treatment costs. These costs can vary widely depending on the specialist contractor's equipment and procedures. In general, the larger the treatment area per penetration, the lower the treatment cost per unit volume. Be on guard to this as a design with the least amount of treatment volume may not be the least costly to satisfy the support or resistance requirements. Performance specifications for the quantities and quality offer the best solution for lowest cost construction.

The design for composite ground DM is further illustrated with the selected case histories in Section 9.7.

9.6 Quality assessment of deep mixing

Quality assessment (QA) of the finished DM product is regarded as one of the pressing issues confronting the implementation of soil mixing (FHWA, 2001; Porbaha, 2002). QA is obtained from the installation records of the columns and from the results of appropriate laboratory and field verification tests. A summary of current testing methods for QA is presented in Table 9.10. Traditionally, the most frequently used testing methods

Table 9.10 Summary of verification methods for quality assessment of deep mixing

Laboratory testing	Wet grab samples	Only for the wet-method: samples taken during construction from different depth along the treated columns
	Core samples	Dry and wet-method, typically 50–100 mm diameter, taken after construction
	Block samples from extracted or exposed columns, or cut column sections	Block sample dimensions ranging from standard cubes 150 × 150 mm to full-diameter columns, taken after construction
In situ testing	Penetration methods	Static cone penetration test (CPT), standard penetration test (SPT), rotary penetration test (RPT), pressuremeter test (PMT), conventional column penetrometer (CCP) test, reverse column penetrometer test (RCP, FOPS), column vane penetrometer (CVP) test, dynamic cone penetration (DCP) test, static–dynamic penetration (SDP) test
	Loading methods	Single column loading test, group column loading test, plate loading test, screw plate test, post-construction monitoring
	Geophysical methods	Seismic methods (inhole, downhole, crosshole), electrical resistivity
	Non-destructive methods	Sonic integrity test
	Drilled shafts or excavations	For visual observation, testing (and sampling)

are coring (Japan), core and wet grab samples (US), and probe testing (Scandinavia). The selection of suitable verification methods should depend on careful analysis of their relevance, accuracy, applicability and cost in relation to the purpose and pattern of soil treatment, strength of stabilised soil and the applied process of soil mixing (dry or wet).

Wet grab samples taken from different depths shortly after construction of wet-mixed columns are used to make cubes or cylinders for later laboratory testing. Typical sampling tools consist of a hollow rod suspending a sample bucket or a tube at its tip, with inside dimensions sufficient to hold the required quantity of treated soil. The sampling device must be able to reach a prescribed depth in order to take a wet sample from a representative soil layer, and must allow it to be retrieved without contamination. Wet grab sampling is relatively easy in operation and permits a large number of specimens to be obtained at low cost. It is common to take wet samples from each days work, e.g. from two columns per shift at different sampling levels, and to check the strength at 3, 7, 14, 28 and 56 days of curing, all depending on project specifications. An early estimate of the field strength is also used for optimisation of the mix design. In the case of less efficient mixing, however, the presence of soil clods (i.e. unmixed soil masses) may prevent the sampler from functioning correctly and/or from obtaining a wet sample whose composition is truly representative of the overall mixed volume. While preparing small specimens for testing, soil clods greater than 10 per cent of the mold diameter should be screened off.

Core drilling is frequently used to obtain field specimens for testing and to inspect continuity and uniformity of DM columns (to less extend practised for the Nordic method). In the case of major Japanese DM applications it is typical to drill one corehole for every 10 000 m^3 of treated soil for marine projects and one for every 3000 m^3 for on-land projects (Okumura, 1996). At Boston's Central Artery/Tunnel Project, involving over 500 000 m^3 of soil–cement, one corehole was drilled per *c.* 2250 m^3. It is important to bear in mind, however, that retrieval of representative core samples from a stabilised soil is difficult and sensitive, both to the sampling device and to the sampling technique. Cracks or micro cracks may occur in the cores during sampling for a variety of reasons, such as bend in the borehole, rigidity of the sampler, wobble of the drill steel, locking of the sampler and rotation of the sampling tube with the outer barrel. Therefore special drilling methods and lengthy field adjustments are often required to improve the quality of core recovery through treated soils of widely varying strength and composition. As an illustration, the coring activities at the CA/T project in Boston, comprising 225 coreholes and more than 7000 linear metres of core, may be cited after Lambrechts and Nagel (2003). At the start of the core-drilling program a double-tube core barrel, equipped with a standard carbide tipped bit, was employed but the core recovery rate in the variable composition soil–cement material was poor (mean 71.5 per cent, standard deviation 16.7 per cent). A significant improvement in core recovery (mean 87 per cent, standard deviation 17.7 per cent) occurred after a number of field trials as a result of the following modifications to coring technique: (1) switch to a triple-tube core barrel, (2) replacement of the carbide bit with a fine diamond step-bit with side discharge waterways to minimise sample washout, and (3) use of synthetic drilling mud and additives. In general, cores with diameters greater than 76 mm are recommended (e.g. 86 mm, 100 mm or more) since they provide less-disturbed core specimens. Moreover, core samples should not be taken exclusively from the middle of the column but rather from all across the radius, including overlapped zones, and also from the weakest soil layers to ensure collection of representative specimen.

The core runs should be visually examined for continuity and uniformity of the soil–binder material. Continuity is evaluated by determining the core recovery rate, as measured similar to a rock quality designation (RQD) value. The recovery rate is defined as the total length of full-diameter core obtained per coring depth, expressed in percentage. In Japan the core RQD should exceed 90 per cent for clayey soils and 95 per cent for sandy soils, while for each individual core run, the RQD may be 5 per cent less than that specified above (Futaki and Tamura, 2002). These requirements are rather high and may be difficult to fulfil even with sophisticated coring techniques, especially in medium to stiff clay. At Boston's CA/T Project, for example, only 38 per cent of the borings drilled with the improved equipment achieved better than 90 per cent overall recovery, while with the double-tube barrel equipment used initially only one-core boring out of 25 achieved 90 per cent RQD. Although much of the recovered core failed to meet the minimum required compressive strength and RQD criteria, the overall DM mass performed adequately during excavation and tunnel construction (Lambrechts and Nagel, 2003). Uniformity is dependent upon the quality of mixing. An assessment of the uniformity with respect to the distribution of binder should preferably be based on the concepts of scale and intensity of segregation (Danckwerts) or some other form of defined mixing indices to avoid subjective visual judgement, as pointed out by Larsson (2001). However, in practice it is difficult to draw any definite conclusions about the homogeneity of soil–binder mix from a limited number of samples, especially if the degree of mixing is low and the sample size is small. Therefore mixing indices and statistical methods for assessing uniformity have been used only sporadically, mainly in connection with R&D projects in Japan and Sweden.

Controversial opinions exist about the true relationship of UCS data from cores to those from wet grab samples (cf. Bruce *et al.*, 2002). It may be expected, however, that core strength is higher than that of wet grab sample, providing good-quality specimens are obtained from careful drilling operations. For instance, Taki and Yang (1991) published data for various soils which show that the core strengths were about twice those obtained by samples made from wet grabs. The scatter of core sample results is, however, typically larger.

The opportunity to expose the treated ground permits block samples to be taken with different shapes, sizes and orientations. Moreover, it is also possible to verify column position and to examine column shape, homogeneity, integrity, nature of column overlap, etc. Single columns can even be fully extracted, while multiple columns can be constructed in a ring or box arrangement to allow a self-supporting excavation to be completed. Drilled observation shafts may serve the same purpose. This kind of QA inspection is often limited due to cost, time and site logistics constraints, but may be very useful to resolve any apparent anomalies identified by coring or penetration testing (e.g. Burke *et al.*, 2001).

Evaluation of the results obtained from testing of wet grab, core and block samples should be based on a statistical approach taking into account the actual number of tested specimen, especially if the number of data is low. Considering, for example, compression testing, the mean value of field UCS, $\overline{q}_{uN}$, evaluated for N randomly distributed sampling points within a specific layer of stabilised soil treated in the same manner, is calculated as follows:

$$\overline{q}_{uN} = \frac{\sum_{i=1}^{N} q_{ui}}{N} \tag{9.11}$$

where q_{ui} represent the average UCS of a series of 3 tests conducted for each sampling point, i.e.:

$$q_{ui} = \frac{(q^i_{u1} + q^i_{u2} + q^i_{u3})}{3}, \quad \text{for } i = 1 \ to \ N \tag{9.12}$$

The corresponding QA criterion can be derived from equations (9.6) and (9.9), and reads:

$$\overline{q}_{uN} \geq \frac{f_{ca}\ F_s}{1 - k_a \nu_d} \tag{9.13}$$

where k_a is the coefficient of acceptance depending on the number of sampling points N, and ν_d represents the designed value of the coefficient of strength variation for a specific DM method. The k_a values currently used in Japan are shown in Table 9.11. Taking for example $\nu_d = 0.35$ and safety factor $F_s = 3$ it can be calculated that for two sampling points ($N = 2$), the mean field UCS should be at least 7.4 times larger than the allowable design

Table 9.11 Number of sampling points, N, and coefficient of acceptance, k_a, used in Japan (after Futaki and Tamura, 2002)

N	1	2	3	4–6	7–8	≥ 9
k_a	1.8	1.7	1.6	1.5	1.4	1.3

compressive stress f_{ca}, while for $N \geq 9$ the ratio drops to 5.5 (assuming $\eta_2 = \eta_3 = 1$, see equation (9.7)).

In the US, the practise of DM simplified evaluation criteria for QA used in the case of a large number of wet grab samples and for relatively high area improvement ratios. For instance, the following acceptance criteria have been developed (Burke, 2002):

- the strength of stabilised soil must achieve an average UCS equal or greater than the minimum design stress value necessary multiplied by a safety factor of 2.5;
- no greater than 5 per cent of the test results shall be less than the minimum design stress value;
- a ceiling value of twice the average required strength shall be used for individual UCS values in calculating the average strength achieved in the field.

Laboratory tests provide verification data only in discrete points of DM columns and may not localise weak zones along the entire depth of treatment. Accordingly, various *in situ* techniques have been developed mainly in Japan, Scandinavia and the US, usually adapting existing geotechnical and geophysical testing methods. Current state-of-the-art summaries of available *in situ* verification techniques have been published by FHWA (2001), Porbaha (2002), Axelsson and Larsson (2003), and Porbaha and Puppala (2003).

Penetration testing is feasible in low-strength DM or young soil mix columns designed for higher strength. For the Nordic method all routine testing on installed columns, comprising usually 0.5–2 per cent of columns, is carried out by some form of penetrometer testing, including specially developed reverse testing methods (e.g. FOPS, limited to column's shear strength of *c.* 600 kPa) to avoid the problem of the cone's tendency to steer out of the column. CPT is most effective when the length of the column is less than 10 m and the shear strength is below 1 MPa. Stepwise pre-boring may be used to keep the cone inside long columns. SPT and DCP tests are also limited to UCS of about 1.5 MPa, while uncertainty exists as to the reliability of the correlation between UCS and the number of blow counts. For hard-treated columns with shear strength lower than 2 MPa static–dynamic (SPD) test may be used, combining the mechanical CPT and dynamic probing. In Japan, rotary penetration tests (RPT) has been developed to allow column testing to a large depth in stratified ground. The cutting resistance during rotary penetration is measured using sensors installed at a special drilling bit. Pressuremeter tests (PMT) were also conducted in Sweden and the US inside a drilled hole at the centre of the treated column to estimate the *in situ* strength and compression modulus (e.g. for the Nordic method).

Various loading tests carried out on the ground surface or at depth in test pits or inside the columns (screw plate test) provide reliable information about the strength and deformation of the treated soil. They include small-diameter plate loading tests as well as full-scale load tests of individual columns or a group of columns. Examples of full-scale tests are presented in Section 9.7. Post-construction monitoring of settlements provides the most objective data of the treated ground's overall performance, and is practised in many projects.

Modified geophysical methods are undergoing extensive testing, especially in Japan and Sweden, as a means of assessing column strength, integrity and homogeneity. Types include seismic methods (inhole P-S logging, downhole logging, crosshole logging), borehole electrical resistivity profiling and low-strain sonic integrity testing. Broadly, each can be described as 'promising', having provided reasonable correlation with cores, but it does not seem that any geophysical method is routinely used (FHWA, 2001). According to Japanese experience, it is possible to obtain reliable information from integrity tests if the column compressive strength exceeds 1 MPa and the column length exceeds 4 m (Futaki and Tamura, 2002).

Experiments have also been made in Finland and Japan with, respectively, 'measurement while drilling' (MWD) and 'factor of drilling energy' tests, which relate the records of various drilling parameters to the strength properties of the treated soil (Bruce *et al.*, 2002).

Finally, it should be emphasised that the QA programs and the adopted control criteria should be dictated by the main purpose of soil treatment, and by careful evaluation of associated design limit states. Even with close controls, significant field variability of the properties of *in situ* treated soil is most probable. This should be understood as an inherent characteristic of SM technology. Therefore, QA control programs need sufficient flexibility to respond to variable characteristics of soil mix, avoiding too restrictive criteria for occasional low strength or existence of soil clods inside the treated ground if the overall performance of stabilised soil is satisfied.

9.7 Selected case histories

Eight case histories, selected from recently completed projects conducted in Malaysia, Japan, Sweden, USA and Poland, are presented in a synthetic manner to illustrate current applications of the DMM, focusing mainly on embankment and foundation support. With respect to the dry-method of DM the applications of the Nordic method and the Japanese DJM method are described, both referring to important infrastructure projects. The third case presents a novel combination of shallow mass stabilisation of organic soils with dry cement and deep stabilisation of the underlying soft clay with dry lime and cement. Four subsequent applications of the wet-method cover mechanical mixing using single axis tools of different diameters (0.8 and 2.1 m) and illustrate column-type and grid-type installation patterns. The last case presents application of hybrid mixing for control of deep excavation near a railway tunnel.

9.7.1 Railway embankment in Malaysia (dry DM)

Source	Raju, Abdullah and Arulrajah (2003)
Location	Railway line between Rawang and Ipoh, Malaysia,
Construction site	800 m long, 20–25 m wide (Figure 9.42)
Soils	Very soft silty clay or clayey silt to loose silty clayey sand, typical CPT log see Figure 9.43a, moisture content $w = 50$–70 per cent, groundwater *c.* 1 m below ground surface
Embankment height and load	1.5–3 m, equivalent traffic load 30 kPa
Design requirements	Train speed 160 km/h, max. settlement 25 mm in 6 month of operation, max. differential settlement 0.1 per cent along the centreline, safety factor for slope failure 1.5 (long-term)
Applied DM method	Nordic method, single shaft (Figure 9.43b)
Column data	Diam. 0.6 m, length 7–14 m, overall 50 000 linear m
Column pattern	Detached columns, square/rectangle, 1–1.3 m c/c under the rails ($a_p = 28$–17 per cent), 1.4–1.5 m c/c remaining area ($a_p = 14$–13 per cent)
Design shear strength	250 kPa under the rails, 150 kPa remaining area
Binder type and factor	Portland cement 100 per cent, 100–150 kg/m^3
Embankment reinforcement	Geotextile 100/50 kN/m (longitudinal/transverse direction)
Observed performance	Settlement below 10 mm for embankment 1–1.5 m, lateral displacement below 15 mm; Loading test see Figure 9.44

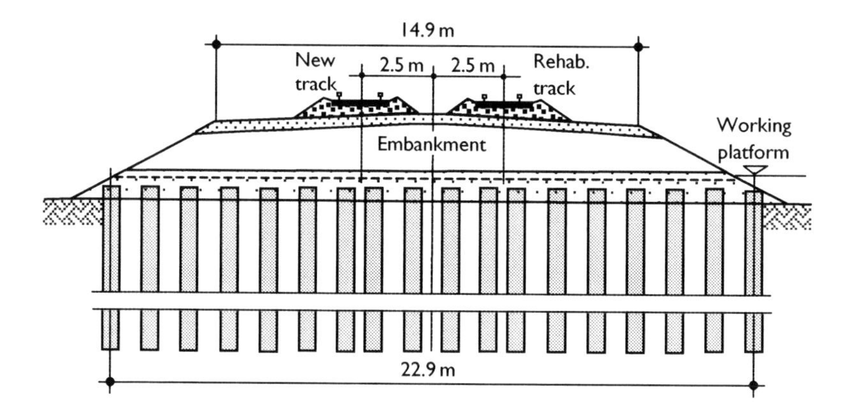

Figure 9.42 Typical cross-section of the railway embankment and treated zone (Raju *et al.*, 2003).

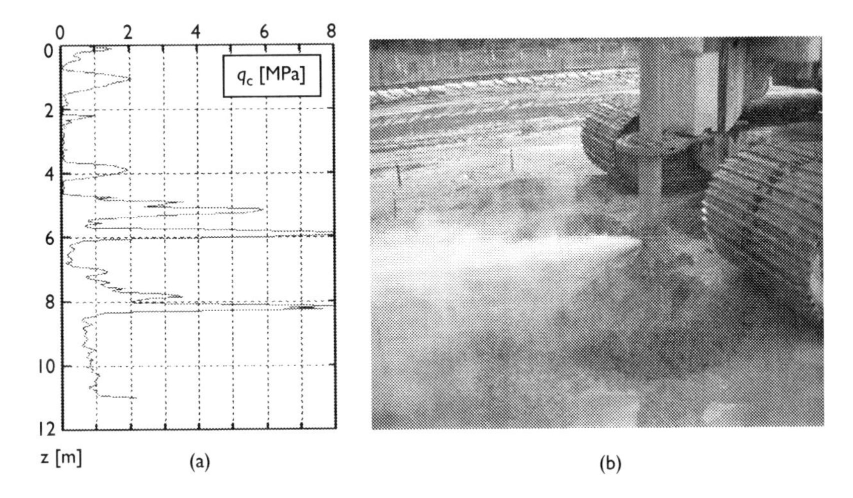

Figure 9.43 (a) Typical CPT log; (b) The LCM machine at work (Raju *et al.*, 2003).

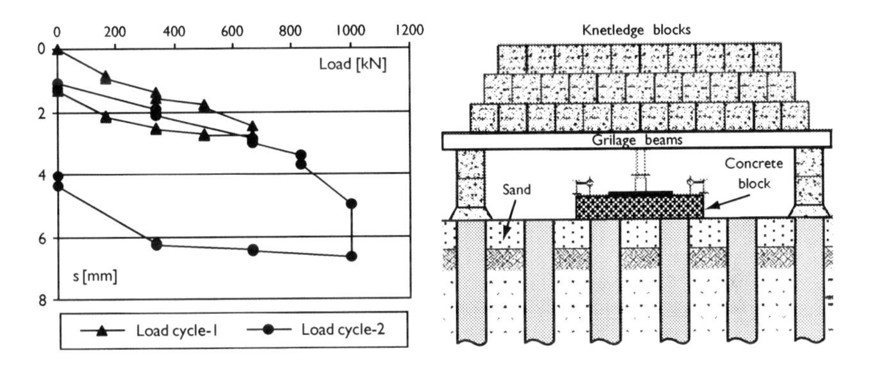

Figure 9.44 Control static loading test over an area of 3×3 m, 4 columns (Raju *et al.*, 2003).

9.7.2 Metropolitan intercity expressway in Japan (dry DM)

Source	Technical site visit 14.10.2002 (courtesy of DJM Association), also Ohdaira *et al.* (2002)
Location	Metropolitan highway at Kawashima, Japan
Construction site/ description	*c.* 100 m long, 47.6–57.4 m wide, case A: settlement reduction for a box-culvert, case B: protection against slope failure (Figure 9.45)
Soils	Soft clay, 5–7 m thick, N = 0, c_u = 18–26 kPa, sand, 3.3–5.4 m thick, N = 10–20, clay, 9–13.5 m thick, N = 4, c_u = 50–70 kPa
Embankment height and load	Average 8 m

Design requirements	Residual settlement 30 cm
Applied DM method	Dry Jet Mixing (DJM), two mixing shafts (Figure 9.46)
Column data	Diam. 1.0 m, length 20–22 m, overall 34 200 linear m
Column pattern	Case A: detached single columns, spacing 1.5 × 1.7 m, $a_p = 31$ per cent Case B: overlapped columns 0.8 m c/c, wall spacing 1.45 m c/c, $a_p = 61$ per cent

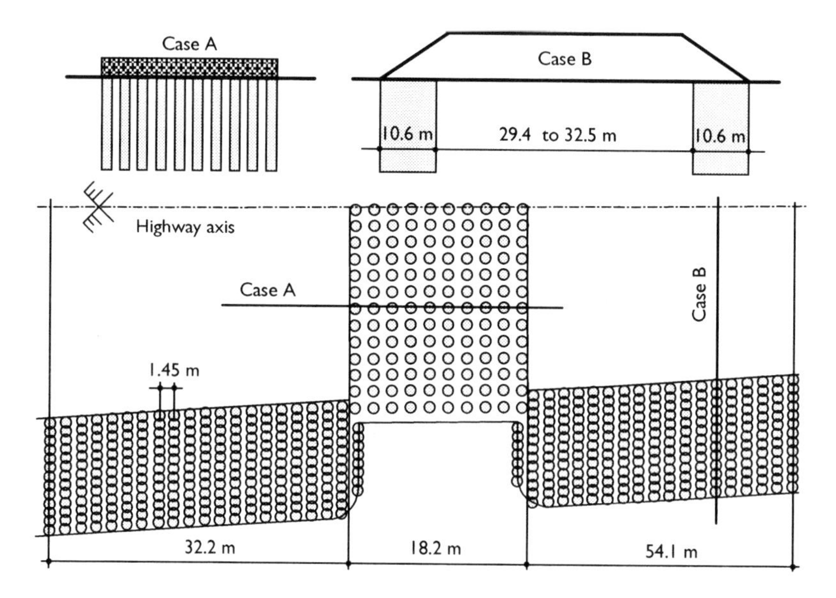

Figure 9.45 Applied column patterns at the Kawashima site (based on the design supplied by the DJM Association).

Figure 9.46 (a) DJM machine at the construction site; (b) Exposed twin columns diam. 1 m.

Design UCS	230 kPa
Binder type and factor	Blast furnace slag cement type B, top clay 150 kg/m^3, sand 110 kg/m^3, bottom clay 125 kg/m^3
Observed performance	Pressure increase on column 130 kPa, maximum excess pore water pressure 80 kPa Settlement after embankment construction *c.* 29 cm, lateral displacement at embankment toe *c.* 2 cm

9.7.3 Road embankment in Sweden (dry mass stabilisation and dry DM)

Source	LCM AB, a Keller Company (2003)
Location	Road 255 in Sweden, between Södertälje and Nynäshamn
Construction site	*c.* 500 m long, 20–25 m wide
Soils	0.5–3 m of superficial organic soil (peat/gyttja), underneath 3 to 15 m of soft silty clay, laying upon moraine; water content: peat up to $w = 1200$ per cent, gyttja $w = 300$–500 per cent; shear strength: gyttja 3–7 kPa, clay 10–25 kPa, increasing with depth
Embankment height and load	Height $h = 1.4$–5.6 m, equivalent traffic load 20 kPa (typical cross-section shown in Figure 9.47)
Design requirements	Max. settlement 30 cm in 12 month of operation, safety factor for slope failure 1.5 (long-term)
Applied Methods	The Nordic method (lime–cement column) for deep stabilisation of clay (Figure 9.48a), LCM system for mass stabilisation of organic soils (Figure 9.48b)
Execution of LC columns (first phase)	Detached columns diameter 0.8 m, square grid of 1.8 m c/c for $h < 4$ m, for $h > 4$ m rows of overlapping columns at c/c 0.7 m spaced 2 m, $a_p = 15.4$–22 per cent, column length 2–10 m, mean length approx. 7 m, total column length *c.* 57 000 m
Execution of mass stabilisation (second phase)	Grid pattern of soil blocks 3 × 4.5 m, 0.3 m vertical overlap with respect to column heads, total volume of stabilised soil 34 000 m^3
Embankment construction (third phase)	Geotextile placed on stabilised soil, crushed fill 0.3–0.5 m thick + geogrid + 0.3–0.5 m crushed fill placed over the stabilised block to pre-load the peat/gyttja layer, subsequent embankment construction after *c.* 1 month

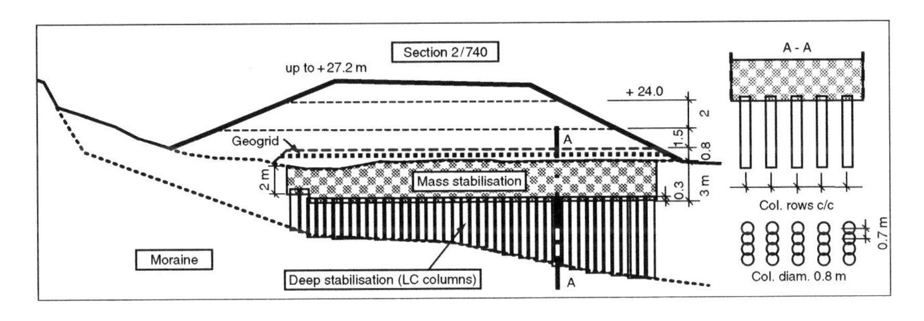

Figure 9.47 Typical cross-section of the road embankment and treated zone.

(a)

(b)

Figure 9.48 (a) First phase: installation of LC columns; (b) Second phase: mass stabilisation (courtesy of LCM AB).

Design shear strength	50 kPa in peat/gyttja, 100 kPa in clay
Binder type and factor	In peat/gyttja: cement Portland CEM II/A-LL 42,5R 100 per cent, 175 kg/m^3; in clay: lime/cement (50 per cent/50 per cent), 80 kg/m^3
Observed strength and performance	Achieved shear strength from 50 kPa in peat/gyttja and 200 kPa in clay, settlement in peat stopped after approximately 4 months

9.7.4 Carriageway Trasa Zielona in Poland (wet DM)

Source	Own project
Location	Lublin, Poland
Construction site/ description	235 m long, 44 m wide, dual three-line urban carriage way
Soils	Weak soils down to 3–8 m, loose anthropogenic fill, underlined by peat ($w = 400$ per cent) and organic clay ($w = 35$ per cent), organic soils 1–4 m thick
Embankment height and load	1.3–2.5 m, equivalent traffic load 30 kPa
Design requirements	Maximum differential settlement 0.2 per cent
Design compressive stress	480–676 kPa, assuming reduced column diameter to 0.7 m
Applied DM method	Wet DM, single axis equipment (Figure 9.49)
Column data	2402 No., diameter 0.8 m, length 3–8.5 m, mean 6.5 m, overall 15 538 lin. m

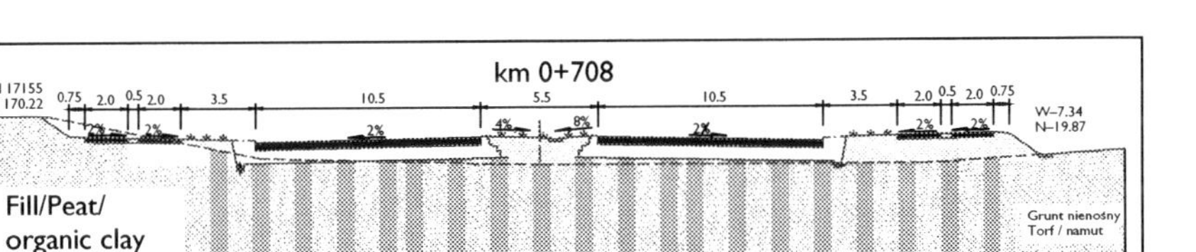

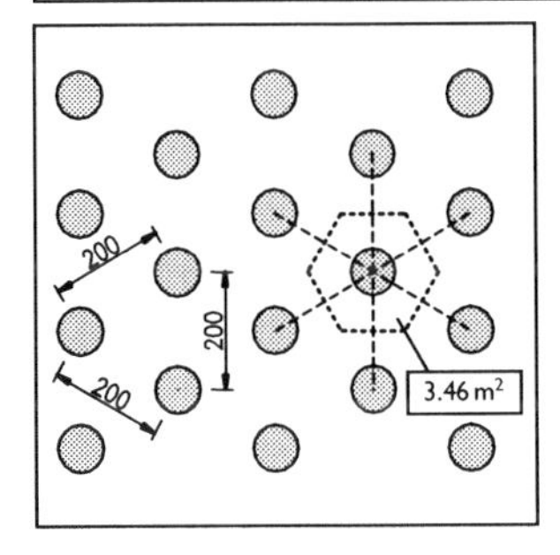

Figure 9.49 Column-type design for a road embankment (single axis mixing tool diameter 0.8 m).

Column pattern	Detached columns, equilateral triangle, side 2 m c/c, $a_p = 14.5$ per cent
Design UCS strength	1.5 MPa (with 90 per cent confidence)
Cement type and factor	Slag cement CEM III/A, 350 kg/m^3
Embankment reinforcement	Two layers of TENSAR geogrid, 30 and 20 kN/m, separated by 30 cm crushed aggregate

9.7.5 *Deepwater bulkhead in USA (wet DM)*

Source	Burke *et al.* (2001)
Location	Ham Marine Inc. Facility, Pascagoula, Mississippi, USA
Construction site	*c.* 610 m long, *c.* 15.5 m wide, outboard soil dredged to −10.7 m upon completion of DM works and the anchored wall
Soils	Loose silty sand, followed by medium stiff to very soft clay (slightly organic), $w = 34$–56 per cent
Client's request	UCS of 96 kPa (1000 psf) at 14 days of cure
Applied DM method	Wet DM, large diameter single axis equipment (Figure 9.51a)
Column data	Diameter 2.13 m (7 ft), length 3.5–14.5 m, overall 72 633 m^3
Column pattern	Cellular grid created with overlapping columns, $a_p = c.$ 85 per cent, stepped arrangement to minimise the quantity of DM works (Figure 9.50)
Design UCS	Average UCS of 689 kPa (100 psi) (obtained laboratory and field strengths depicted in Figure 9.51b)
Binder type and factor	3:1 ratio of GGBFS (ground granulated blast furnace slag) to type I Portland cement, slurry specific gravity 1.42, target binder factor 175 kg/m^3

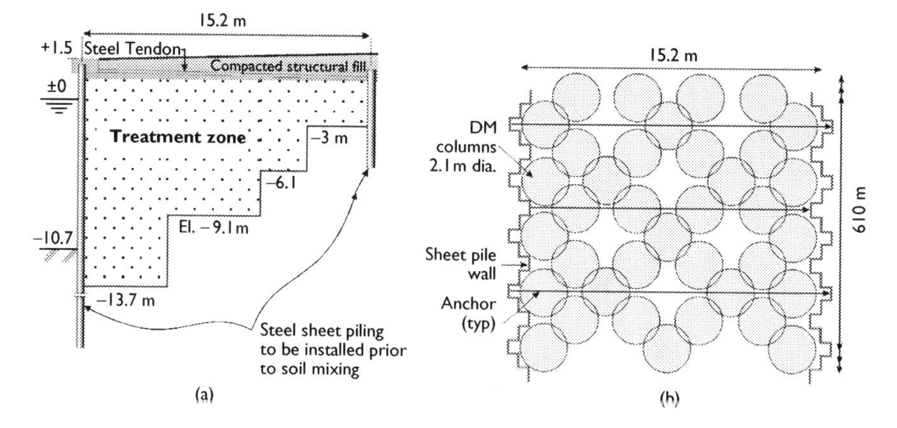

Figure 9.50 Bulkhead design: (a) cross-section; (b) cellular DM column pattern (Burke *et al.*, 2001).

(a)

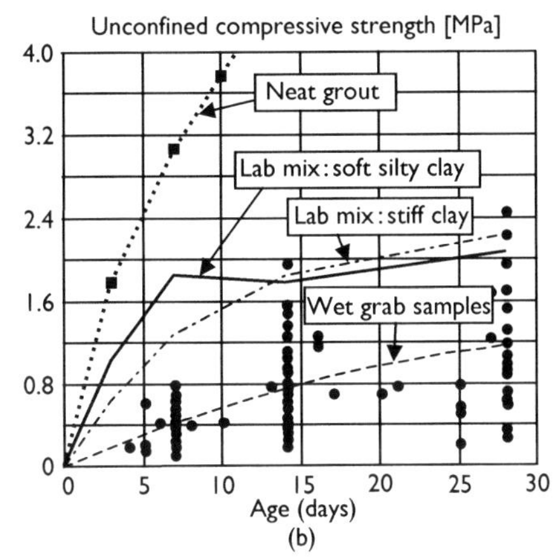

(b)

Figure 9.51 (a) Single axis mixing tool diameter 2.1 m; (b) UCS of laboratory and field sampling (Burke *et al.*, 2001).

Spoil volume	33 per cent of treated ground
Observed performance	63–76 mm of lateral displacement observed during dredging, closely following FEM prediction, subsequent movements during operation of two 250 ton cranes negligible

9.7.6 Foundation of a multistorey building in Poland (wet DM)

Source	Own project
Location	Kielce, Poland
Construction site/description	*c.* 35 × 55 m, foundation slab 1497 m^2, slab thickness 45 cm (60 cm along the edges)
Soils	Heterogeneous soft soils, extending 3–7.5 m below the slab level, including: silt, organic clay, fine sand, peat inclusions 0.5–0.8 m thick, constrained compression modulus 2.1–5.4 MPa
Mean loading pressure	112 kPa (under the foundation slab)
Expected settlement	Without ground improvement: 70–500 mm
Design requirement	Maximum settlement less than 30 mm
Applied DM method	Wet DM, single shaft equipment
Column data	461 No., diameter 0.8 m, length 5–9.2 m (from working level), mean 7.3 m, overall 3370 lin. m
Column pattern	Detached single columns, arrangement adjusted to load distribution and slab–soil-columns interaction, average $a_p = 15.4$ per cent (Figure 9.52)
Cement type and factor	Cement CEM III/A 32.5 NA, 340–380 kg/m^3, slurry specific gravity 1.7–1.75
Design compressive stress	Max 0.86 MPa (factored value), on a single DM column

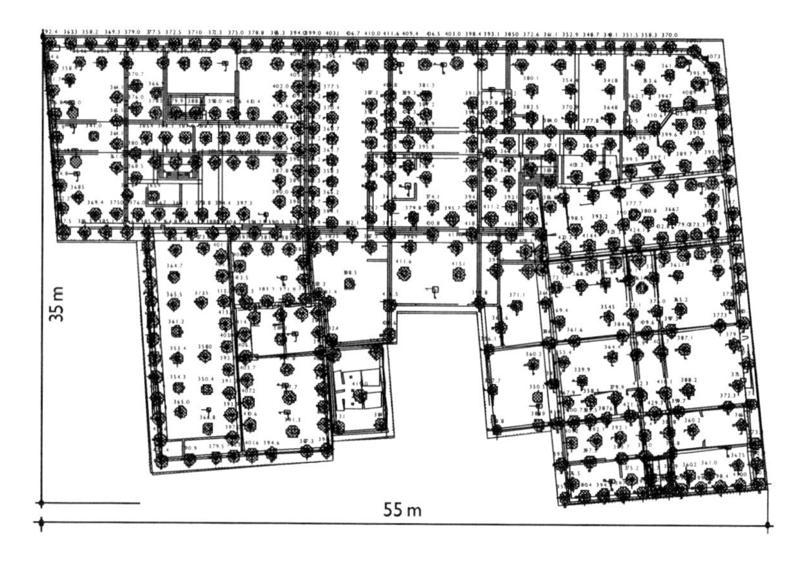

Figure 9.52 Arrangement of DM columns diameter 0.8 m under the foundation slab.

Design UCS — 1.9 MPa (with 90 per cent confidence)
Measured UCS strength — 32 specimens 15 × 15 cm (wet grab), mean UCS at 28-days 5.72 MPa, standard deviation 2.14 MPa, COV 0.38
Observed performance — Below 10 mm

9.7.7 *Foundation of highway bridge supports in Poland (wet DM)*

Source — Own project
Location — A2 highway n/Poznań, Poland, bridge WD-23 with five supports, i.e. P1 and P5 – bridge heads, P2 and P4 – intermediate supports (supports in plan view shown in Figure 9.53)
Construction site/ description — The whole project included construction of 39 new bridges across and along A2 highway, with 2–5 foundation supports each, DM applied for reduction of total and differential settlements
Soils — Boulder sandy clays (CPT log shown in Figure 9.55a)
Loading pressures — P3 (central support): $\sigma_{mean} = 251$ kPa, $\sigma_{max} = 406$ kPa, $\sigma_{min} = 96$ kPa
P5 (bridge head): $\sigma_{mean} = 138$ kPa, $\sigma_{max} = 248$ kPa, $\sigma_{min} = 27$ kPa
Maximum column load — P3: 399 kN, P5: 418 kN
Design requirements — Settlement <2 cm, settlement difference between supports <1 cm
Applied DM method — Wet DM, single axis equipment, diameter 0.8 m (Figure 9.54)
Column length — For P3 at elevation of +87.0 m: l = 3.75 m, for P5 at +89.78 m: l = 6.3 m, overall for five supports: 200 columns and 880 lin. m

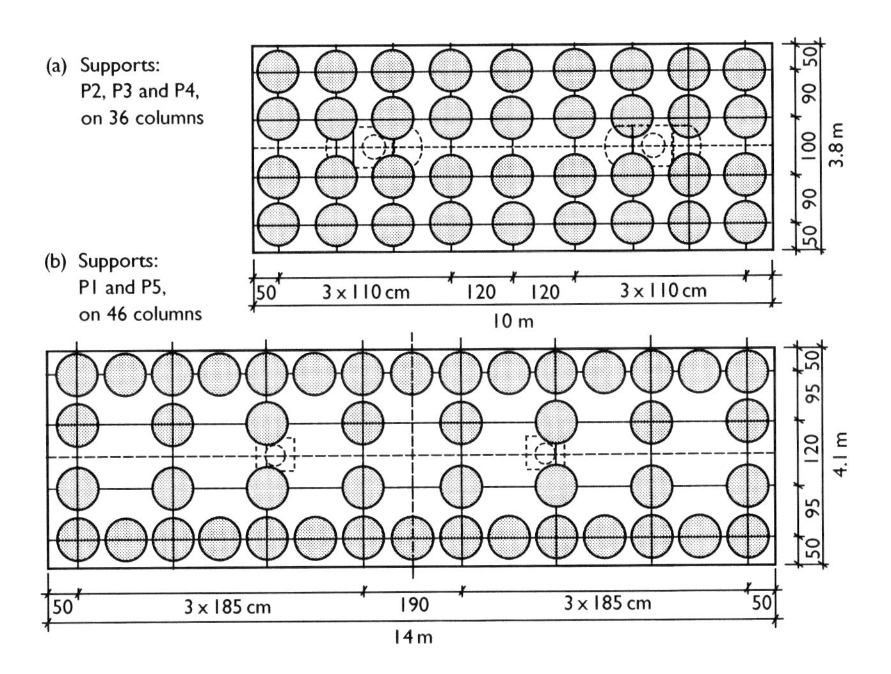

Figure 9.53 Arrangement of DM columns diameter 0.8 m under bridge supports (bridge WD-23).

(a) (b)

Figure 9.54 (a) Single axis mixing tool diameter 0.8 m; (b) Exposed columns (WD-23, P5).

Column pattern	Detached columns, P3: $a_p = 47$ per cent, P5: $a_p = 40$ per cent (Figure 9.53)
Cement type and factor	Slag cement CEM III/A 32.5 NA, 320 kg/m^3, slurry specific gravity 1.65
Design compressive stress	Max. 0.83 MPa (characteristic value) on a single DM column

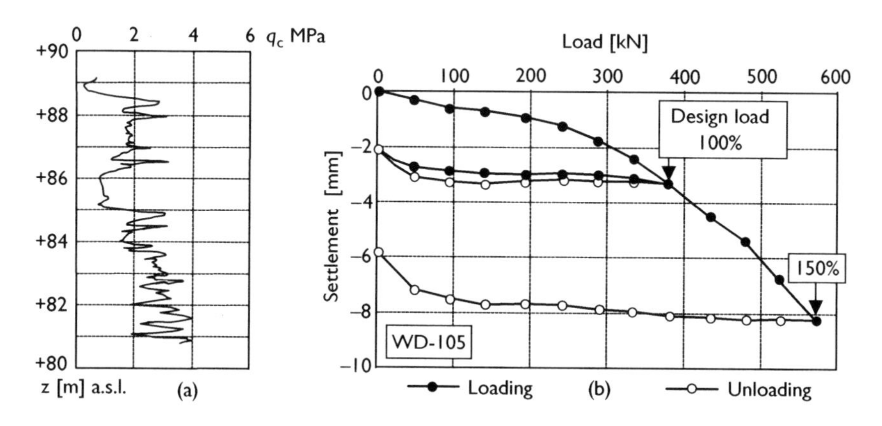

Figure 9.55 (a) Soil conditions (CPT log); (b) Typical result of column static loading test.

Design UCS	2.5 MPa (with 90 per cent confidence), global safety factor 3.0
Measured UCS strength	30 specimens 15 × 15 cm (wet grab), mean UCS at 28 days 8.73 MPa, standard deviation 1.60 MPa, COV 0.18
Observed settlement	P1: 12 mm, P2: 7 mm, P3: 9 mm, P4: 6 mm, P5: 9 mm (Typical result of a single column static loading test is shown in Figure 9.55b for a similar bridge WD-105)

9.7.8 Excavation control in Japan (hybrid wet mixing)

Source	Technical site visit (courtesy of Chemical Grouting Co., Japan)
Location	Tokyo, Japan
Construction site/ description	Four-storey basement, 17.2 m deep, part of the site close (*c.* 6 m) to existing Japan Railway tunnels at depths from 13.5 to 22 m (Figure 9.56)
Purpose of DM works	Reduction of the risk of potential movements resulting from the excavation (movements limited to 9 cm)
Soils	17 m of very soft silts/clays overlying very dense slits and sands
Applied DM method	Wet DM, JACKSMAN method, double-shaft equipment (combined XJET and mechanical mixing)
Column data	Two col. diameter 2.3 m, spaced 1.4 m (JACKSMAN type B), treatment area 7.2 m^2
Column pattern	Block pattern *c.* 12.5 × 56 m, area improvement ratio $a_p = 80$ per cent, column length 9.3 m (from 7.7 to 17 m), total 6423 m^3, 95 No.
Cement factor	$\alpha = 160$ kg/m^3, slurry specific gravity 1.50
Design UCS	0.98 MPa (applied factor of safety 3)
Penetration phase	1.0 m/min, utilising 4 × 75 l/min of water through the jetting system and 2 × 50 l/min of water through the mixing system
Withdrawal phase	0.7 m/min, grout through both XJET nozzles at 4 × 150 l/min and through the mixing system at 2 × 200 l/min
Spoil	Minimal spoil production

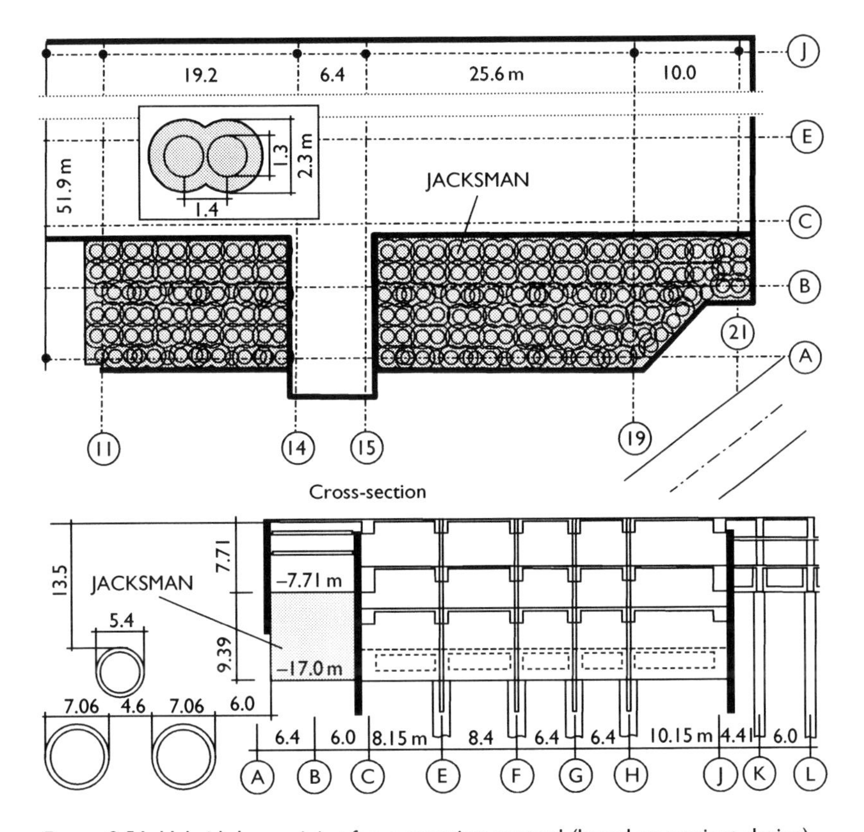

Figure 9.56 Hybrid deep mixing for excavation control (based on project design).

References

Aldridge, C.W. and Naguib, A. (1992) In situ mixing of dry and slurried reagents in soil and sludge using shallow soil mixing, *85th Annual Meeting & Exhibition*, Kansas City, Missouri, 21–26 June (Geo-Cone Inc. home page, paper 92-33.06).

Al-Tabbaa, A. and Evans, Ch. (2002) Geoenvironmental research and applications of Deep Soil Mixing in the UK, *Proceedings of Deep Mixing Workshop 2002 in Tokyo*, Port and Airport Research Institute & Coastal Development Institute of Technology.

Andersson, R., Carlsson, T. and Leppänen, M. (2000) Hydraulic cement based binders for mass stabilization of organic soils, *Proceedings of the Soft Ground Technology Conference*, United Engineering Foundation and ASCE Geo-Institute, Noorwijkerout, Netherlands, 28 May–2 June.

Andromalos, K.B., Hegazy, Y.A. and Jasperse, B.H. (2000) Stabilization of soft soils by soil mixing, *Proceedings of the Soft Ground Technology Conference*, United Engineering Foundation and ASCE Geo-Institute, Noorwijkerout, Netherlands, 28 May–2 June.

Aoi, M. (2002) Execution procedure of Japanese dry method (DJM), *Proceedings of Deep Mixing Workshop 2002 in Tokyo*, Port and Airport Research Institute & Coastal Development Institute of Technology.

Außenlechner, P., Seidel, A. and Girsch, E. (2003) Deichsanierung im Mixed-in-Place-Verfahren (MIP) am Beispiel München, Isarplan, *Proceedings of 18th Christian Veder Colloquium*, Univ. Graz, 24–25 April.

Axelsson, M. and Larsson, S. (2003) Column penetration tests for lime-cement columns in deep mixing – experience in Sweden, *Proceedings of the 3rd International Conference on Grouting and Ground Treatment*, ASCE Geotechnical Special Publication No. 120, pp. 681–694.

Azuma, K., Ohishi, K., Ishii, T. and Yoshimoto, Y. (2002) Typical application to excavation support, Deep Mixing Workshop, *Proceedings of Deep Mixing Workshop 2002 in Tokyo*, Port and Airport Research Institute & Coastal Development Institute of Technology.

Babasaki, R., Terashi, M., Suzuki, T., Maekawa, A., Kawamura, M. and Fukazawa, E. (1996) Japanese Geotechnical Society Technical Committee Report – Factors influencing the strength of improved soil, Grouting and Deep Mixing, *Proceedings of the 2nd International Conference on Ground Improvement Geosystems*, 2:913–918, Balkema.

BCJ (1997) Building Center of Japan. Design and quality control guideline for buildings (in Japanese).

Blackwell, J. (1994) A case history of soil stabilisation using the mix-in-place technique for the construction of deep manhole shafts at Rochdale, *Proceedings of ICE Conference 'Grouting in the Ground'*, Paper 30, Thomas Telford, London, pp. 497–510.

Bredenberg, H. (1999) Keynote lecture: Equipment for deep soil mixing with dry jet mix method, *Proceedings of the International Conference Dry Mix Methods for Deep Soil Stabilization*, Stockholm, Balkema, pp. 323–331.

Bruce, D.A., Bruce, M.E. and DiMillio, A.F. (2002) Deep Mixing: QA/QC and verification methods, Deep Mixing Workshop, *Proceedings of Deep Mixing Workshop 2002 in Tokyo*, Port and Airport Research Institute & Coastal Development Institute of Technology.

Bruce, D.A. and Bruce, M.E.C. (2003) The practitioner's guide to deep mixing, *Proceedings of the 3rd International Conference on Grouting and Ground Treatment*, ASCE Geotechnical Special Publication No. 120, pp. 474–488.

Burke, G.K. (2002) North American single axis wet method of deep mixing, deep mixing workshop, *Proceedings of Deep Mixing Workshop 2002 in Tokyo*, Port and Airport Research Institute & Coastal Development Institute of Technology.

Burke, G.K., Lyle, D.L., Sehn, A.L. and Ross, T.E. (2001) Soil mixing supports a deepwater bulkhead in soft soils, *Proceedings of ASCE Ports Conference 2001*, Norfolk, VA.

CDM Association (1996) Promotional information.

CDM Association (2002) CDM-Cement Deep Mixing Bulletin as of 2002.

CEN TC 288 (2002) Execution of special geotechnical works – Deep mixing, Provisional Version from March, 2002, presented during Deep Mixing Workshop 2002 in Tokyo (prEN 14679).

Coastal Development Institute of Technology (CDIT), Japan (2002) The Deep Mixing Method, A.A. Balkema Publishers.

DJM Association (2002) DJM-Dry Jet Mixing Method Bulletin as of 2002.

Druss, D. (2002a) Boston's Central Artery Project – A Showcase for Ground Treatment Technology, *Proceedings of Deep Mixing Workshop 2002 in Tokyo*, Port and Airport Research Institute & Coastal Development Institute of Technology.

Druss, D. (2002b) Case History: Deep Soil-Cement Mixing at the I-90/I-93 NB Interchange on the Central Artery/Tunnel Project, Contract C09A7 at Fort Point Channel Site, Boston, MA, *Proceedings of Deep Mixing Workshop 2002 in Tokyo*, Port and Airport Research Institute & Coastal Development Institute of Technology.

EuroSoilStab (2002) Development of design and construction methods to stabilise soft organic soils. Design Guide Soft Soil Stabilisation, CT97-0351, European Commission Project BE 96-3177.

FHWA-RD-99-138 (2000) An Introduction to the Deep Soil Mixing Methods as Used in Geotechnical Applications, Prepared by Geosystems (D.A. Bruce) for US Department of Transportation, Federal Highway Administration, p. 143.

FHWA-RD-99-167 (2001) An introduction to the deep soil mixing methods as used in geotechnical applications: verification and properties of treated soil, Prepared by Geosystems (D.A. Bruce) for US Department of Transportation, Federal Highway Administration, p. 434.

Forsberg, T. (2002) Oil tanks on Limix cement columns in Can Tho, Vietnam, *Proceedings of Deep Mixing Workshop 2002 in Tokyo*, Port and Airport Research Institute & Coastal Development Institute of Technology.

Futaki, M. and Tamura, M. (2002) The quality control in deep mixing method for the building foundation ground in Japan, *Proceedings of Deep Mixing Workshop 2002 in Tokyo*, Port and Airport Research Institute & Coastal Development Institute of Technology.

Geo-Con., Inc. (1998) Promotional information.

Herrmann, R., Hilmer, K. and Kaltenecker, H. (1992) Die Entwicklung des Bauverfahrens Mixed-in-Place (MIP) auf der Basis der Rotary-Auger-Soil-Mixing-Methode (RASM), Baugrundtagung 1992, Dresden.

Hidetoshi, Y., Masato, U., Iwasaki, T. and Higaki, K. (1996) Removal of volatile organic compounds from clay layer. Grouting and Deep Mixing, *Proceedings of IS-Tokyo '96, 2nd International Conference on Ground Improvement Geosystems*, Tokyo, 14–17 May, pp. 787–792.

Hioki, Y. (2002) The Construction Control and Quality Control of Dry Method (DJM), *Proceedings of Deep Mixing Workshop 2002 in Tokyo*, Port and Airport Research Institute & Coastal Development Institute of Technology.

HITEC (Highway Innovative Technology Evaluation Center) Evaluation of the Geo-Con Vert Wall System, CERF Report: #40607, April, 2002.

Holm, G. (2002a) Nordic Dry Deep Mixing Method – Execution Procedure, *Proceedings of Deep Mixing Workshop 2002 in Tokyo*, Port and Airport Research Institute & Coastal Development Institute of Technology.

Holm, G. (2002b) Deep Mixing – Research in Europe, *Proceedings of Deep Mixing Workshop 2002 in Tokyo*, Port and Airport Research Institute & Coastal Development Institute of Technology.

Horpibiulsuk, S., Miura, N, Nagaraj, T.S. and Koga, H. (2002) Improvement of Soft Marine Clays by Deep Mixing Technique, *Proceedings of the 12th International Offshore and Polar Engineering Conference*, Kitakyushu, Japan, 26–31 May, pp. 584–591.

Huiden, E.J. (1999) Soil stabilisation for embedment of Botlek Railwaytunnel in the Netherlands, *Proceedings of International Conference on Dry Mix Methods for Deep Soil Stabilization*, Stockholm, Balkema, pp. 45–49.

Jasperse, B.H. and Ryan, Ch.R. (1992) In-situ Stabilization and Fixation of Contaminated Soils by Soil Mixing, ASCE Geotechnical Division Specialty Conference Grouting, Soil Improvement and Geosynthethics, New Orleans, Louisiana, 25–28 February.

Jelisic, N. and Leppänen, M. (2003) Mass stabilization of organic soils and soft clay, *Proceedings of International Conference on Grouting and Ground Treatment*, ASCE Geotechnical Special Publication No. 120, pp. 552–561.

Kamon, M. (1996) Effect of grouting and DMM on big construction projects in Japan and the 1995 Hyogoken-Nambu Earthquake, Grouting and Deep Mixing, *Proceedings of IS-Tokyo '96, 2nd International Conference on Ground Improvement Geosystems*, Tokyo, 14–17 May, pp. 807–823.

Kawanabe, S. and Nozu, M. (2002) Combination mixing method of Jet grout and Deep mixing, *Proceedings of Deep Mixing Workshop 2002 in Tokyo*, Port and Airport Research Institute & Coastal Development Institute of Technology.

Kawasaki, K., Kotera, H., Nishida, K. and Murase, T. (1996) Deep mixing by spreadable wing method, *Grouting and Deep Mixing, Proceedings of IS-Tokyo '96, The 2nd International Conference on Ground Improvement Geosystems*, Tokyo, 14–17 May, pp. 631–636.

Kitazume, M. (2002) Current Design and Future Trends in Japan, *Proceedings of Deep Mixing Workshop 2002 in Tokyo*, Port and Airport Research Institute & Coastal Development Institute of Technology.

Kitazume, M., Okano, K. and Miyajima, S. (2000) Centrifuge model tests on failure envelope of column-type DMM-improved ground, *Soils and Foundations*, Vol. 40, No. 4, 43–55.

Kivelö, M. (1998) Stabilisation of embankments on soft soils with lime/cement columns, PhD Thesis 1023, Royal institute of Technology, Sweden.

Lambrechts, J. and Nagel, S. (2003) Coring Soil-Cement Installed by Deep Mixing at Boston's CA/T Project, *Proceedings of the 3rd International Conference on Grouting and Ground Treatment*, ASCE Geotechnical Special Publication No. 120, pp. 670–680.

Larsson, S. (1999) The mixing process at the dry jet mixing method, dry deep mix methods for deep soil stabilization, *Proceedings of the International Conference on Dry Deep Mix Methods for Deep Soil Stabilization*, Stockholm, Balkema, pp. 339–346.

Larsson, S. (2001) Binder distribution in lime-cement columns, *Ground Improvement*, Vol. 5, No. 3, 111–122.

Lebon, S. (2002) Wet Process Soilmixing – A Review of Central European Execution Practice, *Proceedings of Deep Mixing Workshop 2002 in Tokyo*, Port and Airport Research Institute & Coastal Development Institute of Technology.

Matuso, O. (2002) Determination of Design Parameters for Deep Mixing, *Proceedings of Deep Mixing Workshop 2002 in Tokyo*, Port and Airport Research Institute & Coastal Development Institute of Technology.

Nagata, S., Azuma, K., Asano, M., Nishijima, T., Shiiba, H., Yang, D. and Nakata, R. (1994) Nakajima subsurface dam, Water policy and management: solving the Problems, *Proceedings of the 21st Annual Conference*, Denver, CO., 23–26 May, pp. 437–440.

Ohdaira, H., Hashimoto, H., Gotoh, K. and Nozu, M. (2002) Observation results for the embankment on soft ground improved by DJM, Tsuchi-to-kiso, pp. 31–33, 2002.2. (in Japanese).

Okumura, T. (1996) Deep mixing method of Japan, grouting and deep mixing, *Proceedings of IS-Tokyo '96, the 2nd International Conference on Ground Improvement Geosystems*, Tokyo, Vol. 2, 879–887.

Pampel, A. and Polloczek, J. (1999) Einsatz des FMI- und HZV-Verfahrens bei der DB AG, Der Eisenbahningenieurbau, Nr. 3.

Paviani, A. and Pagotto, G. (1991) New technological developments in soil consolidation by means of mechanical mixing, implemented in Italy for the ENEL power Plant at Pietrafitta, *Proceedings of X ECSMFE*, Vol. 2, 511–516.

Porbaha, A. (1998) State of the art in deep mixing technology. Part I: Basic concepts and overview, *Ground Improvement*, Vol. 2, No. 2, 81–92.

Porbaha, A. (2000) State of the art in deep mixing technology. Part IV: Design considerations, *Ground Improvement*, Vol. 3, 111–125.

Porbaha, A. (2002) State of the art in quality assessment of deep mixing technology, *Ground Improvement*, Vol. 6, No. 3, 95–120.

Porbaha, A. and Puppala, A. (2003) In situ techniques for quality assurance of deep mixed columns, *Proceedings of the 3rd International Conference on Grouting and Ground Treatment*, ASCE Geotechnical Special Publication No. 120, pp. 695–706.

Porbaha, A., Tanaka, H. and Kobayashi, M. (1998) State of the art in deep mixing technology. Part II: Applications, *Ground Improvement*, Vol. 2, No. 2, 125–139.

Porbaha, A., Shibuya, S. and Kishida, T. (2000) State of the art in deep mixing technology. Part III: geomaterial characterization, *Ground Improvement*, Vol. 3, 91–110.

Porbaha, A., Raybaut, J.-L. and Nicholson, P. (2001) State of the art in construction aspects of deep mixing technology, *Ground Improvement*, Vol. 5, No. 3, 123–140.

Raju, V.R., Abdullah, A. and Arulrajah, A. (2003) Ground treatment using dry deep soil mixing for a railway embankment in Malaysia, *Proceedings of the 2nd International Conference on Advances in Soft Soil Engineering and Technology*, Putrajaya, Malaysia, 2–4 July.

Rogbeck, Y., Jelisic, N. and Säfström, L. (1999) Properties of mass- and cell stabilization: Two case studies in Sweden, *Proceedings of the International Conference Dry Mix Methods for Deep Soil Stabilization*, Stockholm, Balkema, pp. 269–274.

Saitoh, S. (1988) Experimental Study of Engineering Properties of Cement Improved Ground by the Deep Mixing Method, PhD Thesis (cited after Terashi 2002b).

Schwarz, W. and Seidel (2003) Das Mixed-in-Place-Verfahren, University Siegen, Symposium "Notsicherung von Dämmen und Deichen".

Shiells, D.P., Pelnik III, T.W. and Filz, G.M. (2003) Deep Mixing: An Owner's Perspective, *Proceedings of the 3rd International Conference on Grouting and Ground Treatment*, ASCE Geotechnical Special Publ. No. 120, pp. 489–500.

Sugiyama, K. (2002) The CDM-LODIC Method – Outline and applications, *Proceedings of Deep Mixing Workshop 2002 in Tokyo*, Port and Airport Research Institute & Coastal Development Institute of Technology.

Takemiya, H., Nishimura, A., Naruse, T., Hosotani, K. and Hashimoto, M. (1996) Development of vibration reduction measure wave impeding block, *Proceedings of 2nd International Conference on Ground Improvement Geosystems*, Balkema, Rotterdam, pp. 753–758.

Taki, O. (2003) Strength Properties of Soil Cement produced by Deep Mixing, *Proceedings of the 3rd International Conference Grouting and Ground Treatment*, ASCE Geotechnical Special Publ. No. 120, pp. 646–657.

Taki, O. and Yang, D.S. (1991) Soil-cement mixed wall technique, ASCE, *Proceedings of Geotechnical Engineering*, Congress, Denver, 298–309.

Tateyama, M., Trauma, H. and Fukuda, A. (1996) Development of a large diameter short reinforced anchor by cement mixing method, *Proceedings of 2nd International Conference on Ground Improvement Geosystems*, IS-Tokyo '96, Balkema, Rotterdam, 759–765.

Terashi, M. (2002a) Development of deep mixing machine in Japan, *Proceedings of Deep Mixing Workshop 2002 in Tokyo*, Port and Airport Research Institute & Coastal Development Institute of Technology.

Terashi, M. (2002b) Long-term strength gain vs. deterioration of soils treated by lime and cement, *Proceedings of Deep Mixing Workshop 2002 in Tokyo*, Port and Airport Research Institute & Coastal Development Institute of Technology.

Terashi, M. (2003) The State of Practice in Deep Mixing Methods, *Proceedings of the 3rd International Conference on Grouting and Ground Treatment*, ASCE Geotechnical Special Publication No. 120, 25–49.

Terashi, M., Tanaka, H. and Kitazume, M. (1983) Extrusion failure of ground improved by the deep mixing method, *Proceedings of the 7th Asian Regional Conference Soil Mechanics and Foundation Engineering*, 1: 313–318.

Topolnicki, M. (2003) Sanierung von Deichen in Polen mit dem Verfahren der Tiefen-Bodenvermörtelung (DMM), Hochwasserschutz, Ernst & Sohn Special 1/2003, 45–53.

Vähäaho, I. (2002) Deep Mixing at Old City Bay in Helsinki, *Proceedings of Deep Mixing Workshop 2002 in Tokyo*, Port and Airport Research Institute & Coastal Development Institute of Technology.

Walker, A.D. (1992) Soil mixing and jet grouting on a hazardous waste site, Pittsburgh, PA, USA. Grouting in the Ground, Institution of Civil Engineers, Thomas Telford, London, 473–486.

Walker, A.D. (1994) A deep soil mix cut-off wall at Lockington dam in Ohio, In situ deep soil improvement, ASCE Geotechnical Special Publication, No. 45, 133–146.

Yang, D. and Takeshima, S. (1994) Soil mixing walls in difficult ground, In-situ deep soil improvement, ASCE Geotechnical Special Publication, No. 45, 106–120.

Yano, S., Tokunaga, S., Shima, M. and Manimura, K. (1996) Centralised control system of CDM Method, Grouting and Deep Mixing, *Proceedings of the 2nd International Conference on Ground Improvement Geosystems*, IS-Tokyo '96, Balkema, Rotterdam, 681–687.

Yu, Y., Pu, J. and Ugai, K. (1997) Study of mechanical properties of soil cement mixture for cutoff walls, Soils and Foundations, Vol. 37, No. 4, 93–103.

Zheng, J. and Shi, P. (1996) Real time monitoring for control of quantity of deep mixing piles, *Proceedings of the 2nd International Conference on Soft Soil Engineering*, Nanjing, 1027–1033.

Index